AF865924

The 15th Anniversary of *JFB*—Functional Biomaterials: Bioactive Properties and Medical Applications

The 15th Anniversary of *JFB*—Functional Biomaterials: Bioactive Properties and Medical Applications

Guest Editors

Cristian Scheau

Andreea Cristiana Didilescu

Constantin Caruntu

MDPI

Basel • Beijing • Wuhan • Barcelona • Belgrade • Novi Sad • Cluj • Manchester

Guest Editors
Cristian Scheau
Department of Physiology
The "Carol Davila" University of Medicine and Pharmacy
Bucharest
Romania

Andreea Cristiana Didilescu
Department of Embryology and Microbiology
The "Carol Davila" University of Medicine and Pharmacy
Bucharest
Romania

Constantin Caruntu
Department of Physiology
The "Carol Davila" University of Medicine and Pharmacy
Bucharest
Romania

Editorial Office
MDPI AG
Grosspeteranlage 5
4052 Basel, Switzerland

This is a reprint of the Special Issue, published open access by the journal *Journal of Functional Biomaterials* (ISSN 2079-4983), freely accessible at: https://www.mdpi.com/journal/jfb/special_issues/UC2HS9Q5KZ.

For citation purposes, cite each article independently as indicated on the article page online and as indicated below:

Lastname, A.A.; Lastname, B.B. Article Title. *Journal Name* **Year**, *Volume Number*, Page Range.

ISBN 978-3-7258-8175-8 (Hbk)
ISBN 978-3-7258-8176-5 (PDF)
https://doi.org/10.3390/books978-3-7258-8176-5

Cover image courtesy of Cristian Scheau

© 2026 by the authors. Articles in this reprint are Open Access and distributed under the Creative Commons Attribution (CC BY) license. The reprint as a whole is distributed by MDPI under the terms and conditions of the Creative Commons Attribution-NonCommercial-NoDerivs (CC BY-NC-ND) license (https://creativecommons.org/licenses/by-nc-nd/4.0/).

Contents

About the Editors . . . . **vii**

Cristian Scheau, Andreea Cristiana Didilescu and Constantin Caruntu
From Scaffolds to Complex Systems: Functional Biomaterials in Regenerative Medicine
Reprinted from: *J. Funct. Biomater.* **2026**, *17*, 51, https://doi.org/10.3390/jfb17010051 . . . . **1**

Nikoleta Ivanova, Stoyan Ivanov, Stefan Peev and Tsanka Dikova
Types of Bone Substitutes and Their Application in Regenerative Medicine: A Systematic Review
Reprinted from: *J. Funct. Biomater.* **2025**, *16*, 341, https://doi.org/10.3390/jfb16090341 . . . . . . **4**

Veronica Manescu (Paltanea), Aurora Antoniac, Maria Cristina Moraru, Iulian Antoniac, Cosmin Mihai Cotrut, Sebastian Gradinaru, et al.
Suitability of Mg-Nd and Mg-Zn Alloys to Obtain Biodegradable Structures for Bone Defects
Reprinted from: *J. Funct. Biomater.* **2025**, *16*, 423, https://doi.org/10.3390/jfb16110423 . . . . . . **53**

Lianjing Yu, Zi Deng, Masato Nakagawa, Shanshan Zheng, Jun-ichiro Jo, Tomonari Tanaka and Yoshitomo Honda
Physicochemical and Antioxidant Alterations of Modified and Free Epigallocatechin Gallate Under Thermal Treatment in Air and Vacuum
Reprinted from: *J. Funct. Biomater.* **2026**, *17*, 18, https://doi.org/10.3390/jfb17010018 . . . . **85**

Todorka G. Vladkova, Dilyana N. Gospodinova, Peter D. Dineff, Milena Keremidarska-Markova, Kamelia Hristova-Panusheva and Natalia Krasteva
Electrospinning of Bovine Split Hide Collagen and Collagen/Glycosaminoglycan for a Study of Stem Cell Adhesion and Proliferation on the Mats: Influence of Composition and Structural Morphology
Reprinted from: *J. Funct. Biomater.* **2025**, *16*, 219, https://doi.org/10.3390/jfb16060219 . . . . . . **105**

Iosif-Aliodor Timofticiuc, Ana Caruntu, Christiana Diana Maria Dragosloveanu, Andreea-Elena Scheau, Ioana Anca Badarau, Argyrios Periferakis, et al.
Head and Neck 3D Bioprinting—A Review on Recent Advancements in Soft Tissue 3D Bioprinting and Medical Applications
Reprinted from: *J. Funct. Biomater.* **2025**, *16*, 240, https://doi.org/10.3390/jfb16070240 . . . . . . **129**

Shota Tojo, Hiromi Miyazaki, Takami Saiki, Yasuyuki Tsunoi, Shingo Nakamura and Ryuichi Azuma
Pre-Vascularized 3-Dimensional Skin Substitutes Promote Angiogenesis and Tissue Repair in a Murine Model of Refractory Skin Ulcers
Reprinted from: *J. Funct. Biomater.* **2025**, *16*, 409, https://doi.org/10.3390/jfb16110409 . . . . . . **173**

Ge Zhu, Su Wang, Zhang Liu, Shengji Gu, Feng Chen and Wangfu Zang
Chlorogenic Acid–Strontium-Containing Dual-Functional Bioresorbable External Stent Suppresses Venous Graft Restenosis via Hippo-YAP Signaling Pathway
Reprinted from: *J. Funct. Biomater.* **2025**, *16*, 259, https://doi.org/10.3390/jfb16070259 . . . . . . **189**

Weronika Radecka, Wiktoria Nogalska and Maria Siemionow
Peripheral Nerve Protection Strategies: Recent Advances and Potential Clinical Applications
Reprinted from: *J. Funct. Biomater.* **2025**, *16*, 153, https://doi.org/10.3390/jfb16050153 . . . . . . **204**

About the Editors

Cristian Scheau

Cristian Scheau is an Associate Professor in the Department of Medical Physiology at Carol Davila University of Medicine and Pharmacy, a primary physician in Radiology and Medical Imaging, and the coordinator of scientific research activity at Foișor Clinical Hospital of Orthopaedics, Traumatology and Osteoarticular TB in Bucharest, with a professional profile that bridges academic physiology, multimodal imaging, and clinically oriented translational research. His main interests include neurophysiology, the interaction between inflammation and carcinogenesis, radiologic and MRI-based investigation, functional biomaterials, and imaging applications in oncology and orthopedics, while his recent projects also include the editorial and scientific coordination of volumes and Special Issues dedicated to biomaterials, epithelial cancers, and advanced imaging in knee surgery. Among his notable achievements are his habilitation in Medicine, his position in the Faculty Council of the Faculty of Medicine, his extensive editorial activity across several international journals, and his strong publication record.

Andreea Cristiana Didilescu

Andreea Cristiana Didilescu is Professor and Dean of the Faculty of Dentistry at Carol Davila University of Medicine and Pharmacy, Bucharest, Romania, where she also holds senior academic leadership responsibilities in scientific research. Her research interests encompass the dentin–pulp complex, oral microbiology, biomaterials, salivary research, and the relationship between oral and systemic disease—areas well represented in her extensive publication profile, which includes original research articles, reviews, academic manuals, and edited scientific volumes addressing topics such as salivary biomarkers, oral health after liver transplantation, and functional biomaterials.

Dr. Didilescu championed the initial accreditation of the English-language Dental Medicine program at Carol Davila University in 2022, expanding access to quality dental education internationally. The following year, she was awarded a Fulbright scholarship for research at the University at Buffalo's School of Dental Medicine. In 2024, she was appointed President of the Dental Medicine Committee for Romania's National Council for Accreditation (CNATDCU) and elected Member of Sigma Xi, The Scientific Research Honor Society. Most recently, in 2025, she and her team received the ADEE–EDSA–Henry Schein Award for their contributions to sustainable dental education and practice.

Constantin Caruntu

Constantin Caruntu is a Full Professor in the Department of Physiology at Carol Davila University of Medicine and Pharmacy and a senior dermatologist at the "N.C. Paulescu" National Institute of Diabetes, Nutrition, and Metabolic Diseases in Bucharest, Romania, where he combines academic leadership, clinical dermatology, and translational research. His scientific interests include in vivo confocal microscopy, the physiology and pathophysiology of the skin and oral mucosa, dermato-oncology, inflammatory skin diseases, neurogenic inflammation, psychodermatology, and the neuroendocrine mechanisms of cutaneous disease. He has coordinated four national research projects and contributed as a member or expert to more than ten others, with a sustained focus on modern investigative techniques for evaluating skin physiological and pathophysiological processes. Among his notable achievements are his habilitation in medicine, his role as Head of the Physiology Discipline, his membership in the CNATDCU Medicine Commission, his authorship of more than 130 Web of Science-indexed papers and over 25 books or book chapters, and his Google Scholar h-index of 52.

Journal of
Functional Biomaterials

MDPI

Editorial

From Scaffolds to Complex Systems: Functional Biomaterials in Regenerative Medicine

Cristian Scheau [1,2,*], Andreea Cristiana Didilescu [3,*] and Constantin Caruntu [1,4,*]

[1] Department of Physiology, The "Carol Davila" University of Medicine and Pharmacy, 050474 Bucharest, Romania
[2] Department of Radiology and Medical Imaging, "Foisor" Clinical Hospital of Orthopaedics, Traumatology and Osteoarticular TB, 021382 Bucharest, Romania
[3] Department of Embryology, Faculty of Dentistry, The "Carol Davila" University of Medicine and Pharmacy, 050474 Bucharest, Romania
[4] Department of Dermatology, "Prof. N.C. Paulescu" National Institute of Diabetes, Nutrition and Metabolic Diseases, 011233 Bucharest, Romania
* Correspondence: cristian.scheau@umfcd.ro (C.S.); andreea.didilescu@umfcd.ro (A.C.D.); costin.caruntu@gmail.com (C.C.)

Regenerative medicine stands at a crossroad between biology and materials science, where functional biomaterials are expected to interact with living tissues, guide repair, and restore functionality [1]. The advancements in material composition, biofabrication, and degradability all work towards the same objective: the design and implementation of regenerative systems capable of actively enhancing healing via interactions with the biological environment [2,3]. This Special Issue brings together diverse papers but with a unified concept, collectively reflecting an honorable ambition of creating biomaterials that do not simply fill defects but actively orchestrate regeneration.

Recent studies have challenged the long-standing paradigm of biomaterials as inert fillers. The systematic review by Ivanova et al. has provided a comprehensive overview of the effectiveness of traditional autografts, allografts, xenografts, and also synthetic materials in various biologic axes [4]. The authors have underlined the trade-off between biological activity and mechanical reliability, which was a recurrent theme in the papers of this Special Issue. This topic was also addressed by another team of authors, which explored biodegradable magnesium-based alloys where Mg-Nd and Mg-Zn are not only studied for their mechanical properties, but also for the kinetics of degradation and tissue responses in vivo [5]. The materials dissolution is reframed here from a limitation to a therapeutic parameter that must be very carefully adapted to synchronize the resorption of the scaffold with bone regeneration. This supports the idea that future biomaterials must be designed as temporally adaptive constructs that evolve together with the surrounding supported tissue [6]. This concept was further expanded by Yu et al. in their paper on the properties of epigallocatechin gallate to retain antioxidant activity when properly processed, thereby bridging molecular bioactivity and scalable biomaterial fabrication [7].

Beyond the topic of bulk composition, multiple authors have emphasized the important role of micro- and nano-architecture in defining tissue interactions. The electrospinning of collagen and collagen-glycosaminoglycan mats demonstrated how minute variations in the morphology of fibers and biochemical composition can influence stem cell adhesion, viability, and proliferation [8]. This reinforces the concept that cells respond not only to physical but also biochemical signals, a concept that resonates with recent work on functional biomaterials [9]. The relevance of architectural design is equally critical on the macrostructural scale, in the field of 3D bioprinting. The extensive review on soft tissue bioprinting by Timofticiuc et al. showed that the current challenge goes beyond material selection and relies on the spatial organization of multiple cell types, bioinks, and mechanical gradients within the construct [10]. The authors show that anatomical fidelity should be doubled by functional integration and that advancements in the field of bioprinting should address not only resolution, but also strategies for vascularization and innervation. These concepts align with recent papers, which have also signaled the necessity of consistent progress in this direction [11,12].

Vascularization appears as a current limiting factor in various classes of materials in different tissues. The study by Tojo et al. demonstrated that embedding vascular networks within the engineered constructs leads to improved angiogenesis and accelerates the wound-healing process [13]. The authors show that host-graft interactions are guided by the microvascular architecture and this underlines the broader theme that the capacity to recruit vasculature is indispensable for clinical translation, be it on a micro- or nano- scale, or even on larger, more complex structures such as entire limbs [14]. Vascular healing may be targeted and controlled on a cellular level by specific biomaterials; this was shown in the paper by Zhu et al. which inhibited vein graft restenosis via developing a chlorogenic acid–strontium bioresorbable stent [15].

Alongside vascularity, preservation and modulation of neural integrity are recognized as highly important factors for functional regeneration and integration [16,17]. Radecka et al. have synthesized the emerging approaches for biomaterial-based neural protection and shown how the control of the microenvironment can promote regeneration and mitigate secondary injury [18].

Overall, across all tissue applications, we are witnessing a maturation of the field of biomaterials, which leads to adaptive, transient, and biologically instructive platforms. Viewed collectively, the manuscripts in this Special Issue reflect and actively advance the coherent evolution and common theme documented in the recent literature: there is a continued pursuit of engineering more sophisticated biomaterials and integrating them into regenerative strategies, appropriate to the complexity of human biology. Advancements are rapid across all domains, and we are eagerly anticipating solid breakthroughs in near-future studies.

Author Contributions: Conceptualization, C.S., A.C.D. and C.C.; writing—original draft preparation, C.S., A.C.D. and C.C.; writing—review and editing, C.S., A.C.D. and C.C. All authors have read and agreed to the published version of the manuscript.

Funding: This research received no external funding.

Data Availability Statement: No new data were created or analyzed in this study. Data sharing is not applicable to this article.

Conflicts of Interest: The authors declare no conflicts of interest.

References

1. Jin, C.; Zhang, X.; Jin, Y.; Chien, P.N.; Heo, C.Y. Acellular Extracellular Matrix Scaffolds in Regenerative Medicine: Advances in Decellularization and Clinical Applications. *J. Funct. Biomater.* **2025**, *16*, 383. [CrossRef] [PubMed]
2. Li, M.; Zhou, S.; Yu, Q.; Wang, C.; Chen, H.; Ma, Y.; Fan, H.; Ni, T.; Lu, M.; Yao, M. A sprayable TQ/Ce6@SAB/F-gel for accelerating wound healing via hypoxia-tolerant photodynamic therapy and immune-metabolic pathway. *Biomaterials* **2026**, *325*, 123602. [CrossRef] [PubMed]

https://doi.org/10.3390/jfb17010051

3. Petrova, V.A.; Dubashynskaya, N.V.; Zhuravskii, S.G.; Poshina, D.N.; Golovkin, A.S.; Mishanin, A.I.; Gofman, I.V.; Ivan'kova, E.M.; Naumenko, M.Y.; Yukina, G.Y.; et al. Design, Characterization, and Enhanced Performance of Electrospun Chitosan-Based Nanocomposites Reinforced with Halloysite Nanotubes and Cerium Oxide Nanoparticles for Wound Healing Applications. *Int. J. Mol. Sci.* **2025**, *26*, 10520. [CrossRef] [PubMed]
4. Ivanova, N.; Ivanov, S.; Peev, S.; Dikova, T. Types of Bone Substitutes and Their Application in Regenerative Medicine: A Systematic Review. *J. Funct. Biomater.* **2025**, *16*, 341. [CrossRef] [PubMed]
5. Manescu, V.; Antoniac, A.; Moraru, M.C.; Antoniac, I.; Cotrut, C.M.; Gradinaru, S.; Dreanca, A.I.; Sevastre, B.; Pop, R.; Tabaran, F.A.; et al. Suitability of Mg-Nd and Mg-Zn Alloys to Obtain Biodegradable Structures for Bone Defects. *J. Funct. Biomater.* **2025**, *16*, 423. [CrossRef] [PubMed]
6. Kumar, R.; Sporn, K.; Prabhakar, P.; Paladugu, P.; Khanna, A.; Ngo, A.; Gowda, C.; Waisberg, E.; Jagadeesan, R.; Zaman, N.; et al. Next-Generation Biomaterials for Load-Bearing Tissue Interfaces: Sensor-Integrated Scaffolds and Mechanoadaptive Constructs for Skeletal Regeneration. *J. Funct. Biomater.* **2025**, *16*, 232. [CrossRef] [PubMed]
7. Yu, L.; Deng, Z.; Nakagawa, M.; Zheng, S.; Jo, J.-i.; Tanaka, T.; Honda, Y. Physicochemical and Antioxidant Alterations of Modified and Free Epigallocatechin Gallate Under Thermal Treatment in Air and Vacuum. *J. Funct. Biomater.* **2026**, *17*, 18. [CrossRef]
8. Vladkova, T.G.; Gospodinova, D.N.; Dineff, P.D.; Keremidarska-Markova, M.; Hristova-Panusheva, K.; Krasteva, N. Electrospinning of Bovine Split Hide Collagen and Collagen/Glycosaminoglycan for a Study of Stem Cell Adhesion and Proliferation on the Mats: Influence of Composition and Structural Morphology. *J. Funct. Biomater.* **2025**, *16*, 219. [CrossRef] [PubMed]
9. Kuznetsova, A.V.; Popova, O.P.; Danilova, T.I.; Latyshev, A.V.; Yanushevich, O.O.; Ivanov, A.A. Effects of ECM Components on Periodontal Ligament Stem Cell Differentiation Under Conditions of Disruption of Wnt and TGF-β Signaling Pathways. *J. Funct. Biomater.* **2025**, *16*, 94. [CrossRef] [PubMed]
10. Timofticiuc, I.-A.; Caruntu, A.; Dragosloveanu, C.D.M.; Scheau, A.-E.; Badarau, I.A.; Periferakis, A.; Dragosloveanu, S.; Didilescu, A.C.; Caruntu, C.; Scheau, C. Head and Neck 3D Bioprinting—A Review on Recent Advancements in Soft Tissue 3D Bioprinting and Medical Applications. *J. Funct. Biomater.* **2025**, *16*, 240. [CrossRef] [PubMed]
11. Elango, J.; Zamora-Ledezma, C. Rheological, Structural, and Biological Trade-Offs in Bioink Design for 3D Bioprinting. *Gels* **2025**, *11*, 659. [CrossRef] [PubMed]
12. Kumar, K.; Nain, A. Emerging 4D Fabrication of Tubular Structures and Clinical Challenges: Critical Perspective. *ACS Mater. Au* **2025**, *5*, 886–895. [CrossRef] [PubMed]
13. Tojo, S.; Miyazaki, H.; Saiki, T.; Tsunoi, Y.; Nakamura, S.; Azuma, R. Pre-Vascularized 3-Dimensional Skin Substitutes Promote Angiogenesis and Tissue Repair in a Murine Model of Refractory Skin Ulcers. *J. Funct. Biomater.* **2025**, *16*, 409. [CrossRef] [PubMed]
14. Timofticiuc, I.-A.; Dragosloveanu, S.; Caruntu, A.; Scheau, A.-E.; Badarau, I.A.; Garofil, N.D.; Didilescu, A.C.; Caruntu, C.; Scheau, C. 3D Bioprinting in Limb Salvage Surgery. *J. Funct. Biomater.* **2024**, *15*, 383. [CrossRef] [PubMed]
15. Zhu, G.; Wang, S.; Liu, Z.; Gu, S.; Chen, F.; Zang, W. Chlorogenic Acid–Strontium-Containing Dual-Functional Bioresorbable External Stent Suppresses Venous Graft Restenosis via Hippo-YAP Signaling Pathway. *J. Funct. Biomater.* **2025**, *16*, 259. [CrossRef] [PubMed]
16. Chu, T.; Xiao, Y.; Lai, H.; Shi, L.; Cheng, Y.; Sun, J.; Pang, Z.; Cheng, S.; Zhao, K.; Gao, Z.; et al. Highly Conductive, Adhesive and Biocompatible Hydrogel for Closed-Loop Neuromodulation in Nerve Regeneration. *ACS Nano* **2025**, *19*, 18729–18746. [CrossRef] [PubMed]
17. Cheng, R.; Liu, Z.; Li, M.; Shen, Z.; Wang, X.; Zhang, J.; Sang, S. Peripheral nerve regeneration with 3D printed bionic double-network conductive scaffold based on GelMA/chitosan/polypyrrole. *Int. J. Biol. Macromol.* **2025**, *304*, 140746. [CrossRef] [PubMed]
18. Radecka, W.; Nogalska, W.; Siemionow, M. Peripheral Nerve Protection Strategies: Recent Advances and Potential Clinical Applications. *J. Funct. Biomater.* **2025**, *16*, 153. [CrossRef] [PubMed]

Disclaimer/Publisher's Note: The statements, opinions and data contained in all publications are solely those of the individual author(s) and contributor(s) and not of MDPI and/or the editor(s). MDPI and/or the editor(s) disclaim responsibility for any injury to people or property resulting from any ideas, methods, instructions or products referred to in the content.

 https://doi.org/10.3390/jfb17010051

Journal of
Functional Biomaterials

Systematic Review

Types of Bone Substitutes and Their Application in Regenerative Medicine: A Systematic Review

Nikoleta Ivanova [1,*], Stoyan Ivanov [2], Stefan Peev [3] and Tsanka Dikova [4,*]

1 Department of Biology, Faculty of Pharmacy, Medical University of Varna, 9002 Varna, Bulgaria
2 Department of Orthopedics and Traumatology, Faculty of Medicine, Medical University of Varna, 9002 Varna, Bulgaria; st.ivanov@mu-varna.bg
3 Department of Periodontology and Dental Implantology, Faculty of Dental Medicine, Medical University of Varna, 9002 Varna, Bulgaria; stefan.peev@mu-varna.bg
4 Department of Dental Material Science and Prosthetic Dental Medicine, Faculty of Dental Medicine, Medical University of Varna, 9002 Varna, Bulgaria
* Correspondence: nikoleta.ivanova@mu-varna.bg (N.I.); tsanka.dikova@mu-varna.bg (T.D.)

Abstract: Background: The growing demand for effective methods of bone tissue regeneration highlights the relevance of studying modern bone substitutes and their applications in regenerative medicine. The aim of this work was to conduct a comprehensive analysis of the biological, mechanical, and clinical characteristics of various types of bone substitutes to determine their potential in regenerative medicine. Methods: The study was performed as a systematic literature review in accordance with PRISMA guidelines, analyzing 68 high-quality scientific sources from 2019 to May 2025, using the PubMed, Scopus, Web of Science, and Google Scholar databases. Results: It was established that autogenous grafts exhibit the highest osteogenic properties due to the presence of growth factors BMP-2, BMP-7, and concentrated growth factors; however, their use is limited by donor site morbidity in 20–30% of patients and the requirement to treat 6% of fractures complicated by non-union. Allogeneic and xenogeneic substitutes provide structural support for large defects but require intensive processing in accordance with European Directives 2004/23/EC and 2006/86/EC to minimize the risk of infection transmission. Synthetic substitutes based on calcium phosphate ceramics with pore sizes ranging from 23 to 210 micrometres demonstrate excellent biocompatibility and controlled degradation, with β-tricalcium phosphate exhibiting optimal characteristics for long-term applications compared to calcium sulphate. Conclusions: The findings of the study highlight the necessity of a personalized approach in selecting bone substitutes, considering the specific requirements of medical specialities, and support the development of hybrid biomaterials to combine structural strength with biological activity.

Keywords: bone tissue; stem cells; clinical efficacy; bone defects; remodelling

1. Introduction

The field of regenerative medicine increasingly requires effective bone regeneration methods, particularly for orthopedic, maxillofacial, and dental procedures. Bone defects resulting from trauma, disease, or surgery have been identified as a significant clinical problem. While minor defects may heal spontaneously, more serious defects necessitate treatment through transplantation or the use of bone substitutes. Traditional transplantation methods, as noted by M.P. Ferraz, include autografts, allografts, and xenografts, which remain routinely used but are associated with complications such as donor site morbidity, immune rejection, and variable integration rates [1]. In response, synthetic and bioengineered bone substitutes have become a central focus in regenerative medicine.

Despite significant advances in tissue engineering and biomaterials, no single, universally accepted optimal bone substitute exists. This is due to the inability to combine, within a single material, all the necessary properties: the biological activity of autografts, the unlimited availability of synthetic materials, controlled degradation, the absence of immune reactions, and adaptability to diverse clinical requirements—from high mechanical load-bearing to rapid resorption. According to M. Heitzer et al., the applicability of these substitutes is determined by their osteoconductivity, osteoinductivity, and mechanical strength, and most importantly, their biocompatibility [2]. Scaffolds offer considerable potential when combined with stem cells, bioactive coatings, and 3D bioprinting; however, challenges remain in optimizing resorption rates, mechanical properties, and clinical relevance.

Overall, H. Fattahian et al. in their review, described the main types of bone substitutes, along with their advantages and limitations [3]. They emphasized that traditional methods such as autografts and allografts possess several shortcomings, thereby prompting the active development of tissue engineering approaches that integrate biomaterials, growth factors, stem cells, and even 3D printing technologies to create ideal bone substitutes. The work of U. Dahiya et al. highlighted the limitations of autologous grafts and examined synthetic substitutes, bioceramics, and polymers with osteoconductive, osteoinductive, and biocompatible properties as promising materials for bone regeneration [4].

In turn, B. Manzini et al. emphasized the importance of biofactors and stem cells in the regeneration process, highlighting the need for further research to develop biomimetic bone substitutes capable of effectively interacting with the body and supporting tissue regeneration [5]. Research represented by the work of M. Miletić et al. which examined the combination of artificial β-tricalcium phosphate with periodontal ligament mesenchymal stem cells, noting a significant increase in regenerative potential compared to cell-free material [6]. This underscores the importance of combined approaches involving cells and substitutes.

Another study by S. Mukhlis et al. highlights the role of pericytes—cells that promote angiogenesis and vascular regeneration in bone tissue [7]. The authors emphasize the potential of using these cells in conjunction with biomaterials to enhance regenerative outcomes, while also noting the need for further investigation into their mechanisms of action and optimal treatment regimens. Other studies, including those by F. Dőri et al. have assessed the combination of biomaterials with various methods for stimulating tissue regeneration, such as the use of plasma, which enhances the integration of membranes and bone substitutes [8]. However, they acknowledge that the correlation between histological changes and clinical outcomes still requires further clarification. At the same time, Q. Zhang addressed the issue of bone cell ageing and osteoporosis, which limit bone tissue regeneration [9]. This study considered the potential use of substitutes containing therapeutic agents to simultaneously address these pathologies, thereby emphasizing the complex and multifactorial nature of the problem.

Overall, current research indicates that successful bone regeneration requires an integrated approach combining innovative biomaterials, cell technologies, growth factors, and vascular growth stimulation. However, many gaps remain, particularly in the optimization of material–cell combinations, the understanding of implant–immune system interactions, and the standardization of clinical protocols to ensure long-term success. Therefore, the aim of this study was to conduct a systematic review of the use of bone substitutes in regenerative medicine, based on an analysis of scientific publications from 2019 to 2025 in accordance with the PRISMA methodology. To achieve this aim, the following research tasks were undertaken: to conduct a comparative analysis of the biological and clinical characteristics of different groups of bone substitutes (autogenous, allogeneic, xenogeneic, and synthetic), with an assessment of their osteogenic, osteoinductive, and osteoconductive potential; to evaluate the clinical and preclinical effectiveness of bioengineered and synthetic bone graft substitutes, with particular attention to their use in bone tissue regeneration and integration into host tissues; and to identify promising directions for development and current challenges in bone substitute technologies, including their limitations and areas requiring further research.

2. Materials and Methods

This study is a systematic review of the literature, conducted in accordance with the PRISMA (Preferred Reporting Items for Systematic Reviews and Meta-Analyses) guidelines to ensure methodological rigour and transparency, registered in PROSPERO under registration number CRD420251117103. The time frame of the study encompassed publications from January 2019 to May 2025, allowing for the analysis of recent advances in the field of bone substitutes and their application in regenerative medicine. As this is a systematic review of published scientific sources and does not involve clinical interventions, ethical approval from the institutional ethics committee was not required.

Data collection was performed through a comprehensive search of the electronic scientometric databases PubMed, Scopus, Web of Science, and Google Scholar. The first stage of the search strategy identified 312 publications in PubMed, 287 sources in Scopus, 254 articles in Web of Science, and 198 additional sources in Google Scholar, resulting in a total of 1051 publications. A combination of keywords and Medical Subject Headings (MeSH) terms was used to ensure maximum coverage of relevant literature, including: "bone substitutes" OR "bone grafts" AND "regenerative medicine," "autogenous bone grafts" OR "autologous bone transplants," "allogeneic bone substitutes" OR "allografts," "xenogeneic bone substitutes" OR "xenografts," "synthetic bone substitutes" OR "bioceramics" OR "calcium phosphate ceramics," "osteoconductivity" OR "osteoinductivity" OR "osteogenesis," "bone tissue engineering" AND "scaffold materials." Search queries were formulated using Boolean operators to optimize search results and exclude irrelevant publications.

Inclusion criteria comprised original studies, systematic reviews, and meta-analyses published in peer-reviewed journals in English within the specified period. The analysis included papers addressing the biological properties, mechanical characteristics, fabrication methods, and clinical applications of autogenous, allogeneic, xenogeneic, and synthetic bone substitutes. Particular attention was paid to studies evaluating the osteoinductive, osteoconductive, and osteogenic properties of materials, as well as their biocompatibility and degradation rate both in vivo and in vitro. The exclusion criteria included publications dated before 2019 (excluding European Directives and documents from the European Medicines Agency), articles in languages other than English, editorial articles, commentaries, and letters to the editor. Studies not directly addressing bone substitutes or their applications in regenerative medicine, papers with insufficient methodological quality, and publications without access to the full text were also excluded. The methodological quality of the included studies was assessed based on the following criteria: a clear description of the methodology, adequate sample size, use of appropriate control groups, statistical analysis of results, and transparency in data presentation. For clinical studies, additional consideration was given to the level of evidence according to the hierarchy of medical evidence and the presence of randomisation.

The selection and analysis of publications were conducted by two independent reviewers, with subsequent discussion of discrepancies to reach a consensus. In cases of disagreement, a third expert was consulted for the final decision.

Systematization and management of bibliographic data were carried out using specialized software: Mendeley Desktop (version 1.19.8) and Zotero (version 6.0.30), which enabled effective organization of the selection and categorisation process for publications. Duplicate records were identified and removed using Mendeley's built-in deduplication algorithm, followed by manual verification. Bibliographic data were exported from databases in RIS and BibTeX formats to ensure compatibility with reference management programmes. Full-text analysis was performed using Adobe Acrobat Reader DC (version 2024.001.20643), employing annotation and highlighting functions to mark key text fragments. Microsoft Excel 365 (version 16.0.17328.20162) was used to construct a data matrix and conduct preliminary analysis, systematizing information on study type, year of publication, type of bone substitute, research methodology, and main results. Visualization of the literature selection process, following PRISMA methodology, was performed using Microsoft Visio 2021 (version 16.0.14332.20447).

The literature review is organized according to the classifications presented in Figure 1.

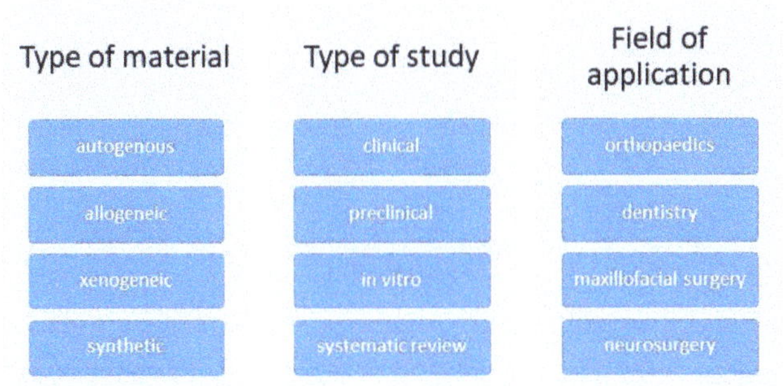

Figure 1. Type of material, type of study, and field of application of bone substitutes included in the review article.

The analysis is conducted across four primary categories of bone substitutes—autogenous, allogeneic, xenogeneic, and synthetic—assessing their biological properties, mechanical characteristics, and clinical efficacy. The studies from which the bone substitute groups are analyzed are clinical, preclinical, in vitro, and systematic reviews. The fields of application in which they are observed are orthopedics, dentistry, maxillofacial surgery, and neurosurgery. In addition, the features of each material type are compared, their indications for use in various medical specialties are systematized, and key areas requiring further investigation are identified. The synthesis of data allows for the identification of major trends in the development of new materials, including the use of 3D printing, the incorporation of growth factors, and the application of stem cells to enhance regenerative potential.

3. Results and Discussion

The selection process was conducted in three consecutive stages in accordance with the PRISMA methodology, as illustrated in the flowchart (Figure 2). As a result, 68 high-quality scientific sources were included in the final systematic analysis.

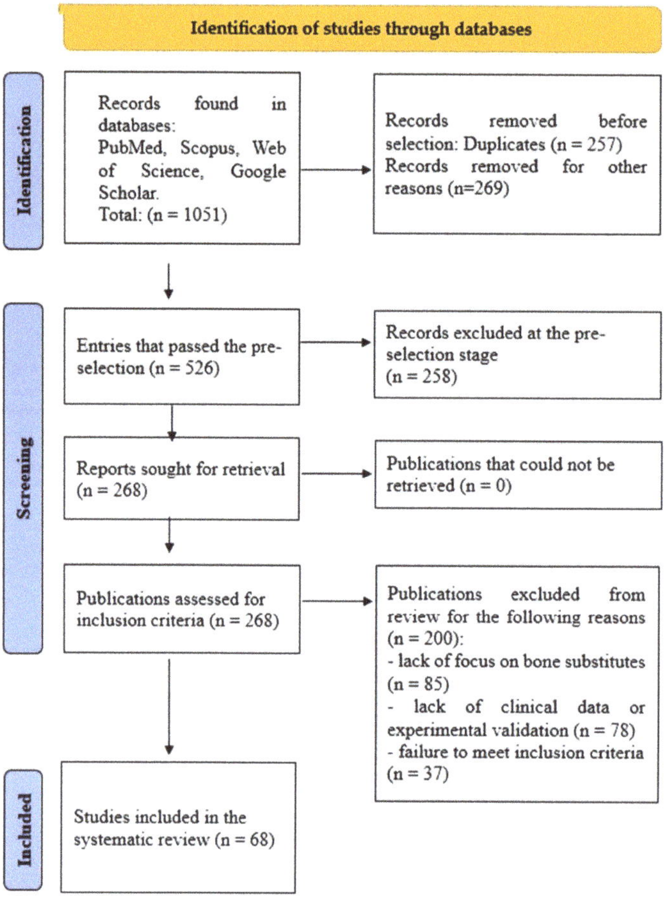

Figure 2. PRISMA flow diagram of the study selection process.

Table 1 presents the characteristics of the studies included in this systematic review.

Table 1. All articles included in the present study.

Reference	Aim of the Study	Type of Material	Field of Application	Type of Study	Outcomes of the Bone Substitute
Ferraz, M.P. [1]	To analyze and compare bone grafting materials used in dental and maxillofacial surgery, focusing on their properties, benefits, and clinical application challenges	Autologous bone, allografts, xenografts, synthetic or xenogeneic scaffolds with growth factors to enhance their osteogenic properties	Dentistry, oral and maxillofacial surgeries	In vivo and clinical research data	Osteogenic properties

Table 1. *Cont.*

Reference	Aim of the Study	Type of Material	Field of Application	Type of Study	Outcomes of the Bone Substitute
Heitzer, M. et al. [2]	To investigate the osteogenic potential of dental pulp stem cells (DPSCs) on three types of bone graft substitutes—alloplastic (BCP), allogeneic (FDBA), and xenogeneic (DBBM)—under in vitro conditions	Alloplastic, allogeneic, and xenogeneic bone graft substitutes	Dentistry, oral and maxillofacial surgeries	In vitro	Osteogenic (bone-forming) potential
Fattahian, H. et al. [3]	To review current materials and methods for the repair and regeneration of bone defects, as well as new tissue engineering approaches to improve the treatment of bone injuries	Autografts, allografts, and xenografts	Orthopedics	Systematic reviews	Biocompatibility, osteoconductivity, osteoinductivity, mechanical strength, resorbability, availability, combinability with biofactors
Dahiya, U. et al. [4]	To analyze several bone substitutes, such as synthetics, bioceramics, and polymers, as a substitute for autologous or allogeneic bone for the treatment of bone defects. These bone substitutes should be biocompatible, bioresorbable, osteoconductive, osteoinductive, and promote new bone growth	Synthetics bone substitutes	Orthopedics, dentistry, oral and maxillofacial surgeries	Review	Biocompatibility, bioresorbability, osteoconductivity, osteoinductivity, support of new bone ingrowth
Manzini, B. et al. [5]	To provide an overview of the bone tissue, including the role of stem cells and some of the bioactive molecules associated with these processes	Synthetic bone substitutes, hydroxyapatite (HA), β-tricalcium phosphate (β-TCP), biphasic calcium phosphate (BCP), combinations of HA and TCP	Orthopedics, dentistry, oral and maxillofacial surgeries	Review	Stem cells and bioactive factors are key for regeneration
Miletić, M. et al. [6]	This in vivo study demonstrates for the superior bone regenerative capacity of CAP-pretreated β-TCP seeded with PDLSCs, highlighting the therapeutic potential of this combined approach in osteoregeneration	Beta-Tricalcium Phosphate	Orthopedics, dentistry, oral and maxillofacial surgeries	In vivo	Periodontal ligament stem cells with beta-tricalcium phosphate treated by cold atmospheric plasma significantly enhanced bone regeneration

Table 1. *Cont.*

Reference	Aim of the Study	Type of Material	Field of Application	Type of Study	Outcomes of the Bone Substitute
Mukhlis, S. et al. [7]	To investigate the therapeutic potential of adipose-derived pericytes, in combination with biomaterials and growth factors, for enhancing bone tissue regeneration	Injectable hydrogels and polymer (gelatin, alginate, collagen, chitosan, poly-L-lactic acid, hyaluronic acid, fibrin, heparin, and polyethylene glycol) scaffolds with pericytes	Dentistry, oral and maxillofacial surgeries	In vivo	Preclinical studies show that pericytes combined with scaffolds or DBM enhance bone healing
Dőri, F. et al. [8]	To evaluate the histological findings of retrieved ePTFE membranes using PRP and bone substitutes, the effect of PRP on graft materials, and the correlation of the findings with the clinical outcomes	PRP + Beta-Tricalcium Phosphate + non-resorbable membrane (GTR)	Dentistry, oral and maxillofacial surgeries	In vivo	Application of β-TCP and PRP may enhance membrane integration and periodontal healing
Zhang, Q. et al. [9]	To analyze tissue engineering and regenerative therapies targeting cell senescence in bone and cartilage, with a focus on treating osteoporosis and age-related degenerative conditions through the use of biomaterials, growth factors, and cell-based strategies	Bone grafts with antiosteoporotic growth factors	Orthopedics	Review	Bone grafts combined with osteoporosis therapies—including scaffolds releasing anti-osteoporotic drugs or implants with therapeutic surface modifications—can improve low bone density and enhance impaired bone regeneration
Rupp, M. et al. [10]	To evaluate the overall use of bone graft substitutes, autografts and allografts, of different types of bone graft substitutes (calcium sulphate, calcium phosphate, calcium phosphate ceramics or polymethyl methacrylate) and of different bone grafts (cancellous vs. cortical), and the use of antibiotic-loading of bone graft substitutes in orthopedic surgery in Germany	Autografts and allografts, synthetic bone substitutes (calcium sulphate, calcium phosphate, calcium phosphate ceramics or polymethyl methacrylate)	Orthopedics	In vivo	Increasing use of bone graft substitutes and antibiotic-loaded bone graft substitutes

Table 1. *Cont.*

Reference	Aim of the Study	Type of Material	Field of Application	Type of Study	Outcomes of the Bone Substitute
Laubach, M. et al. [11]	To review and evaluate current methods for the treatment of bone defects, with a particular focus on the potential and challenges of using personalized 3D-printed implants for long bone regeneration	3D-printed scaffolds	Orthopedics	Review	3D-printed scaffolds provide osteoconductive surfaces, mechanical support, and containment for bone grafts, enhancing bone ingrowth
Santoro, A. et al. [12]	To analyze current natural and synthetic bone substitutes used in orthopedic surgery, with a focus on innovations and challenges in the treatment of complex bone defects	Natural grafts—comprising autologous, allogeneic, and xenogeneic materials; synthetic alternatives, including biodegradable and non-biodegradable biomaterials	Regenerative medicine and orthopedic surgery	Review	Natural grafts (autologous, allogeneic, xenogeneic) offer biological advantages, whereas synthetic substitutes (biodegradable and non-biodegradable) provide structural versatility and lower immunogenicity
Tournier, P. et al. [13]	To evaluate the regenerative potential of a novel partially demineralized allogeneic bone paste as an alternative to traditional allogeneic grafts, focusing on enhanced bone healing and user-friendly application in challenging skeletal defect areas	Demineralized allogenic bone graft in the form of a paste	Orthopedic surgery	In vivo	The bone paste supported bone healing in guided bone regeneration and critical-size defect models
Parikh, S. et al. [14]	To present the ethical and clinical approach in a case of triple bone marrow donation from a single pediatric donor to three HLA-identical siblings with the same primary immunodeficiency, aiming to guide future decision-making in similar cases	Allogeneic bone grafts	Regenerative medicine	In vivo	Multiple harvests from pediatric donors can be performed safely with careful clinical and ethical oversight

Table 1. *Cont.*

Reference	Aim of the Study	Type of Material	Field of Application	Type of Study	Outcomes of the Bone Substitute
Sharifi, M. et al. [15]	To explore the sources, advantages, challenges, and techniques of acellularization and recellularization in allogeneic and xenogeneic bone grafts, aiming to advance future bone defect treatment strategies and support product commercialization	Allogeneic and xenogeneic bone grafts	Regenerative medicine	Review	Allogeneic and xenogeneic acellularized bone grafts mimic native bone structure and support osteoconduction and osteoinduction
Ana, I. [16]	To review the biological aspects, advantages, and disadvantages of various bone graft substitutes used in dental implantology, with a focus on calcium phosphate-based materials	Calcium phosphate-based bone substitutes	Dentistry, oral and maxillofacial surgeries	Review	Advantages of calcium phosphate-based bone substitutes in dentistry
Janjua, O.S. et al. [17]	To investigate the potential and applications of autogenous tooth grafts for regenerating maxillary and mandibular bone defects, including sinus and ridge augmentations and socket preservation before implant placement	Autogenous tooth graft	Dentistry, oral and maxillofacial surgeries	Review	Autogenous tooth bone grafts exhibit osteoconductivity and osteoinductivity, similar to natural bone
Malik, S. et al. [18]	To evaluate the efficacy of adjuvant therapy with Concentrated Growth Factor (CGF) along with xenograft and autograft in mandible fractures, in terms of bone density gain and healing over the period of 6 months	Xenograft and autograft with Concentrated Growth Factor	Dentistry, oral and maxillofacial surgeries	In vivo	CGF enhances bone regeneration with both autograft and xenograft
Black, C. et al. [19]	To evaluate the effectiveness of Laponite™ nanoclay gel as a carrier for localized BMP-2 delivery in bone regeneration, using a relevant large animal preclinical model with femoral condyle defects in aged sheep	Synthetics Smectite nanoclay gel (Laponite™) combined with absorbable collagen sponge and BMP-2 (Bone Morphogenetic Protein-2)	Bone tissue regeneration	In vivo and clinical research data	Autograft showed superior bone formation, nanoclay gels exhibited excellent biocompatibility and potential for delivering bone morphogenetic protein-2 locally

Table 1. *Cont.*

Reference	Aim of the Study	Type of Material	Field of Application	Type of Study	Outcomes of the Bone Substitute
Jin, Y. et al. [20]	To investigate the effectiveness of demineralized bone matrix (DBM) as a carrier for Escherichia coli-derived recombinant human BMP-2 (ErhBMP-2) in bone regeneration	Autograft, demineralized bone matrix (DBM) as a bone graft substitute and growth factor carrier	Orthopedic surgery, bone tissue regeneration	In vivo	DBM as the carrier showed significantly higher bone volume and bone thickness than the groups with HA as the carrier
Tasdemir, U. et al. [21]	To investigate the effect of combining autogenous bone graft with recombinant human vascular endothelial growth factor (rhVEGF) on bone regeneration in a rat mandibular defect model	Autogenous bone graft, gelatin sponge plus rhVEGF, autogenous bone graft plus rhVEGF	Dentistry, oral and maxillofacial surgeries	In vivo	Combining autogenous bone grafts with rhVEGF could potentially improve neovascularization and enhance bone regeneration.
Taşdemir, U. et al. [22]	To evaluate all the tooth layers mixed with simvastatin without any demineralization process effect on bone formation	Autogenous tooth bone grafts (ATGM) with simvastatin, xenogenic bone graft	Dentistry, oral and maxillofacial surgeries	In vivo	Autogenous mineralized tooth bone graft should be mixed with simvastatin for bone regeneration
Kunze, K. et al. [23]	To perform a meta-analysis of RCTs evaluating donor site morbidity after bone-patellar tendon-bone (BTB), hamstring tendon (HT) and quadriceps tendon (QT) autograft harvest for anterior cruciate ligament reconstruction (ACLR)	BTB autograft	Orthopedic surgery	Systematic reviews	Autograft selection should be personalized through considering differential rates of donor-site morbidity
Attia, A. et al. [24]	To evaluate the safety and donor site morbidity of autologous cancellous bone grafts harvested from the distal lower extremity (calcaneus, proximal, and distal tibia) through a meta-analysis of published research	Autologous cancellous bone grafts	Orthopedic surgery	Systematic reviews	Calcaneal, distal tibial, and proximal tibial bone autografts are safe with a low rate of overall and major complications
Khalid, M. et al. [25]	To assess the donor site morbidity in patients having anterior cruciate ligament reconstruction (ACLR) using peroneus longus tendon (PLT) autograft	Autograft	Orthopedic surgery	In vivo	ACLR using the PLT autograft resulted in a good functional outcome, smooth rehabilitation with an early return to sports, and minimal complications at the donor site

Table 1. *Cont.*

Reference	Aim of the Study	Type of Material	Field of Application	Type of Study	Outcomes of the Bone Substitute
Sun, H. et al. [26]	This paper comprehensively explores the composition, mechanisms underlying osteoinductivity, preparation methods, and clinical applications of ADM with the aim of establishing a fundamental reference for future studies on this subject	Autogenous bone	Dentistry, oral and maxillofacial surgeries	Review	ADM has a low rejection rate, possesses osteoinductive properties, and is characterized by easy preparation
Liu, J. et al. [27]	To explore the potential effect of three allogenic bone substitute configurations on the viability, adhesion, and spreading of osteoblasts in vitro	Allogenic bone substitute	Orthopedic	In vitro	Bone powder and bone granule promote cell adhesion and spreading compared to bone fibre group
European Parliament and Council [28]	Setting standards of quality and safety for the donation, procurement, testing, processing, preservation, storage and distribution of human tissues and cells	Allogeneic bone tissue	Orthopedics, dentistry, oral and maxillofacial surgeries, neurosurgery	In vivo	Not applicapable
European Commission [29]	To lay down technical requirements for the coding, processing, preservation, storage and distribution of human tissues and cells	Allogeneic bone tissue	Orthopedics, dentistry, oral and maxillofacial surgeries, neurosurgery	In vivo	Not applicapable
Ilyas, I. et al. [30]	To present the 10-year experience of the bone bank at KFSH&RC, detailing its structure, donor/recipient protocols, graft processing and safety systems, and the clinical use of allografts in surgeries such as joint revision, spine, and tumour procedures	Allografts	Orthopedics, neurosurgery	Clinical, preclinical	Administrative structure, donor and recipient testing protocols, allograft retrieval, processing procedures, and safety arrangements
Regmi, A. et al. [31]	To present the establishment and management of an orthopedic bone bank in India, focusing on the legal, medical, and organizational aspects of using allogenic bone for treating bone defects	Allografts	Orthopedics	Review	Allogenic bone grafts, supported by bone banking systems, provide a safe alternative to autografts, though they require strict legal, medical, and organizational regulation

Table 1. *Cont.*

Reference	Aim of the Study	Type of Material	Field of Application	Type of Study	Outcomes of the Bone Substitute
Asamoto, T. et al. [32]	To analyze the 15-year trends in the supply and use of bone allografts in a regional bone bank in Japan and to assess the effectiveness of the donor screening process	Bone allografts	Orthopedics, dentistry, oral and maxillofacial surgeries, neurosurgery	Preclinical	The study highlights the importance of thorough screening processes for allograft donors to minimize the number of discarded grafts
American Association of Tissue Banks. AATB Standards Rebuild [33]	These regulations and practices underpin modern bone grafting procedures, aiming to optimize clinical outcomes while minimizing associated risks	Allografts	Orthopedics, dentistry, oral and maxillofacial surgeries, neurosurgery	Preclinical	Standards for Tissue Banking
Um, I. et al. [34]	To summarize preclinical studies on the osteoinductive potential and antigenicity of allogeneic demineralized dentin matrix (Allo-DDM), in order to evaluate its viability as a reliable bone graft substitute, particularly at extraskeletal sites	Allogeneic demineral-ized dentin matrix	Dentistry, oral and maxillofacial surgeries	Review	Allo-DDM showed great potential for osteoinductivity in extraskeletal sites with low antigenicity
Waletzko-Hellwig, J. et al. [35]	To evaluate high hydrostatic pressure (HHP) as a gentle sterilization method for allogenic bone substitutes by assessing its effects on the mechanical properties of treated bone granules and trabecular bone, in comparison to untreated samples	Allogenic bone substitutes	Orthopedics	Preclinical	HHP treatment is suitable alternative to current processing techniques for allogenic bone substitutes since with no negative effects on mechanical properties occurred
Shao, A. et al. [36]	To investigate whether GGTA1/iGb3S double knockout (G/i DKO) mice are sensitive to Gal antigen-positive xenoimplants and assess their suitability as a model for studying α-Gal-mediated immunogenicity in xenotransplantation	Xenogeneic bone grafts	Orthopedics, dentistry, oral and maxillofacial surgeries, neurosurgery	Preclinical	Evaluating the α-Gal-mediated immunogenic risk of xenogeneic grafts

Table 1. *Cont.*

Reference	Aim of the Study	Type of Material	Field of Application	Type of Study	Outcomes of the Bone Substitute
Amid, R. et al. [37]	To systematically review and evaluate the structural and chemical properties of various xenograft bone substitutes based on in vitro studies	Xenogeneic bone grafts	Orthopedics, dentistry, oral and maxillofacial surgeries, neurosurgery	Systematic reviews	Xenograft bone substitutes, mainly bovine-derived, show structural similarity and biocompatibility with human bone, but their properties (porosity, crystallinity, Ca/P ratio, and osteogenesis) vary significantly depending on preparation methods. Proper pore size and interconnectivity are crucial for effective bone regeneration
Jerbić Radetić, A.T. et al. [38]	To investigate the biological properties of a new bovine xenogeneic biomaterial enriched with magnesium alloy in a 5 mm critical-sized bone defect (CSBD) model, and to compare its osteoconductive potential with existing xenogeneic biomaterials (Cerabone®, Cerabone® + autologous bone, and OsteoBiol®) for possible use in oral implantology	Xenogeneic bone grafts	Dentistry, oral and maxillofacial surgeries	Preclinical, in vivo	Novel bovine xenogeneic biomaterial enriched with magnesium alloy in a 5 mm CSBD model, compared to Cerabone® and OsteoBiol®, the Cerabone® + Mg group showed higher bone volume, faster biodegradation, and strong osteoinductive properties
Pröhl, A. et al. [39]	To evaluate the inflammatory tissue response and bone healing capacity of a newly developed xenogeneic bone substitute material (BSM) combined with hyaluronate (HY), compared to a control BSM without HY and a sham operation group, using a rat calvaria model over 2, 8, and 16 weeks	Xenogeneic bone substitute	Orthopedics, dentistry, oral and maxillofacial surgeries, neurosurgery	Preclinical, clinical	Both materials proved biocompatible and supported gradual osteoconductive bone regeneration

Table 1. *Cont.*

Reference	Aim of the Study	Type of Material	Field of Application	Type of Study	Outcomes of the Bone Substitute
Denner, J. [40]	To argue that although porcine cytomegalovirus (PCMV) is an important safety concern in xenotransplantation, there are effective methods for its elimination, and therefore, it does not represent an insurmountable obstacle to the advancement of this medical technology	Xenogeneic bone substitute	Orthopedics, dentistry, oral and maxillofacial surgeries, neurosurgery	Preclinical, clinical	To apply methods for eliminating viruses in xenotransplantations
Groenendaal, H. et al. [41]	To develop a risk-based framework for identifying and evaluating porcine microorganisms and parasites (MP) that could pose health risks in xenotransplantation.	Xenogeneic bone substitute	Orthopedics, dentistry, oral and maxillofacial surgeries, neurosurgery	Preclinical, clinical	Provides a structured way to categorize and mitigate the risks posed by porcine microorganisms in xenotransplantation, highlighting that most risks can be addressed with strict biosecurity and monitoring
U.S. Food and Drug Administration [42]	To provide guidance to the industry on the safe development and use of xenotransplantation products in humans	Xenotransplantation products in humans	Orthopedics, dentistry, oral and maxillofacial surgeries, neurosurgery	Preclinical	Guidelines
European Medicines Agency [43]	To provide scientific and regulatory guidance on the quality, safety, and efficacy requirements for xenogeneic cell-based medicinal products intended for human use	Xenogeneic cell-based medicinal products	Orthopedics, dentistry, oral and maxillofacial surgeries, neurosurgery	Preclinical	Guidelines
Hawthorne, W.J. et al. [44]	To support the safe and ethically responsible development of xenotransplantation by promoting international cooperation and coordination	Xenografts	Orthopedics, dentistry, oral and maxillofacial surgeries, neurosurgery	Preclinical	Confirmed the significant progress and the need for global collaboration between science, technology and regulations in the field of xenotransplantation to harmonize practices and rules in different countries

Table 1. *Cont.*

Reference	Aim of the Study	Type of Material	Field of Application	Type of Study	Outcomes of the Bone Substitute
Sánchez-Labrador, L. et al. [45]	To evaluate the feasibility of xenogeneic bone blocks for ridge augmentation compared with autogenous blocks by analyzing block survival rates, block resorption, subsequent implant survival rate, post-surgical complications, and histomorphometric findings	Xenogeneic bone blocks, autogenous blocks	Orthopedics, dentistry, oral and maxillofacial surgeries, neurosurgery	Systematic reviews	Histological and histomorphometric analysis observed more bone formation and less residual bone substitute with autogenous bone blocks than xenogeneic bone blocks
Maevskaia, E. et al. [46]	To evaluate a new design of synthetic bone substitutes based on the ADMS algorithm, which provides high mechanical stability and good osteoconductivity, comparable to traditional lattice structures, and is suitable for clinical use in treating bone defects	Synthetic bone substitutes	Orthopedics, dentistry, oral and maxillofacial surgeries, neurosurgery	Preclinical, in vivo	The study found these substitutes, made from hydroxyapatite (a bone mineral), are highly osteoconductive, meaning they effectively promote the growth of new bone
Ding, Y. et al. [47]	To analyze the advantages and performance requirements of 3D-printed bone scaffolds compared to traditional ones. It reviews key 3D printing technologies used for bone scaffold fabrication, highlighting their benefits and limitations. Although 3D printing shows promise in improving recovery and scaffold performance, material-related challenges remain, and future research directions are suggested	3D-printed bone scaffolds	Orthopedics	Clinical	3D-printed bone scaffolds show clear advantages over traditional ones, offering better performance and faster recovery, though material issues remain
Li, X. et al. [48]	This paper reviews current optical 3D printing methods for tissue engineering scaffold fabrication, highlighting techniques like stereolithography, two-photon polymerization, and others	3D-printed bone scaffolds	Orthopedics, dentistry, oral and maxillofacial surgeries, neurosurgery	Preclinical	Each technique—extrusion-based printing, selective laser sintering, stereolithography, and two-photon polymerization—offers distinct advantages and limitations regarding precision, fabrication speed, material compatibility, and application suitability

Table 1. *Cont.*

Reference	Aim of the Study	Type of Material	Field of Application	Type of Study	Outcomes of the Bone Substitute
Nifant'ev, I. et al. [49]	The study develops bone substitute composites using biodegradable polyesters (PLLA, PCL) and carbonated hydroxyapatite. It shows that adding special compatibilizers improves the composites' strength, thermal stability, and overall performance, making them suitable for bone surgery and orthopedics	Synthetic bone substitutes	Bone surgery and orthopedics	Preclinical	PLLA composites with 25 wt% pCAp exhibited the best mechanical performance and thermal stability
Kang, T.Y. et al. [50]	This study enhanced the mechanical strength of bioactive glass (BAG) by incorporating varying amounts of zirconia (ZrO_2) while preserving its bioactive properties. ZrO_2-BAG showed improved toughness, slower degradation, good biocompatibility, and promoted cell growth and bone formation. In vivo tests confirmed its potential as a strong and effective bone graft substitute	Synthetic bone graft substitute	Bone surgery and orthopedics	Preclinical, In vivo	Incorporation of zirconium oxide (ZrO_2) into bioactive glasses improved their mechanical strength while preserving porosity
Ferraz, M.P. et al. [51]	This review focuses on methods to produce nanophased HA through precipitation and discusses both traditional and recent advances in HA nanoparticle applications	Synthetic bone graft	Orthopedics, dentistry, oral and maxillofacial surgeries, neurosurgery	Review	Nanophased HA obtained via precipitation methods shows strong promise for improving the bioactivity and functionality of synthetic ceramics
Cheng, L. et al. [52]	This article reviews recent advances in 3D printing for bone repair, highlighting its ability to create custom scaffolds using biomaterials and living cells. It covers key printing techniques (SLA, SLS, FDM, inkjet) and their clinical potential in regenerating large bone defects	Synthetic 3D printing for bone repair	Orthopedics, dentistry, oral and maxillofacial surgeries, neurosurgery	Review	Various biomaterials and nanomaterials have been successfully applied in 3D-printed scaffolds, while major printing methods—including SLA, SLS, FDM, and ink-jet printing—have been utilized to create constructs with structural and functional similarities to bone tissue

Table 1. *Cont.*

Reference	Aim of the Study	Type of Material	Field of Application	Type of Study	Outcomes of the Bone Substitute
Liang, D. et al. [53]	To present a simple method for fabricating porous tantalum (Ta) with controlled pore sizes and bone-like mechanical properties, demonstrating its excellent biocompatibility and effectiveness in supporting bone repair in vitro and in vivo	Synthetic bone graft	Orthopedics, dentistry, oral and maxillofacial surgeries, neurosurgery	In vitro, in vivo	The porous tantalum demonstrated excellent biocompatibility in vitro and effective bone repair in vivo in a rat femur defect model, validating it as a viable bone substitute
Kumar, A. et al. [54]	To developed biodegradable bone scaffolds using a blend of polycaprolactone (PCL)	Synthetic bone graft	Orthopedics, entistry, oral and maxillofacial surgeries, neurosurgery	Preclinic	Increasing PLGA content from 25 wt.% to 75 wt.% accelerated the biodegradation rate threefold within 2 weeks in phosphate-buffered saline
Celik, D. et al. [55]	To developed a hybrid hydroxyapatite (HA) scaffold mimicking both cortical and cancellous bone using slip casting and freeze-drying	Synthetic bone graft	Orthopedics, dentistry, oral and maxillofacial surgeries, neurosurgery	Preclinic	Hybrid scaffold demonstrated 70% open porosity, adequate mechanical strength, and structural resemblance to natural bone. Analyses confirmed its suitability for supporting vascularization, osteoinduction, and osteoconduction
Marongiu, G. et al. [56]	To evaluate the effectiveness of cements, bone substitutes, and other augmentation techniques in the surgical treatment of proximal humeral fractures	Synthetic bone graft	Orthopedics, dentistry, oral and maxillofacial surgeries, neurosurgery	Systematic review	Calcium phosphate and calcium sulphate injectable composites provided good biocompatibility, osteoconductivity, and lower mechanical failure rate when compared to non-augmented fractures

Table 1. *Cont.*

Reference	Aim of the Study	Type of Material	Field of Application	Type of Study	Outcomes of the Bone Substitute
Kang, H. et al. [57]	To fabricate and compare multichannel biphasic calcium phosphate (BCP) and β-tricalcium phosphate (TCP) scaffolds in terms of their biodegradation and bone regeneration capabilities over time	Synthetic bone graft-biphasic calcium phosphate, β-tricalcium phosphate	Orthopedics, dentistry, oral and maxillofacial surgeries, neurosurgery	Preclinic	BCP exhibited higher compressive strength, while TCP showed greater calcium and phosphorus ion release in SBF. Both scaffolds demonstrated excellent in vitro biocompatibility
Kowalewicz, K. et al. [58]	To evaluate the in vivo degradation, osseointegration, and biocompatibility of 3D powder-printed calcium magnesium phosphate cement (CMPC) scaffolds (Mg225 and Mg225d) compared to tricalcium phosphate (TCP), assessing their potential as bone substitutes	Synthetic bone graft	Orthopedics, dentistry, oral and maxillofacial surgeries, neurosurgery	In vivo	All scaffolds showed excellent biocompatibility and gradual osseointegration over 6 weeks
Heitzer, M. et al. [59]	To evaluate the osteogenic potential of dental pulp stem cells (DPSCs) and their extracellular vesicles (EVs) on different types of bone grafts (alloplastic, allogeneic, and xenogeneic), and to determine if EVs enhance DPSC proliferation and differentiation for potential use in bone regeneration therapies	Alloplastic, allogeneic, and xenogeneic bone grafts	Dentistry, oral and maxillofacial surgeries, neurosurgery	In vitro	Addition of EVs significantly enhances the biocompatibility of human pulp stem cells with xenogeneic bone grafts, leading to superior cell proliferation, attachment, and osteogenic differentiation compared to other bone grafts tested
Sallent, I. et al. [60]	To highlight successful clinical cases and key factors in the translation of innovative bone substitutes—such as biological glues, stem cell-seeded scaffolds, and gene-functionalized grafts—from research to clinical use, and to inform future strategies for effective bone graft development	Synthetic bone graft-biological glues, stem cell-seeded scaffolds, gene-functionalized grafts	Dentistry, oral and maxillofacial surgeries, neurosurgery	Review	Highlights important ideas and aspects in the field of bone grafting

Table 1. *Cont.*

Reference	Aim of the Study	Type of Material	Field of Application	Type of Study	Outcomes of the Bone Substitute
Massari, L. et al. [61]	To evaluate the safety of porous hydroxyapatite bone substitutes in trauma orthopedic surgery by analyzing adverse events in a retrospective cohort of 114 patients	Synthetic bone graft-porous hydroxyapatite bone substitutes	Orthopedics	Clinical	Porous hydroxyapatite bone substitutes appear safe for use in trauma orthopedic surgery when applied following proper biomechanical principles
Mahardawi, B. et al. [62]	To evaluate the available literature on the Allogenic Demineralized Dentin Matrix (Allo-DDM), revealing its clinical performance when used for implant place-ment procedures	Allogenic demineralized dentin matrix graft	Dentistry, oral and maxillofacial surgeries	Systematic review	Allo-DDM could be a possible alternative to other grafting materials used for bone augmentation and implant placement
Batra, C. et al. [63]	To evaluate the clinical effectiveness of combining recombinant human platelet-derived growth factor-BB (rhPDGF-BB) with xenogeneic bone substitutes in treating severe intrabony periodontal defects	Xenogeneic bone substitutes with platelet-derived growth factor-BB	Dentistry, oral and maxillofacial surgeries	Clinical	Combination of rhPDGF-BB and xenogeneic bone substitutes to be a safe and effective graft for severe periodontal intrabony defects
Zhang, S. et al. [64]	To highlight the potential of autogenous odontogenic materials as a novel and biocompatible bone substitute for jaw restoration, comparing their composition and osteogenic properties with other conventional bone grafts	Autogenous odontogenic materials	Dentistry, oral and maxillofacial surgeries	Review	Autogenous odontogenic materials represent a novel and promising class of bone substitutes, combining scaffold functionality with intrinsic osteogenic stimulation
Jung, K. et al. [65]	To demnstrate the clinical applicability and effectiveness of multichannel granular bone substitutes in treating various orthopedic conditions, including bone tumours, fractures, and defects associated with arthroplasty	Synthetic bone substitutes	Orthopedics	Clinical, in vivo	Bone substitute stabilized the bone defect without any complications, and the defect regenerated slowly during the postoperative period

Table 1. *Cont.*

Reference	Aim of the Study	Type of Material	Field of Application	Type of Study	Outcomes of the Bone Substitute
Baldwin, P. et al. [66]	To review and compare the properties, benefits, and clinical applications of autografts, allografts, and synthetic bone graft substitutes in orthopedic trauma surgery, aiming to guide treatment decisions for bone grafting	Autografts, allografts, and synthetic bone graft substitutes	Orthopedics	Review	Autograft, allograft, and bone graft substitutes all possess their own varying degrees of osteogenic, osteoconductive, and osteoinductive properties that make them better suited for different procedures
Meesuk L. et al. [67]	To evaluate the proliferation and osteogenic differentiation potential of bone marrow and umbilical cord mesenchymal stem cells cultured on 3D-printed hydroxyapatite scaffolds, with or without biomimetic calcium phosphate coating, for applications in bone tissue engineering	Synthetic bone substitutes-3D-printed hydroxyap-atite scaffolds, with or without biomimetic calcium phosphate coating	Orthopedics, dentistry, oral and maxillofacial surgeries	Preclinic	Osteogenic differentiation: Cultivation on HA scaffolds resulted in increased alkaline phosphatase ALP activity and increased expression of osteogenic genes and proteins compared to control 2D culture cells
Adamska, P. et al. [68]	To review the literature on teeth autotransplantation, supported by a case report involving the autotransplantation of a third mandibular molar into the site of an extracted infraoccluded first mandibular molar, as well as the utilization of advanced platelet-rich fibrin (A-PRF) alongside autogenous dentin grafts for bone tissue regeneration	Teeth autotrans-plantation	Dentistry, oral and maxillofacial surgeries	Systematic review	Evaluate the efficacy of autotransplantation, the application of growth factors, and the integration of autogenous dentin grafts in remedying dental deficiencies resulting from reinclusion

Figure 3 shows the publications, included in this literature review, distributed by year of publication. A clear trend of high research interest in bone substitutes and regenerative medicine can be seen. Notably, the number of published studies increased significantly from 2021 onwards, with a peak in 2021 and a subsequent high publication rate in 2023 and 2024. This trend indicates that bone tissue engineering continues to be a dynamic and rapidly evolving field.

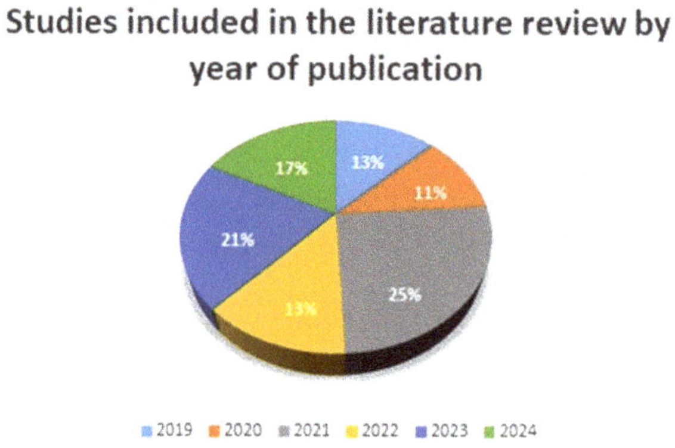

Figure 3. Annual distribution of studies included in the literature review (2019–2024).

3.1. Importance of Bone Substitutes in Regenerative Medicine

Modern regenerative medicine employs bone substitutes as essential tools for the treatment of bone defects caused by trauma, tumours, congenital malformations, and degenerative diseases. The growing demand for effective bone regeneration has directed much research towards the development of biomaterials capable of replacing damaged bone tissue and supporting the body's natural healing processes. A systematic review by M.P. Ferraz [1] outlines the widespread use of bone grafts in dental and orthopedic procedures, including the treatment of fractures, repair of non-union bone tissue, and reconstruction following tumour-related injuries. Evidence confirms that autografts remain the reference standard due to their superior osteogenic potential and biocompatibility. However, their use is limited by donor site morbidity and restricted availability, prompting the ongoing search for alternative bone substitutes.

Current scientific research identifies several key clinical scenarios necessitating the use of bone grafts. Trauma surgery is the leading indication for bone graft application. A clinical study by M. Rupp et al., which analyzed over one million surgical procedures over a ten-year period, demonstrated that bone substitutes are commonly employed to treat bone defects and delayed or non-union fractures—especially in complex cases where an insufficient quantity of bone tissue is available for successful healing [10]. Studies indicate that approximately 6% of all fractures are complicated by non-union, requiring additional surgical intervention using bone substitutes to stimulate the regenerative process. Thus, trauma surgery remains the largest consumer of bone substitutes, where they play a critical role in managing complex fractures that do not respond to standard treatment. The prevalence of fracture non-union highlights the need to develop more effective osteoinductive materials for this field.

Spinal surgery represents another key application area for bone substitutes. A systematic review by M. Laubach et al. reports their use in spinal fusion procedures and the stabilization of vertebral segments [11]. These procedures are essential in the management of degenerative spinal conditions, traumatic injuries, and spinal deformities. The particular demands of spinal surgery require bone substitutes to offer a unique combination of mechanical strength—to support the spine—and biological activity—to ensure reliable vertebral fusion. This makes spinal applications especially demanding in terms of material quality. The treatment of oncological bone lesions also necessitates the use of bone substitutes. A review by A. Santoro et al. highlights the need for bone reconstruction following tumour resection, to restore both the structural integrity of the skeletal system and the functional capacity of affected limbs [12]. In these cases, bone substitutes help preserve anatomical structure and mobility. Oncological surgery presents unique challenges for bone substitute materials, as it requires a balance between radical tumour resection and the preservation of function—necessitating materials with reliable biointegration and the capacity to fill extensive bone defects.

Human bone grafting techniques developed in the past have been successfully refined, leading to the creation of modern synthetic materials that offer improved stability and biocompatibility. Autogenous bone grafts have been referenced in medical literature since the 19th century and have been employed by surgeons to reconstruct bones using material harvested from the fibula or iliac crest. A preclinical study by P. Tournier et al. discussed the historical development of bone grafting techniques and the gradual shift towards bone substitutes [13]. The authors traced the introduction of allografts in the mid-20th century, when they became widely accepted due to their affordability and the reduced surgical burden placed on the patient. However, concerns regarding immune rejection and disease transmission have driven the development of safer alternatives. The study suggested that future advancements in bioprinting, nanotechnology, and tissue engineering will be essential for creating the next generation of bone substitutes, given the currently unpredictable clinical behaviour of these materials.

The ethical considerations associated with the use of various bone graft types, including allogeneic and xenogeneic materials, are complex and multifaceted. As noted in a case study by S. Parikh et al., the use of allogeneic bone grafts raises significant ethical questions, particularly regarding informed consent from donors—especially when the donor is a minor who may not fully comprehend the implications of the procedure [14]. In such cases, decisions are made by parents or legal representatives, which introduces ethical dilemmas surrounding voluntariness and the protection of the donor's interests. This study also underlined the importance of strict material safety monitoring, the prevention of infection transmission, and the maintenance of anonymity, all of which are critical to ethical transplantation practices.

A systematic review by M. Sharifi et al. pointed out that xenografts derived from animals present additional ethical concerns related to animal welfare and humane treatment [15]. The authors emphasized the necessity of accounting for cultural and religious considerations, as some patients may regard the use of animal-derived tissues as unacceptable. In such cases, it is essential to respect patients' beliefs and to offer alternative treatment options.

According to I. Ana, in a chapter of the monograph dedicated to bone substitutes in dental implantology, an ethical approach to the use of bone grafts in clinical practice should also encompass procedural transparency, equitable access to materials, and the avoidance of commercialisation, which may lead to conflicts of interest [16]. The author emphasizes that maintaining a balance between medical efficacy and ethical standards is crucial for the successful and responsible implementation of transplantation procedures. The ethical dimensions of bone substitute use add an additional layer of complexity to the selection of optimal materials, requiring consideration not only of clinical effectiveness but also of cultural, religious, and legal factors. This underscores the need for the development of synthetic alternatives that circumvent the ethical dilemmas associated with donor-derived tissues.

Thus, based on the analysis of the aforementioned studies, it can be concluded that ethical considerations in the use of allogeneic and xenogeneic bone grafts are multifaceted and necessitate a comprehensive approach that includes the protection of donor rights, patient safety, respect for cultural values, and transparency in clinical practice. Bone substitutes have become an essential component of modern regenerative medicine, offering solutions to critical clinical challenges ranging from trauma reconstruction to oncological surgery. Their significance continues to grow, particularly in the context of global population ageing, the rising incidence of osteoporotic fractures, and the expanding indications for complex reconstructive procedures. At the same time, despite the rapid advancement of biomaterials, none of the currently available substitutes fully replicate the biofunctional properties of natural bone tissue, as evidenced by limitations in osteogenic, osteoinductive, and osteoconductive performance. A major challenge remains the absence of universal protocols for material selection tailored to specific clinical scenarios, along with limited long-term clinical data on the efficacy of different types of substitutes. Given the imperative to shorten rehabilitation periods, reduce complication risks, and improve functional outcomes, systematic analysis of the properties and effectiveness of bone substitutes—as well as the development of personalized selection strategies—represents one of the key directions for future advancement in regenerative medicine.

3.2. *Main Categories and Characteristics of Modern Bone Substitutes*

3.2.1. Autologous Bone Grafts

Autologous bone grafts are employed in patients for bone regeneration owing to their highly effective biological characteristics, particularly their osteogenic potential, osteoinductivity, and osteoconductivity. In autologous graft procedures, surgeons harvest the patient's own tissue from four primary donor sites: the iliac crest, fibula, tibia, and mandibular symphysis. A review article by O.S. Janjua et al. analyzed the structure and cellular behaviour of autologous grafts, highlighting the viability of their osteoblasts and the presence of endogenous growth factors that promote bone formation [17]. The study emphasized the natural process of bone remodelling facilitated by living bone cells and bone morphogenetic proteins (BMPs) in such grafts, which is significantly superior to that observed in allografts and synthetic substitutes. However, the review also acknowledged limitations such as the finite quantity of donor bone and the necessity to identify substitutes in cases requiring multiple transplantation procedures.

The effectiveness of autografts in bone tissue regeneration is primarily attributed to the presence of specific growth factors and proteins that play a crucial role in osteogenesis, angiogenesis, and bone remodelling. These biomolecules initiate cellular signalling cascades that activate the proliferation, differentiation, and functional activity of osteoblasts, regulate extracellular matrix formation, and maintain tissue homeostasis. As reported in a randomized clinical trial by S. Malik et al., one of the key contributors to the enhanced effectiveness of autografts is Concentrated Growth Factor (CGF), which comprises various proteins that stimulate osteogenesis and bone healing [18]. The incorporation of CGF into autografts significantly increases bone density and accelerates the regeneration process, as demonstrated by clinical outcomes, particularly in the treatment of mandibular fractures.

An experimental animal study by C. Black et al. supports these findings, showing that BMP-2, in combination with appropriate biomaterials such as collagen sponges, enables effective delivery of growth factors and stimulation of osteogenesis [19]. Furthermore, BMP-7 contributes to maintaining the balance between bone formation and resorption, which is essential for normal remodelling processes. A comparative experimental study by Y. Jin et al. highlighted that autografts also contain proteins that facilitate osteoconductivity, and their effectiveness is highly dependent on the carriers used to deliver recombinant growth factors like BMP-2 [20]. The study demonstrated that demineralised bone matrix (DBM) serves as a more effective carrier for BMP-2 than hydroxyapatite, offering improved bone regeneration and reduced fat infiltration. Thus, the presence of BMP-2 and other growth factors, in combination with optimal carriers, is critical for enhancing the osteoinductive properties of autografts. The effectiveness of autografts stems from the complex interplay of growth factors—including BMP-2, BMP-7, CGFs, and other signalling proteins—that stimulate osteogenesis, alongside the appropriate selection of carriers that ensure their delivery and activate local cellular regeneration mechanisms.

The primary advantage of autogenous bone grafts lies in their excellent biocompatibility, low risk of immune rejection, and strong potential for integration into host tissue. According to a clinical study by U. Tasdemir et al., outcomes of sinus lift and alveolar ridge augmentation surgeries using autogenous bone grafts were evaluated as excellent [21]. The regenerative capacity of autografts surpassed that of synthetic and xenogeneic substitutes, which was partly attributed to their ability to preserve osteocyte viability and stimulate vascularisation. Further enhancement of bone healing outcomes was achieved by combining autografts with growth factors, such as recombinant human vascular endothelial growth factor (rhVEGF).

Despite the superior biological outcomes associated with autogenous bone grafts, their use is limited by donor site morbidity, extended operative time, and postoperative pain. In a pilot study by U. Taşdemir et al., complications related to autogenous bone grafting were assessed, with the most significant being infection, prolonged healing time, and limited bone availability [22]. The authors confirmed that 20–30% of patients experience persistent donor site pain post-healing, with some reporting chronic pain or mild infection at the graft site. The study also highlighted the constraint posed by the limited volume of autografts, which restricts their application in large bone defect reconstructions—necessitating the use of allogeneic or synthetic substitutes in major surgical procedures.

Autograft harvesting is a widely practised technique in orthopedics and reconstructive surgery for the replacement of bone or soft tissue defects. However, this process carries the risk of donor site morbidity, which may manifest as pain, restricted mobility, infectious complications, or other functional impairments. Reducing these complications is essential to improving clinical outcomes and the postoperative quality of life. Scientific efforts are focused on optimizing harvesting techniques, selecting donor sites with lower risk, and identifying alternative sources of autologous tissue to minimize damage and accelerate rehabilitation. Accordingly, as noted in the systematic review and network meta-analysis by K. Kunze et al., minimizing donor site morbidity during autologous graft harvesting depends significantly on the type of graft selected [23]. In anterior cruciate ligament (ACL) reconstruction, the use of hamstring or quadriceps tendon autografts is associated with significantly lower donor site morbidity compared to the traditional bone–patellar tendon–bone graft. The choice of autograft should be individualized based on the patient's level of physical activity and expectations, which reduces the likelihood of postoperative complications such as pain, mobility restriction, and functional loss. Furthermore, a systematic review and meta-analysis by A. Attia et al. emphasized that the technical aspects of bone autograft harvesting—such as from the calcaneus or the proximal and distal tibia—are crucial in reducing complications [24]. Selecting a donor site from the lower limb, rather than the iliac crest, significantly reduces the risk of chronic pain, infection, and neurological complications, which is critical for improving the quality of life of patients following surgery.

As highlighted in the clinical study by M.N. Khalid et al., the use of alternative autograft sources, such as the peroneus longus tendon, also contributes to minimizing donor site morbidity [25]. These methods provide favourable functional outcomes, faster recovery of mobility, and significantly reduced pain symptoms and risk of instability at the harvest site, making them promising candidates for widespread clinical application. Therefore, reducing donor site morbidity is achieved through a comprehensive approach that encompasses the selection of the optimal autograft type, appropriate donor site selection, refinement of surgical techniques, and the utilization of alternative tissues to decrease complication rates and enhance functional outcomes.

Autogenous bone grafts are widely applied in orthopedic, craniofacial, and maxillofacial surgery, where relatively rapid osseointegration and structural support are desired. A review article by H. Sun et al. described the use of guided bone regeneration (GBR) techniques in maxillofacial surgery, reporting predictable preservation of bone volume and excellent implant success rates [26]. In a study investigating autogenous dentin matrix grafts, the authors noted that in sites affected by severe periodontitis and similar conditions, these grafts performed as well as or better than xenogeneic materials. H. Sun et al. further observed that autogenous grafts combined with platelet-rich fibrin resulted in improved soft tissue healing and enhanced implant stability [26]. However, their findings also indicated that the success of autogenous grafts depends heavily on precise surgical handling and the patient's bone metabolism, which may be influenced by factors such as age, systemic health, and lifestyle.

Thus, autogenous bone grafts remain the most effective material for bone regeneration, due to their inherent biocompatibility, the presence of viable cells, and the availability of growth factors that stimulate osteogenesis and remodelling. Nonetheless, their application is constrained by the limited availability of donor tissue and the risk of complications at the harvest site. This underscores the importance of optimizing harvesting techniques and exploring alternative graft sources. The appropriate selection of graft type, along with accurate surgical technique, can minimize complications and facilitate the successful restoration of bone structure.

3.2.2. Allogeneic Bone Substitutes

The need for a large volume of bone is one of the scenarios in which allogeneic bone substitutes—obtained from human donors—serve as a viable alternative to autografts. These grafts are derived from donor tissues processed by bone banks through a screening and preparation process that removes cellular components while preserving elements of the tissue structure. Allogeneic and xenogeneic substitutes have demonstrated structural and osteoconductive advantages; however, they also carry risks of immune response and uncontrolled resorption, necessitating intensive processing methods to optimize their biological behaviour. In an in vitro experimental study, J. Liu et al. conducted laboratory tests to evaluate the effects of various configurations of allogeneic bone substitutes on cell adhesion and proliferation [27]. The researchers confirmed that combinations of bone pellets with bone powder exhibited superior osteoconductivity compared to cortical bone, due to the increased surface area and open architecture of these materials. The study also identified donor age and processing conditions as influential factors affecting graft quality, with bone from younger donors showing better mechanical properties and greater cellular compatibility. However, the findings demonstrated that although allogeneic grafts provide essential structural support and osteoconductivity, their lack of inherent osteogenic potential necessitates the addition of exogenous growth factors or viable cells to achieve optimal regeneration.

Current standards and protocols for donor screening and the processing of allogeneic bone tissue in bone banks are governed by national and international regulations that ensure graft safety, efficacy, and quality. Key regulatory frameworks include European Directives 2004/23/EC and 2006/86/EC, which outline requirements for donor selection, labelling, processing, storage, and transport of tissues—thereby minimizing the risk of infectious transmission and other complications [28,29]. As detailed in the experience report by I. Ilyas et al., these directives mandate comprehensive donor screening, which includes an in-depth medical history, serological testing (for HIV, hepatitis B and C, syphilis), and molecular diagnostics such as nucleic acid amplification tests (NAAT), all of which contribute to a thorough risk assessment [30].

Tissue processing protocols in bone banks involve multiple sequential steps: aseptic tissue procurement, sterile removal of soft tissues, disinfection, preservation (typically via cryopreservation or lyophilisation), and multi-level quality control measures, including microbiological and histological analyses. These procedures are comprehensively described in the review by A. Regmi et al. [31]. A critical component of these protocols is the implementation of a graft tracking system, which ensures full traceability of tissue movement from donor to recipient—an essential measure for transparency and for enabling a prompt response in the event of adverse outcomes.

The feasibility of implementing such stringent standards is supported by evidence demonstrating their effectiveness in significantly reducing the risk of infection transmission and other complications for recipients. These measures also enhance confidence in the use of allogeneic grafts and contribute to greater availability of high-quality materials for surgical interventions, as shown in a survey by T. Asamoto et al. [32]. For instance, the application of NAAT testing in compliance with modern requirements shortens the donor waiting period and decreases the number of rejected grafts caused by the serological window period, thereby increasing the efficiency of bone banks. Likewise, standardized sterilization and storage protocols reduce the risk of microbial contamination, which is particularly crucial for bone grafts due to their porous structure.

The implementation of international standards—such as the recommendations of the American Association of Tissue Banks (AATB)—alongside internal audit and quality control systems, adds another layer of safety and sustains a high level of trust among healthcare professionals and patients [33]. Collectively, these regulations and practices underpin modern bone grafting procedures, aiming to optimize clinical outcomes while minimizing associated risks.

Allogeneic bone undergoes extensive processing, including sterilization, decalcification, and demineralisation, to mitigate the risks of disease transmission and immune rejection. A review by I. Um et al. examined the osteoinductivity and antigenicity of allogeneic demineralised dentin matrix (Allo-DDM), a dentin-derived allograft [34]. The study found that decalcification methods enhanced the release of bone morphogenetic proteins (BMPs), thereby increasing osteoinductivity. Low antigenicity values were achieved through rigorous sterilization protocols that preserved mechanical integrity. The preservation of collagen structures and growth factors through high hydrostatic pressure (HHP) sterilization was shown to be superior to conventional techniques such as gamma irradiation or thermal treatment, which may compromise the structural strength of the bone matrix. However, the authors noted that, despite these optimized processing methods, allografts remain less bioactive than autografts. Consequently, biological augmentation strategies—such as the incorporation of growth factors or the seeding of stem cells—remain important avenues for improving the regenerative potential of allogeneic grafts.

Allogeneic bone substitutes are widely used in fracture healing, spinal fusion, and dental bone grafting—particularly in cases where large volumes of graft material are required. A preclinical study by P. Tournier et al. [13] demonstrated that deproteinised allografts serve as effective scaffolds for stem cells and promote bone formation through natural tissue regeneration mechanisms. The preservation of collagen and mineral structure enhances cell adhesion and proliferation, which is especially valuable in revision surgeries and the reconstruction of extensive bone defects.

However, allogeneic grafts also present notable limitations. An experimental study on mechanical properties conducted by J. Waletzko-Hellwig et al. found that although high-pressure-treated allografts retained structural integrity, they exhibited only limited capacity to induce natural bone regeneration when compared to autografts [35]. The absence of viable osteogenic cells renders these grafts dependent on a slow remodelling process mediated by the host, which may result in the formation of fibrous tissue rather than functional bone. These limitations underscore the need for innovative tissue engineering strategies, such as combining allografts with stem cells and bioactive surface coatings.

Allogeneic bone substitutes derived from human donors therefore offer crucial structural support and osteoconductive properties in the reconstruction of large bone defects. Modern processing protocols—including sterilization and demineralisation—help to minimize the risk of infection transmission and adverse immune responses. Nevertheless, due to their limited osteogenic potential, there is growing interest in enhancing these materials through the incorporation of growth factors and stem cells, with the aim of improving their regenerative capacity and clinical outcomes.

3.2.3. Xenogeneic Bone Substitutes

Xenogeneic bone substitutes—derived from bovine, porcine, and equine sources—are widely used in orthopedics and maxillofacial surgery for various procedures, including alveolar ridge augmentation and sinus lift. An experimental animal study by A. Shao et al. investigated the immunological response to xenogeneic bone matrices [36]. Additionally, a systematic review of in vitro studies by R. Amid et al. found that bovine xenografts possess a mineral structure closely resembling that of human bone, making them effective scaffolds for osteoconduction [37]. Porcine variants, meanwhile, demonstrated superior remodelling properties due to their enhanced resorption capacity. Despite these advantages, xenogeneic materials require thorough processing to remove cellular components and minimize immunogenicity. Furthermore, their prolonged resorption period can delay natural healing, often necessitating the adjunctive use of growth factors or bioactive surface coatings to enhance integration with recipient tissues. To systematize the comparative characteristics of xenogeneic materials of differing origins, a detailed analysis of their structural features, clinical advantages, and limitations is presented in Table 2.

Table 2. Comparative Characteristics of Xenogeneic Bone Substitutes by Source of Origin.

Source of Origin	Structural Features	Clinical Benefits	Main Disadvantages	Indications For Use
Bull	Mineral structure is as similar as possible to human bone; high crystallinity of hydroxyapatite; preserved trabecular architecture	Excellent osteoconductivity; long-term structural support; predictable resorption (12–24 months); broad clinical evidence base	Slow resorption may delay natural remodelling; potential risk of prion disease transmission (theoretical); religious restrictions for some patients	Dental implantology; sinus lift; alveolar ridge augmentation; filling of small bone defects
Pig	Lower density compared to bovine; higher porosity (60–70%); increased resorption capacity	Better remodelling properties; faster integration with native bone (6–12 months); optimal for young patients with active metabolism	Lower mechanical strength; risk of transmission of PERVs and PCMV; cultural and religious restrictions (Islam, Judaism); higher immunogenicity	Periodontal regeneration; filling of defects after tooth extraction; pediatric maxillofacial surgery
Horse	Intermediate characteristics between bovine and porcine; moderate density; balanced porosity	Alternative for patients with religious restrictions; moderate resorption rate (9–18 months); good biocompatibility after processing	Limited market availability; fewer clinical studies; higher cost due to lower prevalence; potential allergic reactions	Orthopedic surgery (limited use); alternative in case of intolerance to other xenografts

Source: created by the author based on R. Amid et al. [37], A.T. Jerbić Radetić et al. [38], A. Pröhl et al. [39].

The presented data demonstrate that the selection of a xenogeneic material source is determined by the balance between structural stability and the speed of biological integration. Bovine-derived materials provide optimal long-term support for reconstructive procedures in dentistry, where volumetric stability is essential. Porcine substitutes offer advantages in clinical situations requiring accelerated remodelling, particularly in periodontology and pediatric applications. Equine materials occupy a niche position as alternatives in cases of cultural or religious restrictions, or in patients with individual intolerance. The key factor in material selection remains not only biological effectiveness, but also the ethical and cultural acceptability for the individual patient.

Xenotransplantation—the transplantation of cells, tissues, or organs between species—carries the risk of transmitting infectious diseases, particularly zoonoses, which may pose serious threats to both recipient health and public safety. Pathogens potentially transmissible through xenografts include viruses such as porcine cytomegalovirus (PCMV/PRV), retroviruses including porcine endogenous retroviruses (PERVs), and various species-specific agents such as hepatitis viruses, as outlined in the expert review by H. Groenendaal et al. [40,41]. The risk of viral transmission from animals to humans, with the possibility of novel infectious disease emergence, requires rigorous oversight and comprehensive risk management. Current protocols for minimizing the risk of zoonotic transmission are highly effective due to a multifaceted approach. This includes mandatory testing of donor animals using polymerase chain reaction (PCR) and nucleic acid amplification tests (NAAT) for a comprehensive panel of known pathogens, following recommendations from regulatory bodies such as the U.S. Food and Drug Administration and the European Medicines Agency [42,43]. Additionally, multi-stage processing of graft materials using combined physical and chemical sterilization techniques, as outlined by Hawthorne et al., is standard practice [44]. Long-term post-operative monitoring of recipients further enhances safety, also in accordance with the protocols established by Hawthorne and colleagues. The effectiveness of these safety measures is evidenced by the absence of documented cases of PERV or other serious zoonotic disease transmission in clinical practice involving properly processed xenogeneic bone substitutes. This supports the adequacy of current safety standards, provided that all processing and monitoring protocols are strictly followed.

Biocompatibility and minimisation of immune response are critical for the clinical success of xenografts. A systematic review by L. Sánchez-Labrador et al. confirmed that hydrothermal and chemical processing techniques effectively remove immunogenic proteins while preserving the structural integrity of the mineralised matrix [45]. This preservation is crucial for ensuring the biocompatibility and osteoconductivity of the material. Furthermore, the review highlighted that the incorporation of osteoinductive agents—such as bone morphogenetic proteins (BMPs) or mesenchymal stem cells—significantly enhances the functionality and integration of xenografts in clinical applications.

An experimental animal study by A.T. Jerbić Radetić et al. highlighted the widespread use of xenografts in orthopedic surgery, dental implantology, and periodontal regeneration, particularly in cases where the application of autografts or allografts is restricted by volume limitations or patient preferences [38]. A preclinical in vivo study by A. Pröhl et al. demonstrated that xenografts suspended in hyaluronic acid achieved excellent integration and bone ingrowth, albeit with slower remodelling compared to autologous materials [39]. The researchers also observed that combining xenografts with bioactive components—such as platelet-rich fibrin or stem cell-based scaffolds—promotes enhanced vascularisation and accelerates healing processes.

Thus, xenogeneic bone substitutes derived from bovine, porcine, and equine tissues are widely applied in orthopedic, maxillofacial, and dental surgical procedures. Bovine-derived materials offer an optimal osteoconductive scaffold due to their structural similarity to human bone, while porcine-derived materials exhibit superior remodelling characteristics. These grafts require complex processing techniques—including hydrothermal, chemical, and genetic engineering methods—to eliminate immunogenic components and reduce the risk of zoonotic disease transmission. The regenerative efficacy of xenogeneic substitutes is significantly enhanced when combined with osteoinductive agents, such as bone morphogenetic proteins or mesenchymal stem cells, which improve tissue integration and vascularisation, compensating for their relatively slow remodelling. Systematic evaluation of clinical applications suggests that xenografts are ideally suited for dental implantology and small bone defects, where their slow resorption offers long-term structural stability. However, they are not recommended for critical load-bearing applications or for large defect reconstruction in younger patients, due to their limited osteogenic potential. The most favourable outcomes are achieved when xenografts are used in conjunction with autogenous bone or growth factors, enabling the combined advantages of both material types to be realized.

3.2.4. Synthetic Bone Substitutes

Synthetic bone grafts play a critical role in regenerative medicine by providing structural support at the defect site while simultaneously serving as a biologically favourable substrate that facilitates osteoblastic adhesion, proliferation, and subsequent induction of new bone formation. Beyond their mechanical and osteoconductive properties, these grafts can be further functionalized as delivery system for bioactive molecules, including growth factors and therapeutic drugs, thereby enhancing their regenerative potential and broadening their clinical applications. Among synthetic bone grafts, the most extensively studied biomaterials include calcium phosphate–based cements, calcium phosphate ceramics, calcium sulphate, bioactive glasses, and polymers [1].

Calcium phosphate cements (CPCs) present several advantages: they are bioactive, suitable for large-scale manufacturing, easy to handle, and injectable, which allows them to adapt to irregularly shaped bone defects. Unlike autogenous and allogeneic grafts, CPCs do not carry risks such as donor site morbidity or infection. Moreover, their biocompatibility and compositional similarity to bone make them highly suitable candidates for bone regeneration.

Synthetic bone substitutes based on calcium phosphate ceramics—particularly hydroxyapatite (HA) and β-tricalcium phosphate (β-TCP)—are characterized by controlled degradation rate, and structural similarity to natural bone. HA is associated with excellent biocompatibility, as it does not provoke local or systemic toxicity, inflammatory responses, or adverse immunological reactions, thereby reinforcing its value as a clinically reliable bone substitute material. Beta-tricalcium phosphate (β-TCP), typically sintered at temperatures below approximately 1125 °C, offers the advantage of thermodynamic stability within the biological environment, while demonstrating greater resorbability than hydroxyapatite under physiological conditions.

In an experimental study by E. Maevskaia et al., the osteoconductive properties of 3D-printed ceramic scaffolds were examined, and it was found that the efficiency of bone tissue adhesion and cell proliferation is directly influenced by the surface microarchitecture and optimal porosity of the material [46]. Structures featuring triply periodic minimal surfaces (TPMS) proved especially promising, offering enhanced mechanical strength, vascularisation, and structural stability. Despite these advantages, synthetic materials possess limited osteoinductive potential, which necessitates their combination with bioactive growth factors or stem cells to achieve optimal integration with host tissues.

3D printing of ceramic frameworks presents significant advantages over traditional manufacturing methods due to its ability to produce complex, patient-specific structures with precisely controlled porosity—thereby promoting improved vascularisation and tissue ingrowth. As noted in the review by Y. Ding et al., this technology also reduces production time and enhances implant precision [47]. However, the technology review by X. Li et al. highlighted that 3D printing of ceramics remains costly and technically demanding, with current limitations in mechanical strength and production scalability [48]. Consequently, although 3D printing represents a highly promising direction, it still requires further development before it can be widely adopted in clinical settings.

Calcium sulphate, one of the earliest synthetic bone graft substitutes, has long been employed in orthopedic and dental applications due to its biocompatibility and ability to completely resorb in vivo. Its rapid degradation profile allows for temporary defect filling and space maintenance, often serving as a carrier for antibiotics or growth factors. However, its primary limitation lies in the imbalance between its fast resorption rate and the relatively slower pace of new bone formation, which can result in incomplete defect regeneration if not combined with other biomaterials. It has been extensively incorporated into composite formulations together with hydroxyapatite, tricalcium phosphate, calcium carbonate, and monocalcium phosphate monohydrate. These mixtures are often designed as injectable pastes that can harden in situ within bone defects, offering a minimally invasive approach to defect stabilization and filling [4].

Biodegradable polymers—such as polylactic acid (PLA), polycaprolactone (PCL), and collagen composites—constitute another promising class of synthetic bone substitutes, owing to their moulding properties and adaptive degradation characteristics. These polymers are readily biodegradable in vivo; however, their limited mechanical strength and inherently low osteoconductivity restrict their application as standalone scaffolds for bone regeneration [1]. In an experimental study by I. Nifant'ev et al., polyester–mineral composites were investigated, revealing that the inclusion of β-tricalcium phosphate or carbonated hydroxyapatite significantly enhanced the mechanical strength and biocompatibility of the materials [49]. The researchers concluded that these composites offer controlled degradation, which is particularly valuable in anatomical areas where gradual bone remodelling is required.

Bioactive glass constitutes a distinct and important class of synthetic bone substitutes, characterized by its ability to establish direct bonds with host bone tissue while releasing biologically active ions that stimulate cellular responses. A critical feature of BGs is their surface reactivity, which closely mimics the chemical and structural properties of the bone mineralization phase, thereby supporting bone reconstruction. Notably, the bioactivity and resorption kinetics of BGs are strongly dependent on their chemical composition, which in turn governs their biological performance and clinical applicability [1]. In an experimental study investigating mechanical properties, T.Y. Kang et al. demonstrated that modification of bioactive glass with zirconium oxide (ZrO_2) significantly enhances its mechanical strength while preserving its bioactivity [50]. The modified materials exhibited improved cellular responses, elevated alkaline phosphatase activity, and accelerated mineralization—making them particularly suitable for applications in bone tissue engineering.

A comparative analysis of the three principal groups of synthetic bone substitutes reveals distinct advantages and limitations relevant to different clinical contexts. Calcium phosphate ceramics—such as hydroxyapatite and β-tricalcium phosphate—exhibit the highest osteoconductivity among synthetic materials due to their chemical resemblance to the mineral phase of natural bone. This makes them particularly well-suited for filling small defects and for augmentation procedures in dentistry. Biodegradable polymers, including polylactic acid (PLA) and polycaprolactone (PCL), offer the benefit of controlled degradation in parallel with new bone formation, which is particularly valuable in pediatric applications and in cases where gradual material replacement with native tissue is desired. Bioactive glass, on the other hand, is notable for its active stimulation of bone formation via the release of silicon and calcium ions. Among synthetic materials, it demonstrates the highest rates of vascularisation, making it especially promising for complex cases involving compromised blood supply. However, a common limitation shared by all three synthetic groups is their insufficient osteoinductive potential. As a result, these materials often require biological augmentation—such as the addition of growth factors or stem cells—to achieve effective integration and regeneration, in contrast to autogenous grafts, which possess inherent biological activity.

In summary, some of the synthetic bone substitutes include calcium phosphate ceramics, biodegradable polymers, and bioactive glass. These materials offer controlled degradation, excellent biocompatibility, and the potential for custom implant fabrication via 3D printing. Despite their favourable osteoconductive properties, their limited osteoinductive capacity necessitates combination with bioactive components for optimal performance in clinical settings. To consolidate and compare the key properties of all bone substitute categories discussed, a systematic overview is provided in Table 3.

Table 3. Comparative Characteristics of the Main Types of Bone Substitutes.

Substitute Type	Source	Osteogenicity	Osteoinductivity	Osteoconductivity	Biocompatibility	Main Advantages	Main Disadvantages
Autogenous	Patient's own tissues	High	High	High	Missing	No rejection, presence of living cells, natural growth factors	Limited availability, donor site morbidity, additional pain
Allogeneic	Human donors	Missing	Limited	Moderate	Good after processing	Large volumes, no donor site morbidity	Risk of infection transmission, need for treatment, slow integration
Xenogenic	Animal tissues	Missing	Low	Moderate	Good after processing	Unlimited availability, structural similarity to human bone	Zoonoses risk, need for careful processing, slow remodelling
Synthetic	Artificial materials	Missing	Missing/Low	Good	Excellent	Controlled properties, no biological risks, 3D printing capabilities	Lack of biological activity, need for additional growth factors

Source: created by the author based on M. Heitzer et al. [2]; M.P. Ferraz et al. [51].

The analysis of the comparative characteristics of bone substitutes reveals a fundamental contradiction between biological effectiveness and the practical availability of materials in modern regenerative medicine. The presented data indicate an inverse relationship between osteogenic potential and the clinical accessibility of different types of substitutes. Autogenous grafts demonstrate the highest values for all three key biological parameters—osteogenicity, osteoinductivity, and osteoconductivity—owing to the preservation of viable osteoblasts and endogenous morphogenetic proteins. However, their clinical use is constrained by the limited volume of available donor tissue and a statistically significant incidence of donor site complications, which, according to the analyzed studies, affects 20–30% of patients.

Allogeneic and xenogeneic substitutes represent the opposite end of this spectrum. While they offer virtually unlimited availability due to standardized donor sourcing and processing, they entirely lack osteogenic properties as a result of decellularisation and sterilization procedures. Their moderate osteoconductivity is retained through the preservation of the mineralised matrix; however, the absence of viable cells renders their efficacy heavily dependent on the host's endogenous regenerative capacity. Synthetic materials occupy a distinct niche, offering high biocompatibility and enabling precise control over physicochemical parameters such as porosity, degradation rate, and mechanical strength. Nevertheless, their biological inertness necessitates functionalisation with osteoinductive molecules or the incorporation of cellular components to achieve satisfactory regenerative outcomes.

This identified heterogeneity in functional characteristics underscores the necessity for a differentiated approach to bone substitute selection. Such an approach must be based on multifactorial analysis, taking into account anatomical and topographic features of the defect, biomechanical loading conditions, the patient's metabolic status, and their potential for endogenous regeneration. A promising direction to address existing limitations is the development of next-generation composite biomaterials. These aim to integrate the structural and mechanical advantages of synthetic scaffolds with the biological activity of natural growth factors and cellular elements. Such a synergistic strategy holds the potential to enhance osteogenesis significantly, thereby improving clinical outcomes in regenerative bone medicine.

3.3. Biological and Mechanical Properties of Bone Substitutes

The porosity of a bone substitute is a determining factor in bone integration and vascularisation and plays a significant role in overall functionality. Porous structures promote osteoconduction by providing a conducive environment for the infiltration, attachment, and proliferation of bone cells within the scaffold, ultimately facilitating bone regeneration and remodelling. L. Cheng et al. investigated the impact of micro- and nanoscale porosity on the performance of bone substitutes and highlighted the necessity of highly porous structures to support nutrient diffusion and osteoblast adhesion [52].

Modern manufacturing methods for bone substitutes aim to achieve maximum precision in creating porous architectures, as pore size and interconnectivity directly influence osteointegration, vascularisation, and the remodelling process. One key approach involves solid-state sintering technologies in combination with porogenic agents, such as ammonium bicarbonate. This method enables the formation of a controlled porous structure with pore sizes ranging from 23 to 210 μm. As emphasized by D. Liang et al., the pore size and shape can be finely tuned by selecting specific porogen granule sizes, achieving a high degree of interconnectivity while maintaining mechanical integrity [53].

Another highly accurate technique for controlling pore geometry involves the use of micro-computed tomography (micro-CT) combined with digital image processing algorithms. In particular, the fuzzy distance transform method allows for detailed assessment not only of pore diameter, but also shape, surface area ratio, and permeability—parameters critical for modelling nutrient diffusion and cellular infiltration. According to M.P. Ferraz et al. [51], this approach facilitates the design of porous structures with precise characteristics without compromising biomechanical performance, thereby enabling more effective, patient-specific implant design. Frameworks produced via 3D printing using triply periodic minimal surfaces and hierarchical pore organization demonstrate significantly enhanced osteoconductivity compared to traditional dense materials. However, excessive porosity leads to reduced mechanical strength, limiting their suitability for load-bearing applications. To address this limitation, L. Cheng et al. [52] proposed composite strategies whereby biodegradable synthetic polymers are used to reinforce porous ceramics, achieving an optimal balance between mechanical strength, porosity, and bioactivity.

To further improve this balance, functionally graded materials combining polymers—such as polycaprolactone (PCL) or poly (lactic-co-glycolic acid) (PLGA)—with β-tricalcium phosphate (β-TCP) have been employed, as demonstrated in the study by A. Kumar et al. [54]. In such systems, porosity gradually increases through polymer degradation while mechanical strength is maintained through concurrent new bone formation. Another approach, proposed by D. Celik and C.B. Ustundag, involves the hybrid design of ceramic frameworks featuring both cortical and trabecular architectures [55]. This configuration provides approximately 70% open porosity while maintaining mechanical strength comparable to natural bone. A comparative analysis of the biomechanical properties of major synthetic bone substitute types reveals a wide spectrum of performance characteristics, supporting the selection of optimal materials for specific clinical applications. These data are summarized in Table 4.

Table 4. Comparative Characteristics of Biomechanical Properties of Synthetic Bone Substitutes.

Material Type	Porosity (%)	Compressive Strength (MPa)	Elastic Modulus (GPa)	Degradation Rate	Clinical Application
β-tricalcium phosphate (β-TCP)	60–80	2–12	0.5–3.5	Moderate (6–24 months)	Filling defects carrying moderate loads
Calcium sulphate	30–50	10–30	1–6	Fast (4–8 weeks)	Short-term applications, drug delivery
Hydroxyapatite	55–75	20–80	7–13	Very slow (>24 months)	Long-term structural support, spinal surgery
Bioactive glass with ZrO_2	40–60	30–140	35–45	Supervised (3–12 months)	Load-bearing areas, dental implantology
3D printed frames with TPMS	65–85	5–30	0.3–2.5	Supervised (6–18 months)	Tissue engineering, individual implants

Source: created by the author based on G. Marongiu et al. [56]; L. Cheng et al. [52]; T.Y. Kang et al. [50].

The presented data confirm an inverse relationship between porosity and the mechanical strength of bone substitute materials. The highest compressive strength is demonstrated by bioactive glass modified with zirconium oxide (up to 140 MPa) at moderate porosity levels (40–60%), whereas highly porous polymer composites (70–85%) show significantly lower strength values (5–25 MPa). Degradation rates also vary widely: from 4 to 8 weeks for calcium sulphate to more than 24 months for hydroxyapatite, defining their clinical application range from short-term drug delivery to long-term structural support, respectively. β-Tricalcium phosphate and hybrid ceramic frameworks demonstrate an optimal balance of properties, combining moderate compressive strength (15–40 MPa) with controlled degradation (6–18 months), making them versatile materials suitable for a broad spectrum of clinical scenarios.

A central challenge in the development of synthetic bone substitutes is achieving an appropriate balance between degradation rate and mechanical properties. Scaffolds must provide sufficient initial structural support while allowing for a gradual transition to regenerated bone. G. Marongiu et al. tested synthetic materials based on calcium sulphate and β-tricalcium phosphate in an osteoporotic bone defect model [56]. Their findings showed that β-TCP exhibited a degradation profile well aligned with new bone formation, supporting long-term structural integrity. In contrast, calcium sulphate resorbs too rapidly to be effective for load-bearing or long-duration applications, making it suitable only for short-term use. The in vivo degradation rate of bone substitutes is influenced by multiple factors, including pH, enzymatic activity, material porosity, and cellular responses. As demonstrated by H. Kang et al., multichannel structures enhance permeability and promote faster degradation through increased ion exchange [57]. Similarly, K. Kowalewicz et al.

found that the level of osteoclast activity in the implantation zone significantly accelerates the resorption of the graft material [58].

Biocompatibility remains a critical requirement for synthetic bone substitutes, as clinical success depends on minimizing immune responses and preventing fibrous encapsulation. M. Heitzer et al. evaluated the biocompatibility of alloplastic, allogeneic, and xenogeneic bone substitutes combined with human dental pulp stem cells (DPSCs) and extracellular vesicles (EVs) [2]. The study showed that EVs derived from stem cells exhibited the greatest potential for promoting osteogenic differentiation, supporting the use of biological enhancements to improve graft integration. Although biphasic calcium phosphate (BCP) is characterized by good biocompatibility, M. Heitzer et al. noted that it lacks intrinsic osteoinductive capacity [59]. Therefore, it requires the addition of growth factors or seeded cells to effectively initiate bone regeneration. The authors concluded that more effective bone healing outcomes are achievable with next-generation bone substitutes that combine biocompatible scaffolds with bioactive molecules, providing a synergistic effect in stimulating osteogenesis.

Therefore, porosity is a critical factor in determining the effectiveness of synthetic bone substitutes, as it directly influences osteoconduction, vascularisation, and cellular infiltration. Modern manufacturing technologies—such as 3D printing and the incorporation of porogenic agents—enable precise control over pore size and interconnectivity, thereby optimizing biological performance. A persistent challenge remains the need to balance porosity with mechanical strength, which is being addressed through the development of functionally graded materials and composite scaffold structures. The degradation rate of the material should be synchronized with the pace of natural bone regeneration, with β-tricalcium phosphate exhibiting particularly favourable characteristics for long-term applications. To enhance biocompatibility and osteoinductive potential, synthetic substitutes are increasingly combined with bioactive molecules, stem cells, or extracellular vesicles—an approach that significantly improves their clinical effectiveness and regenerative outcomes.

3.4. Clinical Applications in Regenerative Medicine

In orthopedic traumatology, bone substitutes are employed for fracture healing, defect repair, and joint reconstruction. I. Sallent et al. demonstrated the clinical efficacy of bioactive scaffolds, ceramics, and engineered allografts, emphasizing the particular utility of synthetic substitutes based on calcium phosphate ceramics and bioactive glass in the management of load-bearing bone defects [60]. The researchers also highlighted the potential of tissue-engineered grafts incorporating mesenchymal stem cells and growth factors, which enhance vascularisation and promote superior regenerative outcomes compared to traditional grafting methods.

In revision surgery, bone substitutes are used to restore bone stock and improve implant integration. L. Massari et al. evaluated the clinical safety of a porous hydroxyapatite substitute in a cohort of 114 patients, confirming excellent biocompatibility and mechanical stability without notable side effects [61]. However, the authors underscored the inherent lack of osteogenic potential in hydroxyapatite-based materials, which necessitates the incorporation of bioactive agents to optimize regenerative performance.

In maxillofacial surgery and dental practice, bone substitutes are commonly applied in alveolar ridge augmentation, sinus lift procedures, and implant site preparation. B. Mahardawi et al. reported that allogeneic demineralised dentin matrix (Allo-DDM) was associated with low postoperative complication rates and sustained long-term functionality, without the morbidity associated with autologous donor sites [62]. C. Batra et al. investigated the combination of recombinant human platelet-derived growth factor-BB (rhPDGF-BB) with xenogeneic substitutes and demonstrated improved soft tissue regeneration and enhanced preservation of bone volume [63].

In neurosurgery, bone substitutes are applied in procedures such as cranioplasty, spinal fusion, and the reconstruction of defects following tumour resection. S. Zhang et al. investigated the use of autogenous dentin material for the reconstruction of cranial defects, confirming its high osteoconductive and osteoinductive potential due to its compositional similarity to natural bone matrix [64]. K.J. Jung et al. demonstrated the clinical success of multi-channel granular calcium phosphate substitutes in spinal fusion procedures, reporting long-term mechanical stability and high union rates [65]. A systematization of clinical experience regarding the use of bone substitutes across different medical specialties is presented in Table 5, which outlines the specific clinical requirements and observed outcomes within each field.

Table 5. Clinical Use of Bone Substitutes by Medical Specialty.

Medical Specialty	Main Indications	Recommended Types of Substitutes	Specific Requirements	Clinical Results
Orthopedic traumatology	Fractures, non-union, large defects	Autogenous, allogenic, synthetic (Ca-phosphate, Ca-sulphate)	High strength, osteoinduction, no immune reaction	High union rates (94% success rate), 6% of fractures are complicated by non-union out of 1,090,167 procedures; donor site morbidity 20–30% with autografts
Spinal surgery	Spondylodesis, segmental stabilization	Allogeneic, synthetic (β-TCP, BMP-2)	Structural support, stimulation of fusion	High fusion rates (85–95% when using BMP-2), β-TCP demonstrates optimal degradation profiles over 6–12 months
Oncological surgery	Tumour resection, restoration of integrity	Combined with growth factors, allogeneic materials	Preservation of anatomy and limb function	Effective restoration of functionality in 9–12 months, multi-channel cylindrical granular substitutes have shown successful regeneration
Maxillofacial surgery and dentistry	Augmentation, sinus lift, implantation	Xenogenic (bovine), Ca-phosphate synthetic, allogenic	Biocompatibility, volumetric stability, implant compatibility	Excellent implantation success rate (92–96%) with autogenous dentin matrices; improved soft tissue healing when combined with PRF
Neurosurgery	Cranioplasty, post-tumour defects	Autogenous dentin, synthetic polymers	Biocompatibility with nervous tissue, controlled degradation	High osteoconductive potential of autogenous dentin (85–90% integration in 6 months); 70% open porosity provides high strength
Revision surgery	Bone restoration, implant integration	Porous hydroxyapatite, combined synthetics	High biocompatibility, long-term stability	Study of 114 patients notes 100% safety with no side effects when using porous hydroxyapatite

Source: created by the author based on I. Ana [16]; A. Santoro et al. [12]; P. Baldwin et al. [66].

3.5. Three-Dimensional Bioprinting and Future Directions in Bone Substitution

Modern advances in 3D bioprinting enable the fabrication of individualized tissue substitutes with precise control over porosity and the incorporation of bioactive coatings. The scaffold, a three-dimensional and highly porous substrate, is a key element in tissue engineering. After expansion in culture, cells are transferred onto the scaffold, where they adhere, proliferate, remain viable, and generate the extracellular matrix composed of structural and functional proteins and saccharides that form living tissue. The material composition of the scaffold, together with its internal architecture (including the dimensions of struts, walls, pores, or channels), regulates and controls the biological behaviour of the cells [69]. Presentation of common bioprinting techniques is provided according to the description of Li et al. [70].

- Extrusion-based bioprinting

Extrusion-based bioprinting relies on the continuous deposition of bioinks via pneumatic, piston, or screw-driven systems. Techniques such as semi-solid extrusion (SSE) and fused deposition modelling (FDM) enable fabrication of both soft tissue analogues and bone-mimetic scaffolds. Natural polymers (alginate, gelatin, collagen, nanocellulose) and synthetic polymers (polycaprolactone [PCL], polyvinyl alcohol [PVA]) can be blended with growth factors or nanoparticles to optimize printability and cellular responses. Advantages include the ability to print high-viscosity bioinks and create mechanically robust structures, whereas lower spatial resolution and shear-induced cell stress remain challenges [71].

Looking forward, future developments are expected to focus on the design of next-generation bioinks with tunable viscoelasticity, the integration of biologically active nanomaterials to guide cellular behaviour, and the advancement of multi-scale fabrication strategies that reconcile structural precision with mechanical robustness. Such innovations are anticipated to expand the applicability of extrusion-based bioprinting to complex, clinically relevant constructs for tissue engineering and personalized medicine.

- Inkjet-Based Bioprinting

Inkjet bioprinting is a non-contact, droplet-based technique that employs thermal, electrostatic, or piezoelectric actuation to deposit cell-laden bioinks with high spatial precision. Its primary advantage lies in rapid, cost-effective fabrication and compatibility with low-viscosity natural and synthetic polymers such as alginate, gelatin, collagen, chitosan, and polyethylene glycol (PEG). While it allows high-throughput printing, limitations include restricted bioink viscosity and lower cell density tolerance, which can affect the physiological relevance of printed constructs [72].

Collectively, inkjet-based strategies hold considerable promise for tissue engineering and regenerative medicine applications, although limitations associated with bioink viscosity and cell density continue to present challenges that warrant further optimization.

- Pressure-assisted bioprinting

Pressure-assisted bioprinting extrudes bioinks under controlled pressure at room temperature, enabling the direct embedding of viable cells. Materials include hydrogels, collagen, chitosan, ceramics, and synthetic polymers such as PEG, PLA, and PCL. The method allows for high-viscosity material printing and customization of construct architecture, though shear stress, bioink optimization, and resolution limitations require careful consideration [70].

Future directions in PAB are likely to focus on engineered bioinks with tunable viscoelastic properties, multi-material printing for hybrid tissue constructs, and integration with post-processing strategies such as crosslinking or mineralization to enhance mechanical and functional properties. These advances aim to broaden the applicability of pressure-assisted bioprinting in tissue engineering, regenerative medicine, and personalized implant fabrication.

- Laser-assisted bioprinting

Laser-assisted bioprinting uses a pulsed laser to transfer bioinks onto a substrate in a droplet-based, non-contact manner. Bioinks comprise natural polymers (collagen, gelatin, alginate, fibrin), synthetic hydrogels (PEG), decellularized extracellular matrix (ECM), nano-hydroxyapatite, and cell-laden formulations. LAB provides high spatial resolution, precise cell placement, and non-contact deposition, but faces challenges related to equipment cost, laser-induced thermal stress, and bioink viscosity constraints [70].

Future directions for LAB involve the development of next-generation bioinks with enhanced optical and rheological properties, integration with multi-material printing strategies, and the exploration of tissue-specific composite formulations for regenerative medicine and personalized tissue engineering applications. Table 6 presents comparative overview of 3D bioprinting modalities.

Table 6. Comparative Overview of 3D Bioprinting Modalities.

Feature	Inkjet Bioprinting	Extrusion-Based Bioprinting	Pressure-Assisted Bioprinting	Laser-Assisted Bioprinting
Mechanism	Droplet ejection via thermal, electrostatic, or piezoelectric forces	Continuous extrusion (SSE/FDM)	Continuous extrusion under pneumatic, plunger, or screw pressure	Laser-induced droplet transfer
Materials	Low-viscosity hydrogels: alginate, gelatin, collagen, chitosan, PEG	Natural (alginate, gelatin, collagen, nanocellulose) and synthetic (PCL, PVA), often blended	Hydrogels, collagen, chitosan, ceramics, PEG, PLA, PCL	Natural polymers (collagen, gelatin, alginate, fibrin), PEG, dECM, nHA, cell-laden bioinks
Cell Compatibility	High	Moderate	High	High
Resolution	High	Moderate	Moderate	Very high
Viscosity Range	Low	High	High	Moderate
Advantages	Low cost, high-throughput, non-contact	High-viscosity bioinks, mechanically robust, multi-material	Cell-friendly, customizable geometry, high-viscosity	Non-contact, precise, complex architectures, high spatial resolution
Limitations	Limited viscosity, low cell density tolerance	Shear stress, lower resolution, slower	Shear stress, bioink optimization required	Equipment cost, thermal stress, viscosity constraints
Applications	Soft tissue models, high-throughput screening	Soft tissue, bone scaffolds, implants	Tissue constructs, high-viscosity bioinks	Complex tissue constructs, bone and vascular models

L. Cheng et al. demonstrated that scaffolds composed of hydroxyapatite, β-tricalcium phosphate (β-TCP), and bioactive polymers exhibit enhanced osteoconductivity and excellent resistance to mechanical loads [52]. Three-Dimensional bioprinting technology offers unprecedented control over scaffold architecture: porosity can be modulated within the range of 30–90%, with pore size precision of ±50 μm. Printing resolution ranges from 100 to 200 μm depending on the technology used (stereolithography, selective laser sintering, or extrusion printing). Additionally, it allows for the creation of complex internal geometries, including gradient structures and interconnected channel systems with diameters of 150–500 μm to facilitate vascularisation. Compared with traditional implantation methods, 3D bioprinting offers critical advantages: personalisation based on patient-specific anatomy derived from computed tomography data, integration of multiple materials within a single construct, controlled spatial distribution of bioactive molecules, and the fabrication of functionally graded scaffolds that mimic the transition from cortical to cancellous bone. However, the technology presents significant limitations, including high equipment costs (ranging from $100,000 to over $1 million for industrial systems), lengthy fabrication times for complex structures (6–24 h depending on size and complexity), a limited range of clinically certified biocompatible materials, and the requirement for post-processing (e.g., sintering, sterilization), which may affect mechanical integrity and bioactivity. The integration of stem cells with bioprinted scaffolds represents a major advancement in regenerative medicine. An experimental study by Meesuk L. et al. reported that both 3D-printed HA and calcium phosphate-coated 3D-printed HA supported the proliferation and osteogenic differentiation of BM-MSCs and UC-MSCs [67]. A clinical study by P. Adamska et al. confirmed the therapeutic efficacy of off-the-shelf stem cell therapies in geriatric patients, offering regeneration without the need for autologous bone harvesting [68].

In 2023 Hao and colleagues reported a clinical case of bone defect repair using a self-developed 3D bioprinted active bone [73]. To restore the left tibia, they fabricated a bone scaffold from a polycaprolactone/β-tricalcium phosphate (PCL/β-TCP) composite integrated with bio-ink derived from the patient's own platelet-rich plasma (PRP).

Scaling and commercializing bioprinted tissues face major challenges. Designing, sourcing cells, and manufacturing are often highly personalized, which drives up costs. In the short term, companies may focus on universal scaffolds or components that reduce the need for individualized materials. Acellular scaffolds seeded with a patient's own cells could lower costs while limiting rejection risks. Although personalized 3D bioprinting requires high upfront investment, it may prove more cost-effective than long-term non-curative treatments such as dialysis or diabetes management. In many cases, a single intervention is economically preferable to lifelong therapies.

One of the central obstacles in bioprinting is achieving an optimal balance between scaffold degradability and mechanical functionality. Controlling the degradation rate is crucial, since overly rapid degradation can compromise structural support, whereas excessively slow degradation may induce inflammatory responses, thereby hindering tissue regeneration [74]. In parallel, ensuring appropriate mechanical performance requires careful evaluation of multiple parameters—including compressive and tensile strength, elasticity, and fatigue resistance—throughout scaffold fabrication [75].

Hydrogels, which represent water-swollen porous networks, have emerged as promising candidates for cell encapsulation, tissue engineering, and 3D bioprinting applications. For successful use in bioprinting, hydrogels are required to exhibit tunable substrate stiffness and the ability to undergo network reconfiguration post-printing, thereby enabling critical cellular processes such as spreading, migration, proliferation, and intercellular interaction. Despite these advantages, their inherently poor mechanical properties remain a major limitation [75].

Stem cells are considered the preferred seed cells in 3D bioprinting owing to their capacity for self-renewal and multilineage differentiation, as well as their genetic compatibility, which reduces the risk of immune rejection. Nevertheless, their declining proliferative ability upon differentiation limits their use in large 3D constructs. Induced pluripotent stem cells (iPSCs) offer an alternative source, but their generation is complex, time-consuming, and associated with risks such as epigenetic abnormalities and tumorigenicity. Another approach is the use of differentiated cell types for tissue and organ modelling; however, this strategy is costly, technically demanding, and hampered by low differentiation efficiency [76].

3D-bioprinted tissues and organs face major regulatory challenges. While the FDA currently applies the same rules to printed and conventional devices, the 21st Century Cures Act allows some regenerative medicine therapies to qualify for RMAT status, including cell and tissue therapies, engineered constructs, and gene-based interventions. Regulating functional bioprinted products is difficult because of their complex mechanisms, multiple active components, and uncertain long-term effects. Even small manufacturing changes may alter product properties unpredictably, underscoring the need for clear and centralized regulatory pathways, especially as hospitals and research institutions produce personalized devices. Current FDA guidelines and international efforts acknowledge these issues, while standardization and technological convergence could help address them. To speed clinical translation, thorough safety and efficacy testing must be combined with clear instructions for use, defining purpose, patient population, and conditions of application. This structured approach could simplify approval and enable faster patient access to bioprinted therapies, improving healthcare outcomes [77].

The analysis of clinical applications reflects an evolution from simple structural substitutes towards sophisticated bioactive systems with osteoinductive capabilities. Each medical specialty presents distinct requirements for bone substitutes, necessitating the design of specialized materials and technologies. The most promising outcomes are observed when synthetic scaffolds are integrated with biological components, achieving an optimal balance between structural support and regenerative functionality. Nonetheless, clinical practice reveals a spectrum of potential complications. With autografts, donor site morbidity ranges from 20 to 30%, including persistent pain and infection risk [22]. Allogeneic and xenogeneic materials, despite advanced processing, still pose risks of infection transmission and immune reaction. Synthetic substitutes may result in fibrous capsule formation due to their lack of intrinsic osteoinductivity. The use of BMP-2 has been associated with ectopic bone formation and local inflammation in 15–20% of cases. Furthermore, 3D-printed implants present risks of mechanical failure under excessive loading and susceptibility to bacterial colonization of porous structures.

In recent years, bone regeneration has benefited from the development of novel biomaterials (such as energetic materials, metabolic regulatory materials, hydrogels, etc.) that go beyond traditional autogenous, allogenic, and synthetic options. A growing body of evidence highlights the crucial role of bioelectric signals and cellular bioenergetics (CBE) in tissue repair.

Electrical stimulation can enhance the healing of nerves, bones, cardiovascular tissues, and wounds, yet conventional metal electrode devices are invasive and rely on external power sources. Consequently, biocompatible electrically active materials have emerged as promising alternatives, offering the potential to modulate bioelectric cues in situ. At the same time, tissue regeneration is tightly linked to CBE, as enhanced metabolic states promote anabolic biosynthesis and mitosis, thereby accelerating repair. Integrating these bioelectric and bioenergetic strategies through advanced biomaterials may provide a synergistic approach for next-generation bone repair therapies. Haoming Liu et al. develop strategy to produce a bioenergetic scaffold that is simple and easy to implement for enhanced bone regeneration [78].

Recent studies suggest that the intrinsic properties of synthetic materials can influence cell metabolism and potentially guide cell behaviour to affect regenerative engineering outcomes. This can occur through the release of soluble metabolic regulatory factors (such as ions, degradation products, and oxygen), the incorporation of antioxidative properties, and the fine-tuning of cell adhesion, chemical composition, topography, and material stiffness [79]. Metabolic regulation of energy, oxygen, and redox homeostasis plays a key role in controlling cell behaviour during differentiation, angiogenesis, and immune responses in regenerative engineering. Cells can also sense and respond to extracellular cues—such as adhesivity, topography, and stiffness—through mechanosensing networks, influencing their metabolic and biosynthetic states. Although still an emerging field, leveraging metabolic regulation in biomaterials design holds promise for both cell therapy and advanced scaffold development. For instance, citrate-based biomaterials with tunable properties could dynamically provide or absorb metabolic cues to optimize regeneration. Integrating material-driven metabolic pathways offers a unique opportunity for interdisciplinary collaboration to create biomaterials that actively guide cell metabolism and behaviour for predictable regenerative outcomes [79].

With the advancement of medical technology, bioactive materials have become a central focus in promoting bone repair, with hydrogels emerging as one of the most prominent biomaterials in bone tissue engineering (BTE). Hydrogels offer high biocompatibility and mimic the structural characteristics of the porous extracellular matrix (ECM), providing an in vivo environment conducive to cell survival and facilitating bone regeneration. Injectable hydrogels are particularly advantageous, as they can conform to irregular or large defects, filling the cavity while maintaining granule cohesiveness during and after delivery. Various hydrogel matrices have been developed in recent years, enhancing osteogenic induction and improving scaffold efficacy in the treatment of complex bone defects arising from trauma, tumour resection, or infection [80,81].

Composite scaffolds, combining hydrogels with organic or inorganic fillers, further enhance mechanical and biological performance. Examples include poly(ethylene glycol) diacrylate with clay, Oligo(poly(ethylene glycol) fumarate) with calcium phosphate, cyclic acetal hydrogels with nano-hydroxyapatite, alginate with 45S5 bioactive glass, gelatin methacrylate with hydroxyapatite, elastin-like polypeptide collagen with 45S5 bioactive glass, and carbon nanotube-based systems. The primary goal of these composites is to maintain structural integrity during injection and implantation, while the in situ gelation ability of hydrogels—triggered by physical, chemical, temperature, or pH stimuli—provides additional functional advantages [81,82].

Despite their benefits, conventional injectable hydrogels face limitations such as gelation timing, biomechanical compatibility, and potential mechanical damage to surrounding tissues, which can compromise long-term function. Recent innovations in shape-memory and self-healing hydrogels address these challenges, enabling restoration of scaffold morphology while preserving cell viability and tissue functionality, thereby paving the way for more effective and durable bone regeneration strategies [82].

3.6. Advanced Bioactive Strategies

Beyond the traditional use of growth factors and stem cells, emerging approaches in bone tissue engineering highlight bioactive platforms that utilize advanced delivery and signalling mechanisms to shape the regenerative microenvironment. The development of smart implants represents an innovative approach in which advanced materials science is integrated with biological mechanisms to deliver structural support while simultaneously enhancing osteogenesis, preventing infection, promoting vascularization, modulating immune activity, and enabling neuromodulatory osteogenesis [83]. These functionalities arise from carefully engineered material designs—such as bioactive coatings, surface nanostructures, and other surface modifications—that dynamically interact with the cellular microenvironment to improve clinical outcomes [84]. In recent years efforts are focused on the design of synthetic smart bone implants, along with the refinement of their functions and fabrication strategies, such as the induction and modification of bioactive complexes including exosomes and functional proteins. At the same time, considerable research has been devoted to the regulation of macrophage polarization (M1 and M2 phenotypes) by tailoring implant surface topographies and incorporating drugs, cytokines, and bioactive metal ions. Collectively, these advances underscore the importance of the immunomodulatory role of implants as a fundamental mechanism facilitating osteointegration and bone defect repair [85].

Liposomes and other lipid-based nanomedicines, protein nanoparticles, polymeric micelles and nanoparticles, polymer–drug conjugates, and various inorganic nanoparticles are among the most widely studied candidates proposed for nanomedical applications [86]. Mesoporous silica nanoparticles (MSNs), among the most studied nanocarriers, have attracted significant attention for smart drug delivery owing to their favourable physicochemical characteristics. Various studies highlight the effectiveness of silica-based mesoporous nanomatrices in transporting antitumoral [87,88] and imaging agents [89] to bone cancer cells. By incorporating phosphorus groups, SBA-15 mesoporous silica nanomatrices achieved increased alendronate loading and induced apatite formation, making them highly promising for applications in osteoporosis therapy [90]. Silica-based mesoporous glasses incorporated into PLGA or MSNs embedded in porous collagen–gelatin scaffolds [91] have shown controlled vancomycin release for bone infection treatment. Moreover, MSN-loaded scaffolds allow the concurrent delivery of multiple therapeutics; for example, cephalexin can be confined within the mesopores while vascular endothelial growth factors are integrated into the scaffold matrix, enabling simultaneous bacterial eradication and bone regeneration [92].

In tissue engineering, combining gene-based therapies with scaffolds typically involves either loading the scaffolds with transfection/transduction vectors (gene- and RNAi-activated scaffolds) to induce in situ modification of host cells, or pre-seeding the scaffolds with cells that have been transfected or transduced with the target gene(s) before implantation. To promote the expression of specific proteins, gene-activated scaffolds can be loaded with various agents, including pDNA, viral vectors, and RNA transcripts. Complexing pDNA with cationic polymers, such as PEI (polyethylenimine), forms nanoplexes that are endocytosed by cells, enabling nuclear delivery and gene expression. While the precise mechanisms of decomplexation and nuclear translocation remain unclear, this non-viral approach has proven effective for inducing gene expression [93]. Khorsand et al. loaded PEI-pDNA nanoplexes encoding BMP-2 and/or FGF-2 (Fibroblast Growth Factor-2) onto collagen scaffolds to evaluate whether co-expression of the two growth factors could synergistically enhance bone regeneration. In vitro, co-delivery of BMP-2 and FGF-2 nanoplexes to human bone-marrow-derived mesenchymal stem cells (BMSCs) doubled BMP-2 expression compared to single-gene treatments. In a diabetic rabbit long bone model, scaffolds loaded with both nanoplex formulations yielded greater bone regeneration (133.7 mm^3 new bone) and higher union rates than scaffolds delivering FGF-2 (96.4 mm^3) or BMP-2 (82.2 mm^3) alone [94]. Malek-Khatabi et al. developed a gene-activated scaffold that releases BMP-2–encoding chitosan-pDNA nanoplexes in response to enzymatic activity from infiltrating cells [95]. They modified electrospun PCL mats with MMP (metalloproteinases) -sensitive peptide sequences and loaded them with homogeneous nanoplexes produced via a microfluidic system. The scaffolds demonstrated MMP-triggered nanoplex release ($\approx$60% over 24 h), enhanced osteogenic differentiation of MSCs in vitro, and superior bone regeneration in a rat calvarial defect model (~18 mm^3 for modified scaffolds vs. ~9 mm^3 and ~4 mm^3 for unmodified scaffolds and controls) [95].

RNAi-activated scaffolds deliver siRNAs, shRNAs, or miRNAs to silence specific genes and modulate protein expression through RNA interference. This regulation induces phenotypic changes in cells and alters the local microenvironment, supporting bone regeneration. Zhang et al. demonstrated that incorporating miR-26a nanoplexes encapsulated within slow-release PLGA microspheres outperformed both fast-release platforms and those lacking miR-26a in a mouse ectopic bone formation model (~4.3 mm^3 vs. ~1.6 mm^3 and 0 mm^3 for fast-release and no miR-26a, respectively) and in a 5 mm calvarial defect model (~100% vs. ~39% and ~22% for fast-release and no miR-26a, respectively) [96].

Collectively, these advanced bioactive strategies represent a shift toward multifunctional and responsive biomaterials, designed not only to provide structural support but also to actively guide and accelerate the complex biological processes underlying bone regeneration.

4. Conclusions

A systematic review of the current state of bone substitutes in regenerative medicine, based on publications from 2019 to May 2025, has identified key patterns in their biological and mechanical properties that determine their clinical application across various medical specialties. The critical role of porosity in the functionality of synthetic bone substitutes has been established, with structures featuring pore sizes from 23 to 210 micrometres providing optimal osteoconduction, vascularisation, and cellular infiltration, while maintaining mechanical strength at up to 70% open porosity.

The review demonstrates the transformation of bone substitute technologies from traditional structural support materials to bioactive systems with osteoinductive properties. A systematization of the main categories of bone substitutes has shown that autogenous grafts possess the highest osteogenic, osteoinductive, and osteoconductive properties, owing to the complex activity of growth factors, including BMP-2, BMP-7, and concentrated growth factors. However, critical limitations of autografts have been identified, with 20–30% of patients experiencing persistent donor site pain and limited material availability for the reconstruction of significant bone defects, which complicate approximately 6% of all fractures with nonunion. Allogeneic and xenogeneic substitutes demonstrate moderate osteoconductive properties and preserve structural integrity, but require intensive processing to minimize the risks of infection transmission and immune reactions, in accordance with European Directives 2004/23/EC and 2006/86/EC. Synthetic substitutes—based on calcium phosphate ceramics, biodegradable polymers, and bioactive glass—offer controlled degradation and excellent biocompatibility, but lack intrinsic osteogenic potential unless combined with bioactive factors. Advanced bioactive strategies illustrate a transition toward multifunctional, responsive biomaterials that not only provide structural support but also actively modulate the cellular microenvironment to enhance and accelerate bone regeneration. Recent advances in bioactive and composite biomaterials, including hydrogels, energetic and metabolic regulatory materials, demonstrate that integrating bioelectric and bioenergetic cues with tailored scaffold designs can actively guide cell behaviour and metabolism, offering a promising strategy for more effective and durable bone regeneration.

The clinical feasibility of specific types of bone substitutes is determined by the particular requirements of individual medical fields. In orthopedic traumatology, autogenous grafts and synthetic calcium phosphate materials with high mechanical strength are preferred for load-bearing applications. In dentistry and maxillofacial surgery, xenogeneic bovine materials are favoured due to their volumetric stability and reliable implant integration. In neurosurgery, materials with controlled degradation and high biocompatibility are essential, with autogenous dentin matrices and synthetic polymers proving effective. The study confirmed the suitability of β-tricalcium phosphate for long-term applications due to its optimal degradation profile, in contrast to calcium sulphate, which exhibits overly rapid resorption. The highest regeneration rates are observed when synthetic scaffolds are combined with mesenchymal stem cells and extracellular vesicles, which significantly enhance the osteogenic potential of these materials.

The future of bone substitution is likely to be shaped by the integration of advanced biomaterials, bioactive cues, and personalized scaffold designs. Emerging materials, including metabolic regulatory biomaterials, electrically active scaffolds, and self-healing hydrogels, offer the potential to modulate cell behaviour and tissue regeneration more effectively than traditional autogenous, allogenic, or synthetic substitutes. Composite scaffolds combining organic and inorganic components can further enhance mechanical stability while providing biologically active environments that promote osteogenesis and vascularization.

Personalized approaches, particularly 3D bioprinting, will enable the fabrication of patient-specific scaffolds that conform to complex anatomical defects, potentially reducing complications associated with donor site morbidity or immune reactions. In parallel, integrating bioelectric and bioenergetic cues into scaffold design may further accelerate tissue repair by regulating cellular metabolism, redox balance, and signalling pathways. Achieving these advances will require interdisciplinary collaboration between materials scientists, bioengineers, and clinicians, as well as the establishment of standardized regulatory frameworks to ensure safety, efficacy, and clinical translatability. Together, these strategies promise a new generation of bone substitutes that are both functionally superior and clinically practical.

A limitation of the study is the absence of long-term clinical outcome analyses concerning the use of different bone substitutes in patients with various comorbidities. Future research should focus on the development of fully biodegradable composite systems with controlled growth factor release, alongside the integration of 3D bioprinting technologies to create individualized tissue-engineered solutions incorporating triply periodic minimal surfaces.

Author Contributions: Conceptualization, N.I. and S.I.; methodology, N.I.; validation, S.I. and S.P.; formal analysis, T.D.; investigation, N.I.; resources, S.I.; data curation, S.I.; writing—original draft preparation, N.I.; writing—review and editing, S.P. and T.D.; supervision, S.P.; project administration, T.D.; funding acquisition, T.D. All authors have read and agreed to the published version of the manuscript.

Funding: This study is financed by the European Union-Next Generation EU through the National Recovery and Resilience Plan of the Republic of Bulgaria, Project No. BG-RRP-2.004-0009.

Institutional Review Board Statement: Not applicable.

Informed Consent Statement: Not applicable.

Conflicts of Interest: The authors declare no conflicts of interest. The funders had no role in the design of the study; in the collection, analyses, or interpretation of data; in the writing of the manuscript; or in the decision to publish the results.

Abbreviations

The following abbreviations are used in this manuscript:

MeSH	Medical Subject Headings
BMP	Bone Morphogenetic Proteins
CGF	Concentrated Growth Factor
BMP-2	Bone Morphogenetic Protein-2
BMP-7	Bone Morphogenetic Protein-7
DBM	Demineralised Bone Matrix
rhVEGF	Recombinant Human Vascular Endothelial Growth Factor
ACL	Anterior Cruciate Ligament
GBR	Guided Bone Regeneration
PRF	Platelet-Rich Fibrin
PRP	Platelet-rich plasma
NAAT	Nucleic Acid Amplification Tests
AATB	American Association of Tissue Banks
Allo-DDM	Allogeneic Demineralised Dentin Matrix
HHP	High Hydrostatic Pressure
PCMV/PRV	Porcine Cytomegalovirus/Porcine Retrovirus
PERVs	Porcine Endogenous Retroviruses
β-TCP	β-Tricalcium Phosphate
TPMS	Triply Periodic Minimal Surfaces
PLA	Polylactic Acid
PCL	Polycaprolactone
ZrO_2	Zirconium Oxide
micro-CT	Micro-Computed Tomography
PLGA	Poly(lactic-glycolic acid)
BCP	Biphasic Calcium Phosphate
DPSCs	Human Dental Pulp Stem Cells
EVs	Extracellular Vesicles
rhPDGF-BB	Recombinant Human Platelet-Derived Growth Factor-BB
CPCs	Calcium phosphate cements
PEG	Polyethylene glycol
ECM	Extracellular matrix
RMAT	Regenerative Medicine Advanced Therapy

References

1. Ferraz, M.P. Bone grafts in dental medicine: An overview of autografts, allografts, and synthetic materials. *Materials* **2023**, *16*, 4117. [CrossRef]
2. Heitzer, M.; Modabber, A.; Zhang, X.; Winnand, P.; Zhao, Q.; Bläsius, F.M.; Buhl, E.M.; Wolf, M.; Neuss, S.; Hölzle, F.; et al. In vitro comparison of the osteogenic capability of human pulp stem cells on alloplastic, allogeneic, and xenogeneic bone scaffolds. *BMC Oral Health* **2023**, *23*, 56. [CrossRef] [PubMed]
3. Fattahian, H.; Mansouri, K.; Mansouri, N. Biomaterials, substitutes, and tissue engineering in bone repair: Current and future concepts. *Comp. Clin. Pathol.* **2019**, *28*, 879–891. [CrossRef]
4. Dahiya, U.; Mishra, S.; Bano, S. Application of bone substitutes and its future prospective in regenerative medicine. In *Biomaterial-Supported Tissue Reconstruction or Regeneration*; IntechOpen: London, UK, 2019. [CrossRef]
5. Manzini, B.; Machado, L.; Noritomi, P.; Da Silva, J. Advances in bone tissue engineering: A fundamental review. *J. Biosci.* **2021**, *46*, 17. [CrossRef]
6. Miletić, M.; Puač, N.; Škoro, N.; Brković, B.; Andrić, M.; Prokić, B.B.; Danilović, V.; Milutinović-Smiljanić, S.; Mitrović-Ajtić, O.; Mojsilović, S. Bone regeneration potential of periodontal ligament stem cells in combination with cold atmospheric plasma-pretreated beta-tricalcium phosphate: An in vivo assessment. *Appl. Sci.* **2023**, *14*, 16. [CrossRef]
7. Mukhlis, S.; Issabekova, A.; Kudaibergen, G.; Mukhambetova, A.; Nurkina, A.; Altayeva, N.; Ashikbayeva, M.; Temirzhan, A.; Baidarbekov, M.; Ogay, V. The role of pericytes in bone tissue regeneration and therapeutic applications. *Eurasian J. Appl. Biotechnol.* **2024**, *50*, 236. [CrossRef]
8. Dőri, F.; Sculean, A.; Takács, D.; Suba, Z. Histological examination of retrieved ePTFE membranes following regenerative surgery of intrabony defects treated with platelet-rich plasma and bone substitutes. *Oral Health Prev. Dent.* **2022**, *20*, 143–148. [CrossRef]

9. Zhang, Q.; Nettleship, I.; Schmelzer, E.; Gerlach, J.; Zhang, X.; Wang, J.; Liu, C. Tissue engineering and regenerative medicine therapies for cell senescence in bone and cartilage. *Tissue Eng. Part B Rev.* **2019**, *25*, 500–526. [CrossRef]
10. Rupp, M.; Klute, L.; Baertl, S.; Walter, N.; Mannala, G.K.; Frank, L.; Pfeifer, C.; Alt, V.; Kerschbaum, M. The clinical use of bone graft substitutes in orthopedic surgery in Germany—A 10-years survey from 2008 to 2018 of 1,090,167 surgical interventions. *J. Biomed. Mater. Res. B Appl. Biomater.* **2022**, *110*, 350–357. [CrossRef]
11. Laubach, M.; Hildebrand, F.; Suresh, S.; Wagels, M.; Kobbe, P.; Gilbert, F.; Kneser, U.; Holzapfel, B.M.; Hutmacher, D.W. The concept of scaffold-guided bone regeneration for the treatment of long bone defects: Current clinical application and future perspective. *J. Funct. Biomater.* **2023**, *14*, 341. [CrossRef]
12. Santoro, A.; Voto, A.; Fortino, L.; Guida, R.; Laudisio, C.; Cillo, M.; D'Ursi, A.M. Bone defect treatment in regenerative medicine: Exploring natural and synthetic bone substitutes. *Int. J. Mol. Sci.* **2025**, *26*, 3085. [CrossRef]
13. Tournier, P.; Guicheux, J.; Paré, A.; Maltezeanu, A.; Blondy, T.; Veziers, J.; Vignes, C.; André, M.; Lesoeur, J.; Barbeito, A.; et al. A partially demineralized allogeneic bone graft: In vitro osteogenic potential and preclinical evaluation in two different intramembranous bone healing models. *Sci. Rep.* **2021**, *11*, 4907. [CrossRef]
14. Parikh, S.; Pentz, R.; Haight, A.; Adeli, M.; Martin, P.; Driscoll, T.; Page, K.; Kurtzberg, J.; Prasad, V.K.; Barfield, R.C.; et al. Ethical considerations of using a single minor donor for three bone marrow harvests for three HLA-matched siblings with primary immunodeficiency. *Pediatr. Blood Cancer* **2019**, *66*, e27602. [CrossRef]
15. Sharifi, M.; Kheradmandi, R.; Salehi, M.; Alizadeh, M.; Hagen, T.; Falahati, M. Criteria, challenges, and opportunities for acellularized allogeneic/xenogeneic bone grafts in bone repairing. *ACS Biomater. Sci. Eng.* **2022**, *8*, 3199–3219. [CrossRef] [PubMed]
16. Ana, I. Bone substituting materials in dental implantology. In *Bone Management in Dental Implantology*; Springer: Berlin/Heidelberg, Germany, 2019; pp. 101–119. [CrossRef]
17. Janjua, O.S.; Qureshi, S.M.; Shaikh, M.S.; Alnazzawi, A.; Rodriguez-Lozano, F.J.; Pecci-Lloret, M.P.; Zafar, M.S. Autogenous tooth bone grafts for repair and regeneration of maxillofacial defects: A narrative review. *Int. J. Environ. Res. Public Health* **2022**, *19*, 3690. [CrossRef] [PubMed]
18. Malik, S.; Shaikh, A.; Ayub, T.; Rehman, A.; Khan, H.; Haque, A. Efficacy of concentrated growth factor with autograft and xenograft in mandible fractures: A randomized clinical trial. *J. Univ. Coll. Med. Dent.* **2024**, *4*, e3519. [CrossRef]
19. Black, C.; Gibbs, D.; McEwan, J.; Kanczler, J.; Fernández, M.; Tozzi, G.; Dawson, J.; Oreffo, R. Comparison of bone formation mediated by bone morphogenetic protein delivered by nanoclay gels with clinical techniques (autograft and InductOs®) in an ovine bone model. *J. Tissue Eng.* **2022**, *13*, 20417314221113746. [CrossRef]
20. Jin, Y.; Zheng, G.; Lee, J.; Han, S. Comparison of demineralized bone matrix and hydroxyapatite as carriers of Escherichia coli recombinant human BMP-2. *Biomater. Res.* **2021**, *25*, 20. [CrossRef]
21. Tasdemir, U.; Iyilikçi, B.; Aktürk, M.C.; Ozmen, O.; Kizildağ, A.; Elmali, Z. The effect of autogenous bone graft mixed with recombinant human vascular endothelial growth factor on bone regeneration. *J. Craniofac. Surg.* **2021**, *32*, 2233–2237. [CrossRef]
22. Taşdemir, U.; Kirtay, M.; Keleş, A.; Çil, N.; Abban, G.; Dodurga, Y. Autogenous tooth bone graft and simvastatin combination effect on bone healing. *J. Craniofac. Surg.* **2020**, *31*, 2350–2354. [CrossRef]
23. Kunze, K.; Moran, J.; Polce, E.; Pareek, A.; Strickland, S.; Williams, R. Lower donor site morbidity with hamstring and quadriceps tendon autograft compared with bone-patellar tendon-bone autograft after anterior cruciate ligament reconstruction: A systematic review and network meta-analysis of randomized controlled trials. *Knee Surg. Sports Traumatol. Arthrosc.* **2023**, *31*, 3339–3352. [CrossRef]
24. Attia, A.; Mahmoud, K.; ElSweify, K.; Bariteau, J.; Labib, S. Donor site morbidity of calcaneal, distal tibial, and proximal tibial cancellous bone autografts in foot and ankle surgery. A systematic review and meta-analysis of 2296 bone grafts. *Foot Ankle Surg.* **2021**, *27*, 729–735. [CrossRef]
25. Khalid, M.; Janjua, S.; Sheraz, M.; Kanwal, S.; Ghouri, Q.; Shaheen, U. Quantifying donor site morbidity in anterior cruciate ligament reconstruction using peroneus longus tendon autograft. *J. Musculoskelet. Surg. Res.* **2024**, *8*, 165. [CrossRef]
26. Sun, H.; Yin, X.; Yang, C.; Kuang, H.; Luo, W. Advances in autogenous dentin matrix graft as a promising biomaterial for guided bone regeneration in the maxillofacial region: A review. *Medicine* **2024**, *103*, e39422. [CrossRef] [PubMed]
27. Liu, J.; Yang, L.; Zhang, H.; Zhang, J.Y.; Hu, Y.C. Effects of allogeneic bone substitute configurations on the cell adhesion process in vitro. *Orthop. Surg.* **2023**, *15*, 579–590. [CrossRef] [PubMed]
28. European Parliament and Council. Directive 2004/23/EC of the European Parliament and of the Council of 31 March 2004 on setting standards of quality and safety for the donation, procurement, testing, processing, preservation, storage and distribution of human tissues and cells. *Off. J. Eur. Union* **2004**, *L102*, 48–58. Available online: https://eur-lex.europa.eu/legal-content/EN/TXT/?uri=CELEX%3A32004L0023 (accessed on 7 April 2004).

29. European Parliament and Council. Commission Directive 2006/86/EC laying down technical requirements for the coding, processing, preservation, storage and distribution of human tissues and cells. *Off. J. Eur. Union* **2006**, *L294*, 32–50. Available online: https://eur-lex.europa.eu/legal-content/EN/TXT/?uri=CELEX%3A32006L0086 (accessed on 25 October 2006).
30. Ilyas, I.; Al-Rabiah, A.; Alhussainan, T.; Alrumaih, H.; Fallatah, A.; Alsakran, S.; Al-Mohrej, O.A. Principles of bone and tissue banking in Saudi Arabia: 10-year experience report. *Cell Tissue Bank* **2021**, *22*, 93–101. [CrossRef]
31. Regmi, A.; Niraula, B.; Maheshwari, V.; Nongdamba, H.; Karn, R.; Bondarde, P.; Anand, U.; Dhingra, M.; Kandwal, P. Establishing a bone bank within a hospital setting in India: Early insights from a tertiary care center in Northern India—A review article. *Cell Tissue Bank* **2024**, *25*, 873–882. [CrossRef]
32. Asamoto, T.; Osawa, Y.; Takegami, Y.; Takemoto, G.; Komatsu, D.; Seki, T.; Imagama, S. The survey of bone allograft transplantation in a Japanese regional bone bank. *J. Orthop. Sci.* **2024**, *29*, 804–809. [CrossRef]
33. American Association of Tissue Banks. AATB Standards Rebuild Project: Draft 15th Edition Standards For Tissue Banking. 2023. Available online: https://www.aatb.org/news/aatb-standards-rebuild-project-draft-15th-edition-standards-tissue-banking (accessed on 18 December 2023).
34. Um, I.; Lee, J.; Kim, J.; Kim, Y.; Bakhshalian, N.; Jeong, Y.; Ku, J.K. Allogeneic dentin graft: A review on its osteoinductivity and antigenicity. *Materials* **2021**, *14*, 1713. [CrossRef]
35. Waletzko-Hellwig, J.; Saemann, M.; Schulze, M.; Frerich, B.; Bader, R.; Dau, M. Mechanical characterization of human trabecular and formed granulate bone cylinders processed by high hydrostatic pressure. *Materials* **2021**, *14*, 1069. [CrossRef] [PubMed]
36. Shao, A.; Ling, Y.; Chen, L.; Wei, L.; Fan, C.; Lei, D.; Xu, L.; Wang, C. GGTA1/iGb3S double knockout mice: Immunological properties and immunogenicity response to xenogeneic bone matrix. *Biomed. Res. Int.* **2020**, *2020*, 9680474. [CrossRef] [PubMed]
37. Amid, R.; Kheiri, A.; Kheiri, L.; Kadkhodazadeh, M.; Ekhlasmandkermani, M. Structural and chemical features of xenograft bone substitutes: A systematic review of in vitro studies. *Biotechnol. Appl. Biochem.* **2021**, *68*, 1432–1452. [CrossRef] [PubMed]
38. Jerbić Radetić, A.T.; Zoričić Cvek, S.; Tomas, M.; Erjavec, I.; Oguić, M.; Perić Kačarević, Ž.; Cvijanović Peloza, O. CSBD healing in rats after application of bovine xenogeneic biomaterial enriched with magnesium alloy. *Int. J. Mol. Sci.* **2021**, *22*, 9089. [CrossRef]
39. Pröhl, A.; Batinic, M.; Alkildani, S.; Hahn, M.; Radenkovic, M.; Najman, S.; Jung, O.; Barbeck, M. In vivo analysis of the biocompatibility and bone healing capacity of a novel bone grafting material combined with hyaluronic acid. *Int. J. Mol. Sci.* **2021**, *22*, 4818. [CrossRef]
40. Denner, J. The porcine cytomegalovirus (PCMV) will not stop xenotransplantation. *Xenotransplantation* **2022**, *29*, e12763. [CrossRef]
41. Groenendaal, H.; Costard, S.; Ballard, R.; Bienhoff, S.; Challen, D.; Dominguez, B.; Kern, D.R.; Miller, D.; Noordergraaf, J.; Rudenko, L.; et al. Expert opinion on the identification, risk assessment, and mitigation of microorganisms and parasites relevant to xenotransplantation products from pigs. *Xenotransplantation* **2023**, *30*, e12815. [CrossRef]
42. U.S. Food and Drug Administration. Source Animal, Product, Preclinical, And Clinical Issues Concerning the Use of Xenotransplantation Products in Humans: Guidance for Industry. 2016. Available online: https://www.fda.gov/regulatory-information/search-fda-guidance-documents/source-animal-product-preclinical-and-clinical-issues-concerning-use-xenotransplantation-products (accessed on 22 April 2024).
43. European Medicines Agency. Guideline on Xenogeneic Cell-Based Medicinal Products (EMEA/CHMP/CPWP/83508/2009). 2009. Available online: https://www.ema.europa.eu/en/xenogeneic-cell-based-medicinal-products-scientific-guideline (accessed on 1 December 2019).
44. Hawthorne, W.J.; Cowan, P.J.; Ayares, D.; Bührer-Sékula, S.; Buhler, L.H.; Bottino, R.; Pierson, R.N., 3rd; Ahn, C.; Azimzadeh, A.; Cozzi, E.; et al. Third WHO global consultation on regulatory requirements for xenotransplantation clinical trials. *Xenotransplantation* **2019**, *26*, e12513. [CrossRef]
45. Sánchez-Labrador, L.; Molinero-Mourelle, P.; Pérez-González, F.; Saez-Alcaide, L.M.; Brinkmann, J.C.; Martínez, J.L.; Martínez-González, J.M. Clinical performance of alveolar ridge augmentation with xenogeneic bone block grafts versus autogenous bone block grafts: A systematic review. *J. Stomatol. Oral Maxillofac. Surg.* **2021**, *122*, 293–302. [CrossRef]
46. Maevskaia, E.; Khera, N.; Ghayor, C.; Bhattacharya, I.; Guerrero, J.; Nicholls, F.; Waldvogel, C.; Bärtschi, R.; Fritschi, L.; Salamon, D.; et al. Three-dimensional printed hydroxyapatite bone substitutes designed by a novel periodic minimal surface algorithm are highly osteoconductive. *3D Print. Addit. Manuf.* **2023**, *10*, 905–916. [CrossRef]
47. Ding, Y.; Pan, H.; Wang, Z.; Xiao, B. Progress and prospects of 3D printing human skeletal scaffold. *Highlights Sci. Eng. Technol.* **2024**, *70*, 123–130. [CrossRef]
48. Li, X.; Lu, W.; Xu, X.; Wang, Y.; Chen, S. Advanced optical methods and materials for fabricating 3D tissue scaffolds. *Light Adv. Manuf.* **2022**, *3*, 26. [CrossRef]
49. Nifant'ev, I.; Tavtorkin, A.; Komarov, P.; Kretov, E.; Korchagina, S.; Chinova, M.; Gavrilov, D.; Ivchenko, P. Dispersant and protective roles of amphiphilic poly(ethylene phosphate) block copolymers in polyester/bone mineral composites. *Int. J. Mol. Sci.* **2023**, *24*, 11175. [CrossRef] [PubMed]

50. Kang, T.Y.; Seo, J.Y.; Ryu, J.H.; Kim, K.M.; Kwon, J.S. Improvement of the mechanical and biological properties of bioactive glasses by the addition of zirconium oxide (ZrO_2) as a synthetic bone graft substitute. *J. Biomed. Mater. Res. A* **2021**, *109*, 1196–1208. [CrossRef]
51. Ferraz, M.P.; Monteiro, F.J.; Manuel, C.M. Hydroxyapatite nanoparticles: A review of preparation methodologies. *J. Appl. Biomater. Biomech.* **2020**, *2*, 74–80.
52. Cheng, L.; Suresh, K.S.; He, H.; Rajput, R.S.; Feng, Q.; Ramesh, S.; Wang, Y.; Krishnan, S.; Ostrovidov, S.; Camci-Unal, G.; et al. 3D printing of micro- and nanoscale bone substitutes: A review on technical and translational perspectives. *Int. J. Nanomed.* **2021**, *16*, 4289–4319. [CrossRef]
53. Liang, D.; Zhong, C.; Jiang, F.; Liao, J.; Ye, H.; Ren, F. Fabrication of porous tantalum with low elastic modulus and tunable pore size for bone repair. *ACS Biomater. Sci. Eng.* **2023**, *9*, 2192–2205. [CrossRef]
54. Kumar, A.; Zhang, Y.; Terracciano, A.; Zhao, X.; Su, T.; Kalyon, D.; Katebifar, S.; Kumbar, S.G.; Yu, X. Load-bearing biodegradable polycaprolactone-poly (lactic-co-glycolic acid)- beta tri-calcium phosphate scaffolds for bone tissue regeneration. *Polym. Adv. Technol.* **2019**, *30*, 1189–1197. [CrossRef]
55. Celik, D.; Ustundag, C. Fabrication of biomimetic scaffold through hybrid forming technique. *Int. J. Ceram. Eng. Sci.* **2024**, *6*, e10210. [CrossRef]
56. Marongiu, G.; Verona, M.; Cardoni, G.; Capone, A. Synthetic bone substitutes and mechanical devices for the augmentation of osteoporotic proximal humeral fractures: A systematic review of clinical studies. *J. Funct. Biomater.* **2020**, *11*, 29. [CrossRef] [PubMed]
57. Kang, H.; Makkar, P.; Padalhin, A.; Lee, G.; Im, S.; Lee, B. Comparative study on biodegradation and biocompatibility of multichannel calcium phosphate based bone substitutes. *Mater. Sci. Eng. C Mater. Biol. Appl.* **2020**, *110*, 110694. [CrossRef] [PubMed]
58. Kowalewicz, K.; Vorndran, E.; Feichtner, F.; Waselau, A.; Brueckner, M.; Meyer-Lindenberg, A. In-vivo degradation behavior and osseointegration of 3D powder-printed calcium magnesium phosphate cement scaffolds. *Materials* **2021**, *14*, 946. [CrossRef] [PubMed]
59. Heitzer, M.; Zhao, Q.; Greven, J.; Winnand, P.; Zhang, X.; Bläsius, F.M.; Buhl, E.M.; Wolf, M.; Neuss, S.; Hildebrand, F.; et al. Evaluation of in vitro biocompatibility of human pulp stem cells with allogeneic, alloplastic, and xenogeneic grafts under the influence of extracellular vesicles. *Sci. Rep.* **2023**, *13*, 12475. [CrossRef]
60. Sallent, I.; Capella-Monsonís, H.; Procter, P.; Bozo, I.Y.; Deev, R.V.; Zubov, D.; Vasyliev, R.; Perale, G.; Pertici, G.; Baker, J.; et al. The few who made it: Commercially and clinically successful innovative bone grafts. *Front. Bioeng. Biotechnol.* **2020**, *8*, 952. [CrossRef]
61. Massari, L.; Saracco, A.; Marchesini, S.; Gambuti, E.; Delorenzi, A.; Caruso, G. Safety of a porous hydroxyapatite bone substitute in orthopedics and traumatology: A multi-centric clinical study. *J. Funct. Morphol. Kinesiol.* **2024**, *9*, 71. [CrossRef]
62. Mahardawi, B.; Jiaranuchart, S.; Dhanesuan, K.; Arunjaroensuk, S.; Mattheos, N.; Pimkhaokham, A. The clinical efficacy of the allogenic demineralized dentin matrix graft for implant placement: A systematic review. *Oral Maxillofac. Surg.* **2024**, *28*, 585–593. [CrossRef]
63. Batra, C.; Goel, A.; Daneshparvar, N.; Hamada, Y. Clinical evaluation of the combination of rhPDGF-BB and xenogeneic bone substites for the treatment of severe periodontal intrabony defects: A case series. *Int. J. Periodontics Restor. Dent.* **2023**, *43*, 192–200. [CrossRef]
64. Zhang, S.; Li, X.; Qi, Y.; Ma, X.; Qiao, S.; Cai, H.; Zhao, B.C.; Jiang, H.B.; Lee, E.S. Comparison of autogenous tooth materials and other bone grafts. *Tissue Eng. Regen. Med.* **2021**, *18*, 327–341. [CrossRef]
65. Jung, K.J.; Sarkar, S.K.; Kim, W.J.; Kim, B.R.; Park, J.S.; Lee, B.T. Bone regeneration by multichannel cylindrical granular bone substitutes for regeneration of bone in cases of tumor, fracture, and arthroplasty. *Int. J. Environ. Res. Public Health* **2022**, *19*, 8228. [CrossRef]
66. Baldwin, P.; Li, D.; Auston, D.; Mir, H.; Yoon, R.; Koval, K. Autograft, allograft, and bone graft substitutes: Clinical evidence and indications for use in the setting of orthopaedic trauma surgery. *J. Orthop. Trauma* **2019**, *33*, 203–213. [CrossRef]
67. Meesuk, L.; Suwanprateeb, J.; Thammarakcharoen, F.; Tantrawatpan, C.; Kheolamai, P.; Palang, I.; Tantikanlayaporn, D.; Manochantr, S. Osteogenic differentiation and proliferation potentials of human bone marrow and umbilical cord-derived mesenchymal stem cells on the 3D-printed hydroxyapatite scaffolds. *Sci. Rep.* **2022**, *14*, 19509. [CrossRef] [PubMed]
68. Adamska, P.; Pylińska-Dąbrowska, D.; Stasiak, M.; Sobczak-Zagalska, H.; Jusyk, A.; Zedler, A.; Studniarek, M. Tooth autotransplantation, autogenous dentin graft, and growth factor application: A method for preserving the alveolar ridge in cases of severe infra occlusion—A case report and literature review. *J. Clin. Med.* **2024**, *13*, 3902. [CrossRef] [PubMed]
69. Derby, B. Printing and Prototyping of Tissues and Scaffolds. *Science* **2012**, *338*, 921–926. [CrossRef] [PubMed]
70. Li, J.; Chen, M.; Fan, X.; Zhou, H. Recent Advances in Bioprinting Techniques: Approaches, Applications and Future Prospects. *J. Transl. Med.* **2016**, *14*, 271. [CrossRef]

71. Rasheed, A.; Azizi, L.; Turkki, P.; Janka, M.; Hytönen, V.P.; Tuukkanen, S. Extrusion-Based Bioprinting of Multilayered Nanocellulose Constructs for Cell Cultivation Using In Situ Freezing and Preprint $CaCl_2$ Cross-Linking. *ACS Omega* **2021**, *6*, 569–578. [CrossRef]
72. Vanaei, S.; Parizi, M.S.; Vanaei, S.; Salemizadehparizi, F.; Vanaei, H.R. An Overview on Materials and Techniques in 3D Bioprinting Toward Biomedical Application. *Eng. Regen.* **2021**, *2*, 1–18. [CrossRef]
73. Hao, Y.; Cao, B.; Deng, L.; Li, J.; Ran, Z.; Wu, J.; Pang, B.; Tan, J.; Luo, D.; Wu, W. The first 3D-bioprinted personalized active bone to repair bone defects: A case report. *Int. J. Bioprint* **2022**, *9*, 654. [CrossRef]
74. Zan, J.; Qian, G.; Deng, F.; Zhang, J.; Zeng, Z.; Peng, S.; Shuai, C.J. Dilemma and Breakthrough of Biodegradable Poly-L-Lactic Acid in Bone Tissue Repair. *J. Mater. Res. Technol.* **2022**, *17*, 2369–2387. [CrossRef]
75. Li, G.; Yang, H.; Zheng, Y.; Chen, X.-H.; Yang, J.-A.; Zhu, D.; Ruan, L.; Takashima, K. Challenges in the Use of Zinc and Its Alloys as Biodegradable Metals: Perspective from Biomechanical Compatibility. *Acta Biomater.* **2019**, *97*, 23–45. [CrossRef]
76. Ma, Y.; Deng, B.; He, R.; Huang, P. Advancements of 3D Bioprinting in Regenerative Medicine: Exploring Cell Sources for Organ Fabrication. *Heliyon* **2024**, *10*, e24593. [CrossRef] [PubMed]
77. Mladenovska, T.; Choong, P.F.; Wallace, G.G.; O'Connell, C.D. The Regulatory Challenge of 3D Bioprinting. *Regen. Med.* **2023**, *18*, 659–674. [CrossRef] [PubMed]
78. Liu, H.; Du, Y.; St-Pierre, J.-P.; Bergholt, M.S.; Autefage, H.; Wang, J.; Cai, M.; Yang, G.; Stevens, M.M.; Zhang, S. Bioenergetic-Active Materials Enhance Tissue Regeneration by Modulating Cellular Metabolic State. *Sci. Adv.* **2020**, *6*, eaay7608. [CrossRef] [PubMed]
79. Ma, C.; Kuzma, M.L.; Bai, X.; Yang, J. Biomaterial-Based Metabolic Regulation in Regenerative Engineering. *Adv. Sci.* **2019**, *6*, 1900819. [CrossRef]
80. Nallusamy, J.; Das, R.K. Hydrogels and Their Role in Bone Tissue Engineering: An Overview. *J. Pharm. Bioallied Sci.* **2021**, *13* (Suppl. S2), S908–S912. [CrossRef]
81. Utech, S.; Boccaccini, A.R. A Review of Hydrogel-Based Composites for Biomedical Applications: Enhancement of Hydrogel Properties by Addition of Rigid Inorganic Fillers. *J. Mater. Sci.* **2016**, *51*, 271–310. [CrossRef]
82. Shen, C.; Han, Y.; Xiong, H.; Wang, Y.; Tan, Z.; Wei, H.; Ding, Q.; Ma, L.; Ding, C.; Zhao, T. Multifunctional Hydrogel Scaffolds Based on Polysaccharides and Polymer Matrices Promote Bone Repair: A Review. *Int. J. Biol. Macromol.* **2025**, *294*, 139418. [CrossRef]
83. Sun, W.; Ye, B.; Chen, S.; Zeng, L.; Lu, H.; Wan, Y.; Gao, Q.; Chen, K.; Qu, Y.; Wu, B. Neuro–Bone Tissue Engineering: Emerging Mechanisms, Potential Strategies, and Current Challenges. *Bone Res.* **2023**, *11*, 65. [CrossRef]
84. Zhou, H.; Ye, S.; Xu, M.; Hao, L.; Chen, J.; Fang, Z.; Guo, K.; Chen, Y.; Wang, L. Dynamic Surface Adapts to Multiple Service Stages by Orchestrating Responsive Polymers and Functional Peptides. *Biomaterials* **2023**, *301*, 122200. [CrossRef]
85. Wang, T.; Bai, J.; Lu, M.; Huang, C.; Geng, D.; Chen, G.; Wang, L.; Qi, J.; Cui, W.; Deng, L. Engineering Immunomodulatory and Osteoinductive Implant Surfaces via Mussel Adhesion-Mediated Ion Coordination and Molecular Clicking. *Nat. Commun.* **2022**, *13*, 160. [CrossRef]
86. Sun, T.; Zhang, Y.S.; Pang, B.; Hyun, D.C.; Yang, M.; Xia, Y. Engineered Nanoparticles for Drug Delivery in Cancer Therapy. *Angew. Chem. Int. Ed. Engl.* **2014**, *53*, 12320–12364. [CrossRef] [PubMed]
87. Zhang, Y.; Hu, M.; Wang, X.; Zhou, Z.; Liu, Y. Design and Evaluation of Europium Containing Mesoporous Bioactive Glass Nanospheres: Doxorubicin Release Kinetics and Inhibitory Effect on Osteosarcoma MG-63 Cells. *Nanomaterials* **2018**, *8*, 961. [CrossRef] [PubMed]
88. Ravanbakhsh, M.; Labbaf, S.; Karimzadeh, F.; Pinna, A.; Houreh, A.B.; Nasr-Esfahani, M.H. Mesoporous Bioactive Glasses for the Combined Application of Osteosarcoma Treatment and Bone Regeneration. *Mater. Sci. Eng. C* **2019**, *104*, 109994. [CrossRef] [PubMed]
89. Lu, Y.; Li, L.; Lin, Z.; Li, M.; Hu, X.; Zhang, Y.; Peng, M. Enhancing Osteosarcoma Killing and CT Imaging Using Ultrahigh Drug Loading and NIR-Responsive Bismuth Sulfide@Mesoporous Silica Nanoparticles. *Adv. Healthc. Mater.* **2018**, *7*, 1800602. [CrossRef]
90. Colilla, M.; Izquierdo-Barba, I.; Vallet-Regí, M. Phosphorus-Containing SBA-15 Materials as Bisphosphonate Carriers for Osteoporosis Treatment. *Microporous Mesoporous Mater.* **2010**, *135*, 51–59. [CrossRef]
91. Zhou, X.; Weng, W.; Chen, B.; Feng, W.; Wang, W.; Nie, W.; Chen, L.; Mo, X.; Su, J.; He, C. Mesoporous Silica Nanoparticles/Gelatin Porous Composite Scaffolds with Localized and Sustained Release of Vancomycin for Treatment of Infected Bone Defects. *J. Mater. Chem. B* **2018**, *6*, 740–752. [CrossRef]
92. Paris, J.L.; Lafuente-Gómez, N.; Cabañas, M.V.; Román, J.; Peña, J.; Vallet-Regí, M. Fabrication of a Nanoparticle-Containing 3D Porous Bone Scaffold with Proangiogenic and Antibacterial Properties. *Acta Biomater.* **2019**, *86*, 441–449. [CrossRef]

93. Benjaminsen, R.V.; Mattebjerg, M.A.; Henriksen, J.R.; Moghimi, S.M.; Andresen, T.L. The Possible "Proton Sponge" Effect of Polyethylenimine (PEI) Does Not Include Change in Lysosomal pH. *Mol. Ther.* **2013**, *21*, 149–157. [CrossRef]
94. Khorsand, B.; Nicholson, N.; Do, A.-V.; Femino, J.E.; Martin, J.A.; Petersen, E.; Guetschow, B.; Fredericks, D.C.; Salem, A.K. Regeneration of Bone Using Nanoplex Delivery of FGF-2 and BMP-2 Genes in Diaphyseal Long Bone Radial Defects in a Diabetic Rabbit Model. *J. Control Release* **2017**, *248*, 53–59. [CrossRef]
95. Malek-Khatabi, A.; Javar, H.A.; Dashtimoghadam, E.; Ansari, S.; Hasani-Sadrabadi, M.M.; Moshaverinia, A. In situ bone tissue engineering using gene delivery nanocomplexes. *Acta Biomater.* **2020**, *108*, 326–336. [CrossRef]
96. Zhang, X.; Li, Y.; Chen, Y.E.; Chen, J.; Ma, P.X. Cell-Free 3D Scaffold with Two-Stage Delivery of miRNA-26a to Regenerate Critical-Sized Bone Defects. *Nat. Commun.* **2016**, *7*, 10376. [CrossRef]

Disclaimer/Publisher's Note: The statements, opinions and data contained in all publications are solely those of the individual author(s) and contributor(s) and not of MDPI and/or the editor(s). MDPI and/or the editor(s) disclaim responsibility for any injury to people or property resulting from any ideas, methods, instructions or products referred to in the content.

Journal of
Functional Biomaterials

MDPI

Article

Suitability of Mg-Nd and Mg-Zn Alloys to Obtain Biodegradable Structures for Bone Defects

Veronica Manescu (Paltanea) [1,2], Aurora Antoniac [1], Maria Cristina Moraru [3], Iulian Antoniac [1,4,*], Cosmin Mihai Cotrut [1], Sebastian Gradinaru [5,6], Alexandra Iulia Dreanca [3], Bogdan Sevastre [3], Romelia Pop [3], Flaviu Alexandru Tabaran [3], George Mihail Vlasceanu [7,8], Mariana Ionita [7,8] and Marius Manole [9]

1 Faculty of Material Science and Engineering, National University of Science and Technology Politehnica Bucharest, 313 Splaiul Independentei, District 6, 060042 Bucharest, Romania; veronica.paltanea@upb.ro (V.M.); aurora.antoniac@upb.ro (A.A.); cosmin.cotrut@upb.ro (C.M.C.)
2 Faculty of Electrical Engineering, National University of Science and Technology Politehnica Bucharest, 313 Splaiul Independentei, District 6, 060042 Bucharest, Romania
3 Faculty of Veterinary Medicine, University of Agricultural Sciences and Veterinary Medicine of Cluj-Napoca, 3-5 Calea Manastur, 400372 Cluj-Napoca, Romania; maria.moraru@usamvcluj.ro (M.C.M.); alexandra.dreanca@usamvcluj.ro (A.I.D.); bogdan.sevastre@usamvcluj.ro (B.S.); romelia.pop@usamvcluj.ro (R.P.); alexandru.tabaran@usamvcluj.ro (F.A.T.)
4 Academy of Romanian Scientists, 54 Splaiul Independentei, 050094 Bucharest, Romania
5 Department of Medical-Clinical Discipline, Faculty of Medicine, Titu Maiorescu University, 67A Gheorghe Petrascu, 031593 Bucharest, Romania; sebastian.gradinaru@prof.utm.ro
6 Department of General Surgery, County Hospital Ilfov, 050474 Bucharest, Romania
7 Faculty of Medical Engineering, National University of Science and Technology Politehnica Bucharest, 1-7 Gheorghe Polizu, 011061 Bucharest, Romania; george.vlasceanu@upb.ro (G.M.V.); mariana.ionita@polimi.it (M.I.)
8 Advanced Polymer Materials Group, National University of Science and Technology Politehnica Bucharest, 1-7 Gheorghe Polizu, 011061 Bucharest, Romania
9 Department of Prosthetics and Dental Materials, Faculty of Dentistry, Iuliu Hatieganu University of Medicine and Pharmacy, 8 Victor Babes, 400012 Cluj-Napoca, Romania; marius.manole@umfcluj.ro
* Correspondence: antoniac.iulian@gmail.com

Abstract: Mg-based alloys are one of the most promising materials used in regenerative medicine for bone tissue engineering. Considering the increasing prevalence of a continuously aging population, as well as the high incidence of accidents and bone cancers, it is crucial to explore biomaterials that can serve as bone substitutes. After carefully analyzing the literature in the introduction section, we proposed two Mg-based alloys as suitable for obtaining biodegradable structures for bone defect treatment. To achieve trustworthy results, the alloys' microstructure was investigated using microscopic techniques coupled with energy-dispersive spectroscopy and X-ray diffraction. The obtained results were comparable with those described in references on similar Mg alloys. Then, the mechanical compression properties were highlighted, and the *in vitro* corrosion behavior proved that Mg-Zn exhibited a reduced corrosion rate compared to the Mg-Nd alloy, as tested using electrochemical methods. However, the *in vivo* tests showed good biocompatibility for both magnesium alloys. In conclusion, both alloys are suitable for use as potential bone substitute applications, but it must be taken into consideration that Mg-Zn alloys present lower biodegradation and mechanical properties. For future investigations, we aim to develop bone substitutes made from these materials, specifically designed for small bone defect treatment and with patient-adapted geometry. Due to the differences mentioned above, various designs will be tested.

Keywords: magnesium alloys; microstructure; corrosion; biocompatibility

1. Introduction

Nowadays, modern lifestyle changes correlated with an increased life duration have made orthopedics one of the most challenging medical domains. Due to car or motorcycle accidents or even trauma caused by sports practice, there is an important need to address the bone defects that usually occur [1,2]. However, in medical science, it is stated that bone injuries are generated also caused by tumors, infections, congenital issues, and even surgeries [3,4]. Small bone defects exhibit the ability to spontaneously heal, but large bone defects usually require supplementary care, and various solutions are applied to alleviate the patient's suffering [5–7]. A very realistic and generally accepted classification of bone defects was made by Schemitsch [8], who defined small bone defects as those with a size below 2 cm and a maximum of 50% cortical loss circumference. In his opinion, intermediate defects have a size between 2 and 6 cm correlated with a circumference loss higher than 50%. Last but not least, the largest defects with a size greater than 6 cm are considered a major health concern, being very serious and leading in some cases to limb amputation and to an important decrease in the patient's quality of life with a high social burden [9].

To fix these bone defects, one can identify the continuous need to develop new and innovative solutions for bone substitutes, which is a major concern for scholars. The use of the autograft was long considered the gold standard, but problems such as limited patient bone quantity and donor site morbidity prevented it from being applied on a large scale [10,11]. Another solution met in practice consists of using allografts or xenografts, which are harvested from human or animal donors, respectively. Unfortunately, these interventions are also associated with a high risk of graft rejection or disease transmission, with unanticipated evolution due to bacteria or pathogen adaptation from passing through the organisms of different individuals or species [12–15]. Another major drawback of the grafting techniques is their high cost, which is associated with a continuous increase in graft demand that has led to the necessity of developing synthetic bone substitutes.

Georgeanu et al. [1] underlined the most important attributes of the ideal biomaterial used as a bone substitute. The authors stated that it must exhibit osteoinductive, osteogenic, and osteoconductive properties. An adequate bone substitute should stimulate both angiogenesis and neurogenesis, and it must not induce any teratogenic, allergic, antigenic, carcinogenic, or other adverse reactions. In addition, its surface must be rough and exhibit good wettability to facilitate cell adhesion and proliferation. Also, the mechanical properties of an ideal biomaterial should have values close to those of the human bone, and last but not least, it must not be degraded by clinical handling and must offer the possibility of being easily sterilized. Other characteristics that have lately been underlined are bone substitute bioactivity and biodegradability, which concomitantly help the bone regeneration process with the bone substitute's gradual dissolution [16,17]. One of the most modern solutions should incorporate all these requirements.

The synthetic materials that are today used as bone substitutes are classified as ceramics, polymers, or composites. Bongio et al. [16] identified the first material generation as inert and tolerant, providing material fibrous encapsulation as a host tissue response. The second generation constitutes responsive biomaterials such as bioglasses, being characterized by degradation and osteoconductivity, while the third generation comprises inductive materials that are able to develop a strong link between natural bone and substitute, as well as to enhance osteoblast production simultaneously with blood vessel formation and, where necessary, nerve regeneration.

From the first generation of biomaterials, one can name thermoplastic polymers and ceramics including zirconia or alumina with important drawbacks such as inadequate mechanical properties and the lack of ability to connect with the hard tissue of the host [6,16–19]. From the second generation, bioglasses are considered osseous replacements with high biocompatibility and increased osteoconductivity. When adequate processing conditions are chosen, a porous structure can be achieved. This is, in particular, beneficial for bone development. Usually, bioglasses are not associated with inflammatory reactions or systemic toxicity, being resorbable in a certain amount of time. The degradation rate could be controlled through adjustments in chemical composition. Some of the most common glasses in practice are those based on silica, borate, or phosphate [20–22]. The last ones are especially characterized by a bioactive character, and many studies [23–25] proved that they could form a strong connection with natural bone. The main disadvantage of glasses is that they are brittle, have low mechanical strength, and decreased fracture resistance. Regarding the third biomaterial group, one can mention different types of calcium phosphate such as β-tricalcium phosphate (β-TCP), hydroxyapatite (HA), or a combination between them in the shape of biphasic calcium phosphate (BCP) ceramic [26,27]. These types of materials are characterized by increased osteoinduction that can be enhanced through different methods such as bone-forming cells, growth factors, and bioactive protein addition [28–30]. Some studies [31,32] showed that the slow release of vascular endothelial growth factor (VEGF) from BCP determined an increase in blood vessel formation. Polymers represent another solution adopted by practitioners because they are very versatile regarding their physical or chemical properties [33–35]. In the case of these biomaterials, an important drawback, which can be immediately foreseen, consists of the fact that they are inert, and osteoinduction must be controlled through different approaches because hydrogels used as simple templates do not sustain cell adhesion and viability in a high percentage of cases [36,37]. If polymers are associated with bioactive factors, highly bioactive bone substitutes can be obtained with adequate degradation kinetic process but with poor mechanical properties such as reduced stiffness and strength.

A response to the current limitations, which was adopted in many studies, consists of using biodegradable metal in bone substitute manufacture (Figure 1). From this class of biomaterials, magnesium (Mg) and its alloys stand out as the most promising candidates due to their excellent physical, chemical, and mechanical properties. This biodegradable and lightweight metal can be used as a bone substitute in two forms: as a powder to fill the bone gap directly and as a block inserted into the defect. One of the most vital mechanical properties of Mg is its Young's modulus, which is between 41 and 45 GPa, a value considered very close to that of human bone ($3 \div 20$ GPa) [38–40]. Regarding the yield strength of Mg ($65 \div 100$ MPa), this mechanical quantity is slightly lower than that of natural bone ($104 \div 121$ MPa). However, one can immediately notice that the stress shielding effect is estimated to occur very rarely due to similar values between Mg and human bone in terms of their mechanical properties. Another key feature in bone substitute design is the degradation process of the alloys. In the case of Mg, the main secondary products are magnesium hydroxide ($Mg(OH)_2$) and hydrogen (H_2), which should be carefully controlled to avoid generating harmful effects such as reactive oxidative species formation and necrosis due to hydrogen pockets that usually occur in the vicinity of the Mg bone substitute [41]. In addition, the concentration of Mg^{2+} ions must be kept in the following interval: $5 \div 10$ mmol/L to avoid adding supplementary toxicity [42–44]. Some studies [45–47] showed that Mg^{2+} ions could exhibit a beneficial effect on osteoblast differentiation due to the medium alkalinization that reduces osteoclast activity and accelerates mineralization, concomitantly with a stimulation of the calcitonin gene-related peptide (CGRP) that is well-known to sustain the differentiation of stem cells into osteoblasts [48–50]. Usually, the most common solutions to control the biodegradation rate of Mg-based alloys consist of Mg alloying with different chemical elements to obtain new Mg-based alloys or developing biocompatible coatings based on ceramics, polymers, or composites for existing Mg-based alloys [6,7,51,52].

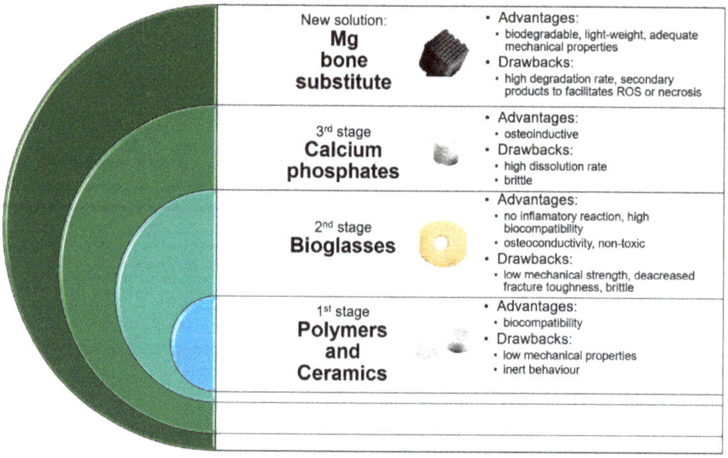

Figure 1. Main bone substitutes used in clinical practice and new solutions with advantages and drawbacks.

The main binary Mg-based systems used in developing medical applications are Mg-Ca, Mg-Zr, Mg-Zn, Mg-Ag, and Mg- rare earths (REs) such as Mg-Nd and Mg-Y [6,39,53–56]. We will discuss only the importance of alloying Mg with neodymium (Nd) and zinc (Zn) from all of these systems because they are directly related to our experimental investigation. Neodymium is a rare earth and is usually added to improve the alloy's mechanical properties and increase the corrosion resistance due to the modifications induced in the material microstructure [57]. Different studies demonstrated that the Nd percent should be carefully selected because it forms a phase (Mg_3Nd) that can lead to an increase in the corrosion process [58–60]. In addition, a shielded oxide layer was noticed at the surface of the Mg-Nd binary alloy. Regarding the alloying of Mg with Zn, the tolerance limit of Zn in Mg-Zn alloys was established to be around 3 wt.% [61,62]. In the Mg-Zn alloys, the Zn percent should be controlled because it plays an important role in biocorrosion by reducing the emission of hydrogen, which typically occurs during the Mg degradation process. Otherwise, Zn, being a trace element in the human body, is highly biocompatible and is present in muscles and bones. Kirkland mentioned that the corrosion rate increases when the Zn content varies from 1 to 3 wt.% due to crack formation [62].

Mg-based alloys have addressed almost all the limitations of traditional bone substitutes (e.g., second surgery, stress shielding, inert character) because they degrade and are replaced by new bone, have a Young's modulus with values very close to those of human bone, prevent the stress shielding effect, and last but not least, promote bone regeneration.

The current study aims to evaluate two binary Mg-based alloys from the systems Mg-Nd and Mg-Zn to be applied as bone substitutes in different shapes. We will investigate their microstructural characteristics based on optical microscopy, scanning electron microscopy (SEM) coupled with energy dispersive spectroscopy (EDS), and X-ray diffraction

(XRD). The mechanical properties of the two binary alloys will be determined based on the compression test. Also, the corrosion of the experimental magnesium alloys will be assessed with electrochemical analysis in simulated body fluid (SBF). The biocompatibility will be evaluated by *in vivo* testing on rats, followed by histological analyses and microcomputed tomography (μCT).

2. Materials and Methods

Two magnesium-based alloys, from the binary systems Mg-Nd and Mg-Zn, were chosen to be investigated and tested to establish if these are suitable for bone substitutes in orthopedics. The Mg-Nd alloy was produced using sand mold and chill casting, while the Mg-Zn alloy was fabricated via the stir casting method. Both alloys were synthesized from high-purity raw materials (Mg, Nd, Y, Zn, Ca, and Mn—99.99% purity). Their chemical composition is provided in Table 1. Experimental samples for material characterizations were prepared as parallelepipedal blocks with edge lengths of 10 mm, obtained through mechanical cutting.

Table 1. Elemental composition of the two Mg-based alloys.

Alloy	Composition (Mass Percents %)					
	Zn	Y	Nd	Ca	Mn	Mg
Mg-Nd	0.4	0.3	2.6	-	-	Bal.
Mg-Zn	1.4	-	-	0.3	0.6	Bal.

2.1. Microstructural Characterization of Experimental Samples

The microstructural characterization of the samples was conducted using a Nikon Eclipse MA100N inverted microscope (Nikon Metrology Europe NV, Leuven, Belgium) and an ESEM Quattro S microscope, which has an EDS UltraDry 60M module (Thermo Fisher Scientific Inc., Waltham, MA, USA). For the optical observations, the Mg-based alloy sample surface was metallographically prepared by chemical etching using a Nital reagent. The phase identification was performed using a PANALYTICAL x-Pert PRO diffractometer (Malvern Panalytical, UK) equipped with CuKα radiation. The acquired information was processed in PDXL Software 1.2.0.1 (ICDD 1999).

2.2. Mechanical Properties

The compression test was conducted based on a Walter + Bai LFV 300 (Walter + Bai AG, Löhningen, Switzerland) device in accordance with the ASTM E9-09 standard [63]. The measurements were performed as a function of displacement control with a speed of 5 mm/min. The ending criterion was set when an applied force decrease of about 75% from the maximum value was reached. After the compression tests were finished, the stress–strain diagram was obtained as the main result.

2.3. In Vitro Corrosion Behavior of the Experimental Samples

The Mg-Nd and Mg-Zn corrosion process was investigated based on electrochemical analysis. In the case of the above-mentioned method, we chose simulated body fluid (SBF) because this electrolyte has a chemical and ionic composition similar to that of human blood plasma [64]. The SBF was prepared in our laboratory using the Kokubo protocol [65] and characterized by a physiological pH of 7.4.

The electrochemical analysis was performed based on PARSTAT 4000 Potentiostat/Galvanostat equipment (Princeton Applied Research, Oak Ridge, TN, USA) with a three-electrode cell according to ASTM G5-94 (2011) standard [66]. The testing protocol used to determine the corrosion parameters was similar to that described in [57,67,68]. Briefly, based on a CW-05G (Jeio Tech, Seoul, Republic of Korea) heating and recirculation bath, the electrochemical measurements were performed at 37 °C. Three samples prepared from the two Mg alloys, respectively, were inserted into a dedicated Teflon support to expose only a surface area of 1 mm^2 to the SBF action. The Tafel dependencies were determined in a potential range of ± 250 mV vs. EOC (potential rich at quasi-equilibrium) at a speed of 1 mV/s. The EOC was measured for 1 h. The main corrosion parameters were computed based on potentiodynamic variations plotted in Versa Studio 2.63.3 Software.

2.4. *In Vivo Animal Study*

2.4.1. Surgical Procedure and Histopathological Analysis Route

The amount of material for *in vivo* tests was measured precisely employing a calibrated analytical balance (Kern ABJ 220-4NM, Merk, Germany). Specifically, 140 mg of material was applied subcutaneously, and 40 mg was used to fill the bone defects. Before implantation, both Mg-Nd and Mg-Zn underwent a wet sterilization procedure by autoclaving with a Trade Raypa Steam Sterilizer (R. Espinar S.I., Spain, A.E-75 Dry) at 105 °C and 1 Bar for 5 min.

Seventy-two specific pathogen-free (*SPF*), 40-week-old female Sprague Dawley rats originated from the "Cantacuzino" National Institute for Medical-Military Research and Development (Bucharest, Romania) were involved in the study. Animals were housed under SPF laboratory conditions with free access to food and water.

For the subcutaneous compatibility investigation, 32 animals were used, 16 for each material. Four animals from the two alloy-based groups were euthanized at 1, 2, 4, and 8 weeks.

The bone implantation was conducted on 40 animals, randomly allocated into two experimental groups (n = 20), of which samples of Mg-Nd and Mg-Zn alloys were used. Five animals from each group were euthanized at the same intervals of 1, 2, 4, and 8 weeks for the collection of the bone tissue. The rationale for choosing an investigation time of a maximum of 8 weeks to collect bone tissue was in accordance with many literature studies that involved bone defects in Sprague Dawley rats and evaluated local and systemic biocompatibility in both soft and hard tissues, rather than long-term bone regeneration and complete material degradation. Therefore, the selected time point corresponds to an acute phase assessment, in accordance with ISO 10993-6 [69] recommendations for biological evaluation of medical devices [70,71]. This time was considered sufficient for the natural bone healing process with substantial but not complete remodeling or bone formation [72]. The study of Nowak et al. [73] recommended 12 weeks for complete healing. However, an 8-week approach is a more adequate choice when biodegradable metals such as Mg alloys are used, as it allows for the study of early bone mineralization and new bone formation in conjunction with alloy degradation [74]. This process is synchronically connected with its replacement by animals' natural hard tissue, and by choosing 8 weeks as a correct time, an *in vivo* evaluation before complete bone maturation could be conducted [74–76]. Most of the research studies opted for a maximum of 8 weeks for the investigation time based on practical considerations because at this point, important effects of the bone substitute are usually still present [77,78]. In addition, at the 1-week time point, early molecular and cellular mechanisms such as the onset of osteogenesis and inflammation could be studied [79]. At 2 weeks, osteogenesis and angiogenesis are enhanced due to the action of Mg^{2+} ions [6,80,81], while at 4 weeks, early bone formation and osteoblast activity can be monitored [79].

All procedures were approved by the Bioethics Committee of the University of Agricultural Sciences and Veterinary Medicine Cluj-Napoca (Ethical approval 386/15 June

2023), and the project was approved by the regional state veterinary authority DSVSA-CN (Project authorization no. 382/25 August 2023).

General anesthesia was induced by intraperitoneal injection of ketamine (80 mg/kg) and dexmedetomidine (0.5 mg/kg). Analgesia was provided using intramuscular tramadol (15 mg/kg). Corneal lubrication was applied to prevent desiccation. Anesthesia depth was monitored continuously, and body temperature was maintained with a heating pad. Atipamezole was administered at the end of the procedure to reverse dexmedetomidine.

Under aseptic conditions, a cranio-lateral skin incision was made over the right femur. Blunt dissection was carried out between the *biceps femoris* and *fascia lata* to expose the femoral diaphysis without damaging the periosteum. A standardized cortical bone defect, 3 mm in length, was created using a 1.2 mm diamond burr, with continuous saline irrigation to prevent thermal injury. Then, the defect was filled with either Mg-Nd or Mg-Zn powder. The wound was closed in three layers using 4-0 absorbable monofilament sutures for fascia and subcutaneous tissue, and 4-0 non-absorbable nylon in an interrupted pattern for the skin (Figure 2).

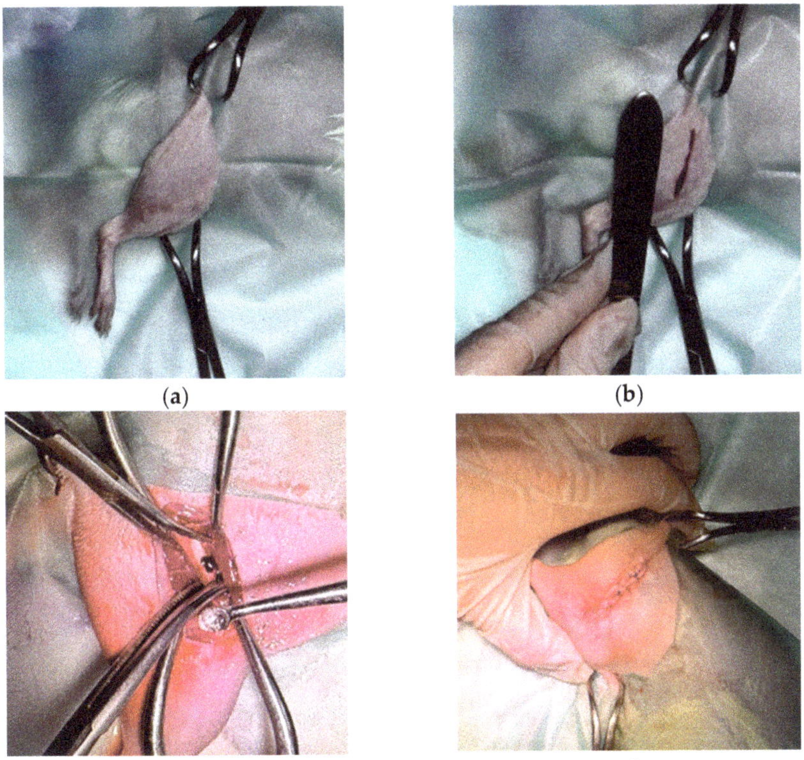

(**a**) (**b**) (**c**) (**d**)

Figure 2. Surgical images during the medical procedure performed on rat bones: (**a**) Placement for surgery; (**b**) Incision site; (**c**) Bone defect and Mg alloy samples; (**d**) Sutured surgical wound.

For subcutaneous implantation, the rats were positioned in sternal recumbency. One small dorsal incision was made under aseptic conditions. In each group, 140 mg of powdered magnesium alloy (Mg-Nd or Mg-Zn) was inserted into the subcutaneous space using a Volkmann spoon. The skin was closed with 4-0 nylon sutures in a simple interrupted pattern (Figure 3).

Post-operative supportive therapy included subcutaneous 0.9% NaCl for 2–3 days. Analgesia was maintained with tramadol (15 mg/kg IM) and oral metamizole (150 mg/kg, dose-adjusted based on the rat grimace scale). Enrofloxacin (10 mg/kg SC) was administered once daily for 5 days as antibiotic prophylaxis. The rats were euthanized at 1, 2, 4, and 8 weeks post-implantation. Skin and femoral bone samples were collected for further evaluation.

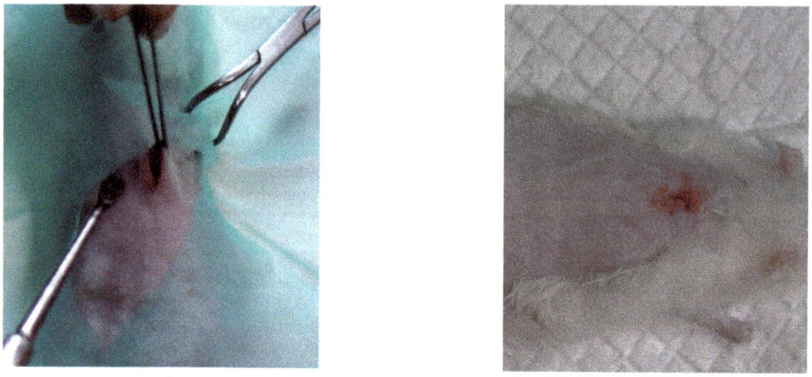

(a) (b)

Figure 3. Surgical images during the medical procedure performed on the rats' skin: (**a**) Subcutaneous placement of Mg alloy samples; (**b**) Surgical site.

After harvesting, skin and bone tissues were fixed in 10% neutral-buffered formalin for 48 h. For bone samples, fixation was followed by 24 h decalcification in a rapid solution and storage in 70% ethanol. All samples were dehydrated in graded ethanol, cleared in xylene, and embedded in paraffin at 58 °C. Sections of 2 μm were cut using a rotary microtome. Slides were deparaffinized in xylene, rehydrated through descending ethanol concentrations, and rinsed in tap water. Standard hematoxylin–eosin (HE) staining was applied: hematoxylin (5 min), rinsing (5 min), eosin (2 min), followed by dehydration in graded ethanol and clearing in xylene. Coverslips were mounted with Permount™ toluene-based synthetic resin.

Histological evaluation was performed under an Olympus BX51 microscope (Olympus Corporation, Tokyo, Japan). Images were acquired using an Olympus UC30 (Olympus Corporation, Tokyo, Japan) digital camera, processed with Olympus cellSens imaging software version 3.1.

Subcutaneous implant-associated lesions were scored according to ISO 10993-6:2021 [69] guidelines for the biological evaluation of medical devices. The average score per parameter and the overall biological response were calculated based on three representative microscopic fields per sample, from three independent, blinded observations. The scoring criteria included the evaluation of parameters such as inflammatory cell infiltration, fibrosis, necrosis, neovascularization, and new bone formation for bone samples, and epithelial integrity, inflammatory response, and fibrosis. Each criterion was graded semi-quantitatively from 0 (no reaction) to 4 (severe reaction).

2.4.2. Micro-Computed Tomography (μCT) Acquisition and Analysis

Femoral specimens were imaged on a Bruker SkyScan 1272 (Kontich, Belgium) at 70 kV and 140 μA, with a 0.5 mm aluminum filter. Projections (2452 × 1640 px) were collected every 0.2° through 360° using 1400 ms exposures and three-frame averaging. Data were reconstructed in NRecon 1.7.1.6 software with ring- and beam-hardening corrections to an isotropic voxel size of 3.5 μm. A volume of interest (VOI) was drawn slice by slice in CTAn to match the original oval cortical window in the mid-diaphysis. The outline followed the adjacent cortex, excluding mineral that migrated into the medullary canal or formed beyond the defect. Irregular margins due to remodeling and variable cortical thickness were accommodated.

Within the VOI, thresholding rendered bone voxels white (bone volume, BV); all remaining voxels were treated as porosity and omitted from BV calculations (Figure 4).

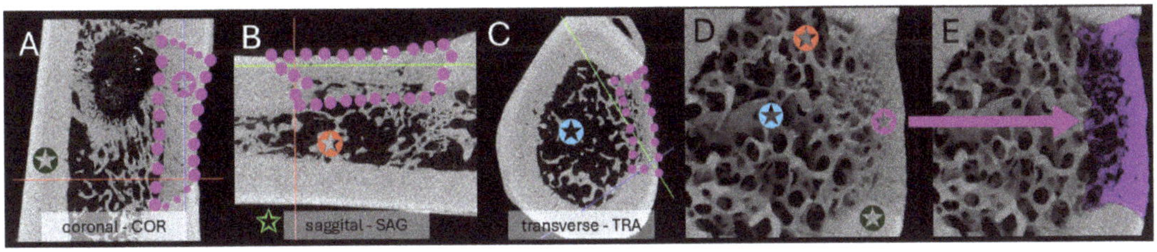

Figure 4. Illustration of manually defined limits of volume of interest (VOI) in orthogonal planes ((**A**) coronal COR, (**B**) sagittal SAG, (**C**)) transverse TRA extracted from DataViewer 1.5.4.6 software (magenta dots), as part of the segmentation process—exemplified for sample Mg-Nd_W8. The tomogram slice highlights key anatomical and pathological features, including intact femoral cortex (✪), medullary channel (✪), mineralized tissue within both the medullary cavity (✪) and the cortical defect (✪), and surrounding soft tissue (☆). The (**D**,**E**) subsets depict a TRA cross-section of the femur tomogram (CTVox 3.3.0.0 software) at the Mg-Nd_W8 (8 weeks) implantation site. The segmented bone volume (BV) selected for analysis is shown in magenta, representing the region of interest within that specific slice.

Bone mineral density (BMD) was measured ex vivo from micro-CT scans of explanted samples collected at 1, 2, 4, and 8 weeks post-implantation. Values (g/cm^3) were calculated in CTAn 1.17.7.2 software using calibrated attenuation coefficients, with hydroxyapatite (CaHA powder) embedded epoxy resin calibration rod phantoms (ø 4 mm) serving as reference standards.

3. Results and Discussion

3.1. *Microstructural Characterization of the Mg-Based Alloys*

3.1.1. Optical Microscopy

The Mg-Nd and Mg-Zn sample surfaces were metallographically prepared for optical observations by chemical etching with Nital 2.5%. Figure 5 presents the optical images obtained for both Mg-based alloys.

In the case of the sample made of Mg-Nd, one can clearly observe polyhedral grains of α-Mg, in which lamellar and globular morphologies are found. The grains are big and exhibit non-uniform size, with a reduced degree of intergranular precipitation. The lamellar compounds tend to form groups into the grains, and this fact induces better mechanical properties. Our findings confirm the fact that Mg-based alloys containing rare earths, like Nd, exhibit good mechanical strength and corrosion behavior [82–85].

For the Mg-Zn experimental samples, the optical microscopy images show that the material microstructure presents uneven and larger α-Mg polyhedral grains compared to the Mg-Nd experimental samples. Inside the grains, spheroidal and irregularly contoured compounds are visible, which will be further investigated based on SEM-EDS analysis. Microstructure refinement and improvement of mechanical properties are usually performed by adding different alloying elements such as Mn and Zn. Mn is usually used to reduce the grain size concomitantly with an improvement in the tensile strength of the alloy [86], while Zn has beneficial effects on increasing the tensile strength and corrosion resistance [87–89].

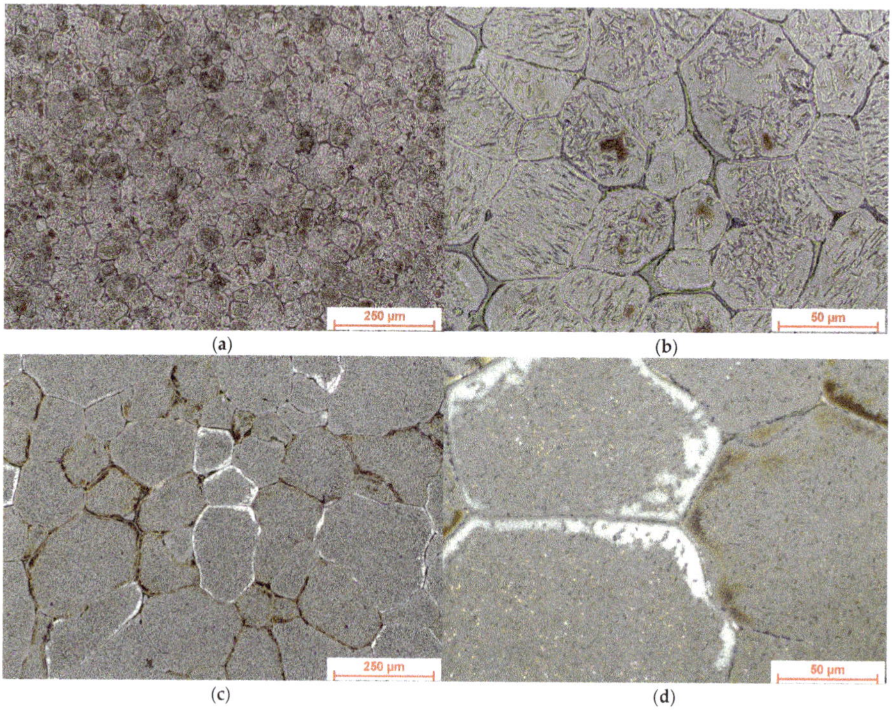

Figure 5. Optical microscopy images showing the microstructural aspects of Mg-Nd alloy: (a) magnification 10×; (b) magnification 50×; and Mg-Zn alloy: (c) magnification 10×; (d) magnification 50×.

3.1.2. X-Ray Diffraction Analysis (XRD)

As one can observe from Figure 6, in the case of Mg-Nd alloy, the identified peaks represent α-Mg and ternary phases (Mg-Zn-Nd and Mg-Zn-Y). Yang et al. [90] noticed in the case of Mg-4.5Zn-*x*Nd (x = 0, 1, and 2 wt%), which contains Zn and Nd as in our investigated magnesium alloys, the presence of the ternary Mg-Zn-Nd phase. They identified this phase as T-phase (Mg_4Zn_2Nd), which exhibits a *c*-centered orthorhombic crystal structure, as presented in [91,92]. In the case of our Mg-Nd alloy, we found that the ternary phase corresponds to the T-phase, as it was noticed in the literature [90,93].

Additionally, in accordance with the research performed by Zengin and Turen [94], considering the fact that the Y concentration is about 0.3 wt%, we can observe a supplementary ternary phase, Mg-Zn-Y, denoted in the literature as I-phase (Mg_3Zn_6Y). Other phases reported are W-phase ($Mg_3Zn_3Y_2$) and X-phase ($Mg_{12}ZnY$) [95–97], but these were not existent in our Mg-Nd alloy, being generally associated with an increased ratio of Y and Zn [98,99].

Regarding the other investigated alloy, Mg-Zn, with a Zn mass percent of 1.4 wt%, the diffraction peaks are associated with α-Mg and Mg-Zn-Ca ternary phase. The existence of the ternary Mg-Zn-Ca ($Mg_6Zn_3Ca_2$) and a supplementary binary Mg-Ca (Mg_2Ca) phases was evidenced by the studies of Farahany et al. [100], Jiang et al. [101], and Schäublin et al. [102] and was conditioned by a Ca content more elevated than 0.5 wt%. In addition, for a Zn/Ca ratio with values higher than 1.23, the α-Mg and Mg-Zn-Ca ternary phases are usually identified in the XRD spectra [103,104]. For our samples, we did not have evidence of the presence of the binary phase Mg-Ca because the value of Zn/Ca is about 4.66, and the Ca content is 0.3 wt%. Some studies [105–107] showed the existence of a supplementary α-Mn phase precipitate, but in our case, this phase was not identified. This observation is in agreement with the analyses performed by Cho et al. [108] and Bakhsheshi-Rad [88], who reported the absence of Mn intermetallic compounds and α-Mn phase in the case of Mg-4Zn-0.5Ca-(0, 0.4, 0.8 wt%)Mn and Mg-2Ca-0.5Mn-(2-7 wt%)Zn, respectively.

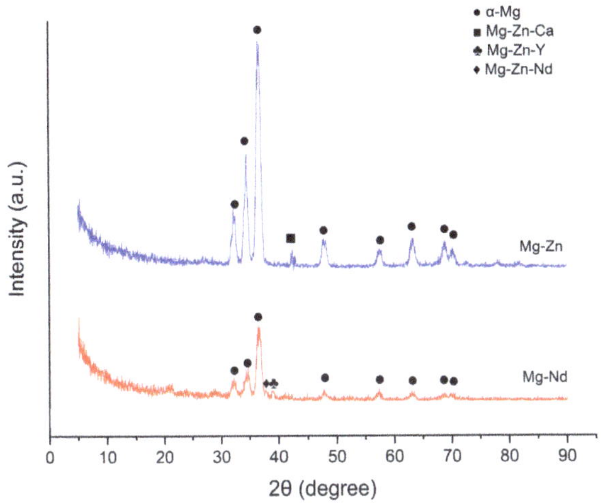

Figure 6. X-ray diffraction patterns of the Mg-Nd and Mg-Zn alloys.

The SEM findings will fulfill the investigation presented above very well by clarifying the chemical and physical processes that occurred during the material manufacturing steps, while XRD clearly showed the phases present in the microstructure.

3.1.3. Scanning Electron Microscopy

It is well-known that material microstructure, grain shape, and size significantly influence alloy properties such as corrosion resistance, plasticity, and strength. In the case of the Mg-Nd and Mg-Zn samples, we obtained microstructure images similar to those from optical microscopy (Figure 7).

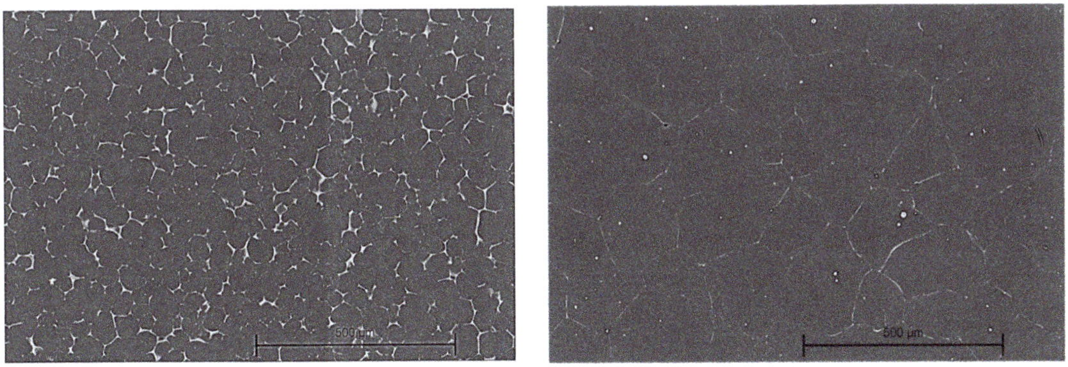

(**a**) (**b**)

Figure 7. SEM images of samples made of: (**a**) Mg-Nd alloy, magnification 100×; and (**b**) Mg-Zn alloy, magnification 100×.

Based on EDS analysis, in the case of both investigated magnesium alloys, the main chemical elements were identified (Table 2). This fact is directly correlated with the chemical composition of the material described in Table 1.

Table 2. EDS chemical composition of the two Mg-based alloys.

Element	Mg-Nd					
	O	Zn	Y	Nd	Mg	Total
Weight (%)	1.56 ± 0.15	0.38 ± 0.04	0.32 ± 0.03	2.58 ± 0.12	95.16 ± 0.34	100.00
Atom (%)	2.08 ± 0.23	0.14 ± 0.02	0.09 ± 0.02	0.41 ± 0.02	97.28 ± 0.35	100.00
Element	Mg-Zn					
	O	Zn	Ca	Mn	Mg	Total
Weight (%)	0.69 ± 0.09	1.20 ± 0.13	0.35 ± 0.06	0.54 ± 0.07	97.22 ± 0.33	100.00
Atom (%)	1.02 ± 0.13	0.51 ± 0.05	0.31 ± 0.02	0.23 ± 0.03	97.93 ± 0.32	100.00

In addition, in Figure 8a, the EDS mapping images of Mg-Nd sample with the chemical element spatial distribution are presented. It can be noticed that the alloy matrix consists mainly of Mg and Nd, while some traces of Zn and Y can be also observed (point 1, Table 3). In [109–111], it is stated that when the Nd quantity is higher than 2 wt.%, only 1.4 ± 0.1 wt.% Nd is stable, proving saturation of the solid solution of Nd in the Mg matrix. For the investigated Mg-Nd alloy, the supplementary Nd quantity can be localized at the grain boundary (Figures 8a and 9a and Table 3). The EDS analysis confirms that a phase Mg-Zn-Nd is present at the grain boundary (point 2, Table 3), and in the lamellar compounds identified based on optical microscopy and SEM in α-Mg matrix (points 3, 5, 6, 7, and 8, Table 3). Regarding the white compound placed at the grain boundary, an intermetallic phase Mg-Zn-Y was detected (Table 3, point 4).

Table 3. Chemical composition for phase identification.

Point	Mg-Nd				
	O	Zn	Y	Nd	Mg
1	1.69 ± 0.16	0.21 ± 0.05	0.18 ± 0.05	1.20 ± 0.10	96.72 ± 0.31
2	2.17 ± 0.16	1.51 ± 0.13	0.00	16.45 ± 0.32	79.87 ± 0.26
3	1.70 ± 0.15	0.53 ± 0.05	0.00	2.25 ± 0.15	95.52 ± 0.28
4	1.35 ± 0.14	0.70 ± 0.06	3.02 ± 0.30	0.00	94.93 ± 0.30
5	1.87 ± 0.16	0.55 ± 0.05	0.00	2.55 ± 0.11	95.03 ± 0.29
6	1.25 ± 0.17	0.61 ± 0.10	0.00	3.94 ± 0.11	94.20 ± 0.30
7	1.28 ± 0.10	0.46 ± 0.05	0.00	2.41 ± 0.11	95.85 ± 0.30
8	1.44 ± 0.16	0.41 ± 0.05	0.00	2.26 ± 0.10	95.89 ± 0.30
Point	Mg-Zn				
	O	Zn	Ca	Mn	Mg
1	0.67 ± 0.09	1.2 ± 0.11	0.22 ± 0.02	0.35 ±0.07	98.23 ± 0.31
2	0.82 ± 0.10	12.71 ± 0.18	11.05 ± 0.11	0.00	75.42 ± 0.27
3	-	0.61 ± 0.05	0.12 ± 0.02	29.90 ± 0.15	69.37 ± 0.22
4	-	0.76 ± 0.09	0.19 ± 0.04	38.97 ± 0.16	60.08 ± 0.24
5	0.81± 0.09	13.73 ± 0.18	12.72 ± 0.10	0.00	72.74 ± 0.25
6	0.96 ± 0.11	15.16 ± 0.18	10.04 ± 0.09	0.00	73.84 ± 0.26

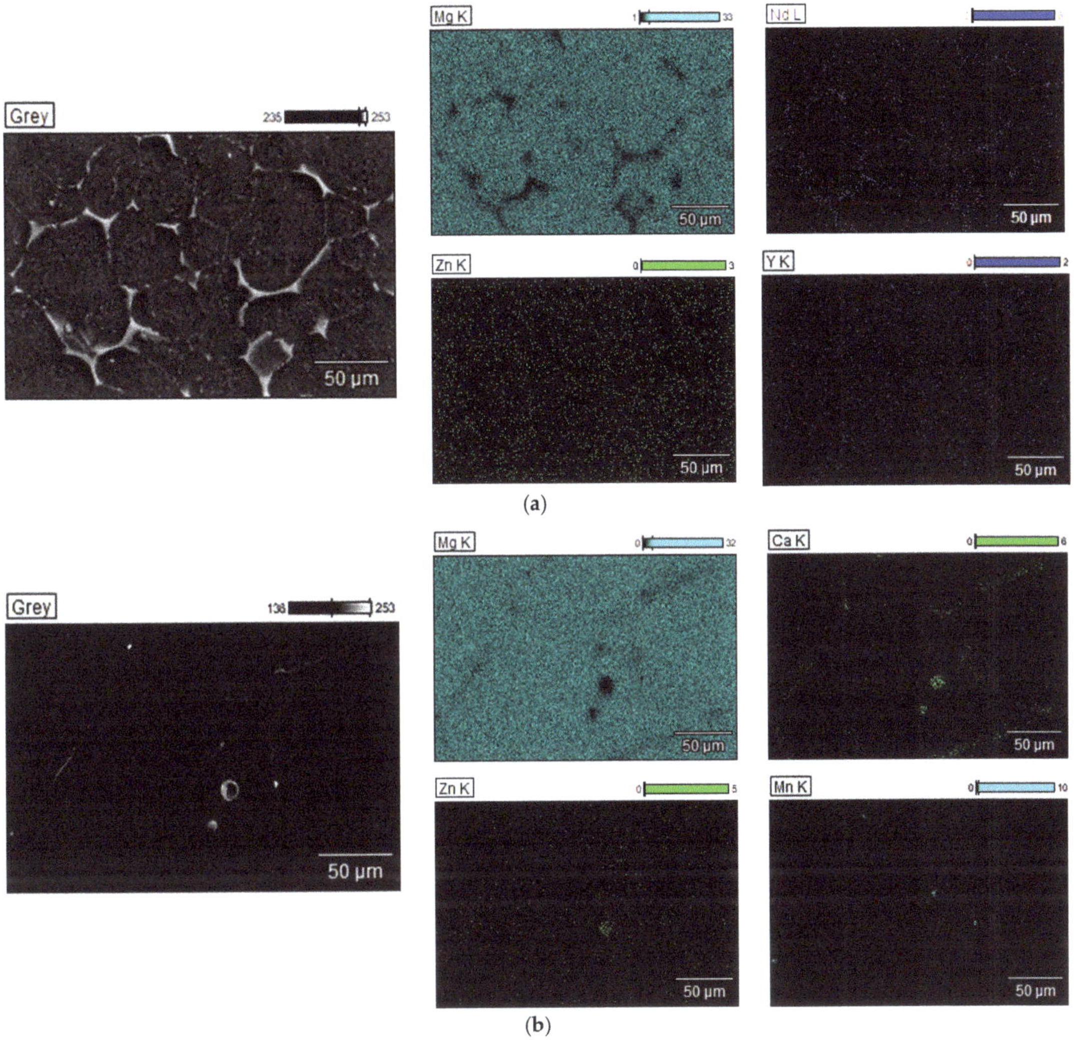

Figure 8. EDS elemental mapping results (magnification 500×) for (**a**) Mg-Nd alloy; (**b**) Mg-Zn alloy.

Elkaiam et al. [112] analyzed the microstructure of Mg-5%Zn-0.13%Y-0.35%Zr, adding up to 3% Nd. SEM images evidenced that α-Mg matrix was prevalent in the alloy, and the Nd additions were linked to precipitations around the magnesium grains. The authors noticed that modifying the Nd concentration led to different stoichiometry of Mg-Zn-Nd ternary phases that were localized based on EDS analysis at the grain boundaries. In addition, in our case, the Y concentration was higher, leading in the Y presence in the α-Mg matrix. Gong et al. [113] identified in Mg-2Zn-0.5Y-1.0Nd a quasi-crystalline phase rich in Zn with a chemical formula of $Zn_{60}Mg_{30}RE_{10}$, in accordance with other studies [94,114].

In the case of the second investigated alloy, Mg-Zn, SEM images showed an α-Mg matrix with secondary phases localized mainly at the grain boundaries in a strip-like structure. EDS analysis revealed a chemical composition (Figure 8b) for these secondary phases with elements such as Mg, Ca, and Zn, and a localized distribution of Mn.

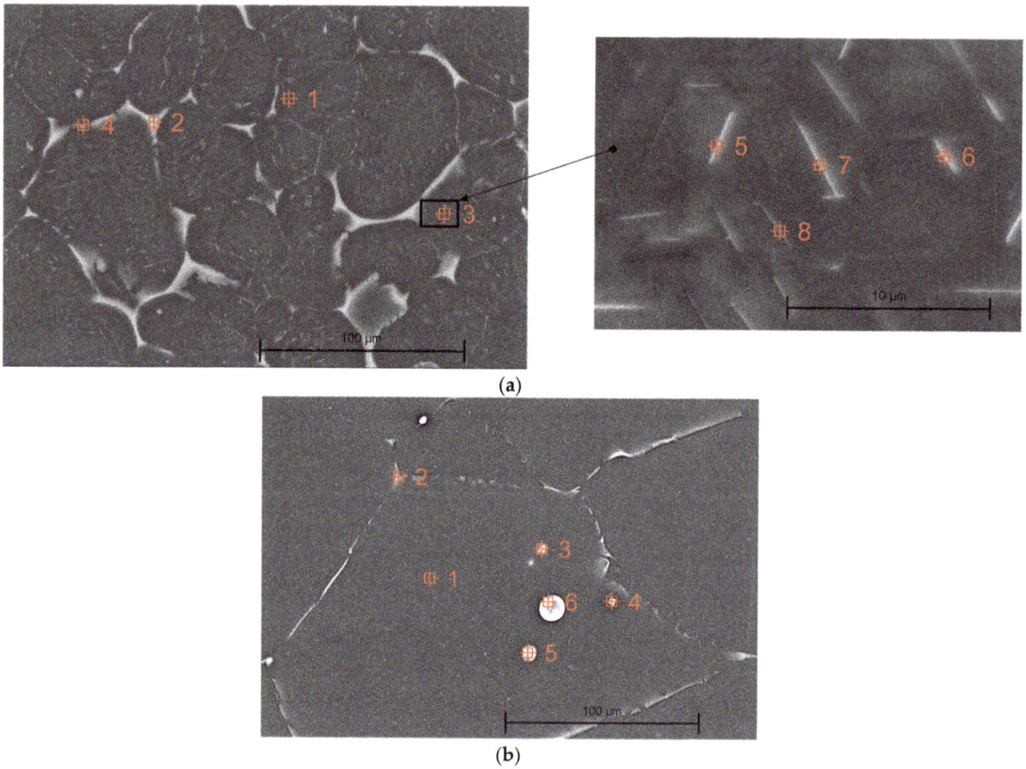

Figure 9. EDS analysis conducted for identification of phases' chemical composition: (**a**) Mg-Nd alloy—magnification 500× (Point 3—Inset (**a**) magnification 6000×); (**b**) Mg-Zn alloy—magnification 500×.

Figure 9b and Table 3 present some representative points set in different zones of the Mg-Zn microstructure. Point 1, Table 3 is localized inside the grain and exhibits mainly α-Mg as well as Zn, Ca, and Mn that are uniformly distributed into the whole matrix. Regarding the grain boundary investigation, we set point 2, Table 3, which was characterized by the presence of Mg, Zn, and Ca. The small round compounds (points 3 and 4, Table 3) are rich in Mn, as reported in the literature [106]. Regarding the large round compounds (points 5 and 6, Table 3), the EDS analysis revealed a similar chemical composition of Mg, Zn, and Ca as that found at the grain boundary. The XRD analysis supports this observation. In good agreement with our results extracted from EDS analysis and XRD investigation, some studies [100,102,106] proved the formation of $Mg_6Zn_3Ca_2$ phase instead of MgZn because the diffusivity of Ca in Mg is superior to that of Zn in the same matrix.

SEM investigations are of utmost importance when the material microstructure is analyzed because they can clarify the observations obtained after optical microscopy.

3.2. Mechanical Characterization of the Mg-Based Alloys

The manufacturing technology of Mg-based alloys has a significant impact on the elastic modulus and compressive strength of the material. For instance, in the case of our cast alloys, when compared to the other technologies, less optimized mechanical properties could be achieved because the coarse grain structure leads to lower values of Young's modulus and compressive strength [115,116]. The wrought processes typically result in a reduced grain size by promoting the precipitation process and enhancing the aforementioned mechanical properties [117,118]. Regarding the additive manufacturing technology, high-temperature working conditions are directly linked to differences in mechanical performance, while low-temperature methods lead to higher mechanical properties [119]. For Mg-based alloys manufactured through powder metallurgy, the compressive strength value is influenced by the powder content and production conditions [120]. To summarize these findings, one can notice that the choice of manufacturing technology influences the material microstructure, with direct implications for Young's modulus and compressive strength.

The compression tests were made on five samples prepared from each Mg-based alloy following the indications of ASTM E9-09 standard [63]. The Young's modulus was calculated by dividing the applied stress by the corresponding measured strain in accordance with ASTM E111 standard [121]. Figure 10 presents the variations in the stress–strain curves and the compressive strength values for Mg-Nd and Mg-Zn alloys. From Figure 10, one can conclude that the strongest strain hardening occurred in the case of the Mg-Nd, while the Mg-Zn exhibited the highest deformation capacity. We could mention that the Nd and Y additions for Mg-Nd alloys generated high compressive strength. Additionally, the differences in mechanical behavior could be linked to grain size, alloying elements, precipitate formation, as well as all material microstructure characteristics. As stated before, the presence of Nd and Zn is linked to various strengthening mechanisms, and their combinations with chemical elements such as Y or Ca generate complex interactions and could modify the material properties. Both alloying elements, Nd and Zn, respectively, produce a grain refining effect.

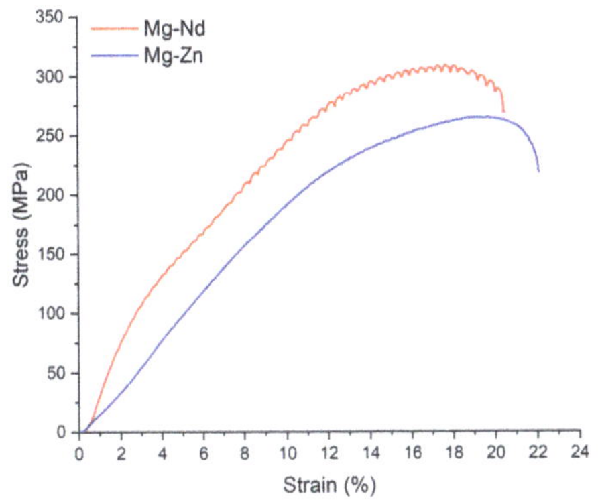

Figure 10. Stress and strain variations obtained after the compression test of the experimental magnesium-based alloys.

The compressive strength is defined as the property of a material to oppose a load that tends to reduce its size. For the Mg-Nd alloy, we have obtained a higher value of about 320.21 ± 13.40 MPa compared to the Mg-Zn alloy, which was characterized by a value of 289.12 ± 12.15 MPa. It is well-known that the ultimate compressive strength of natural bone was determined to be equal to 205 MPa along the longitudinal axis and 131 MPa in the transverse direction [122,123]. The values obtained for the two investigated alloys could be considered close to those of human bone, proving that the selected materials are suitable for bone substitutes. Regarding the Young's modulus, in the case of Mg-Nd alloy, it was determined to be a value of 59.44 ± 0.5 GPa, and for the Mg-Zn alloy, a much lower value of about 28.08 ± 0.5 GPa was obtained. The elasticity moduli obtained in our study are similar to those of human bone (trabecular bone 14.800 ± 1.40 GPa, cortical bone 20.700 ± 1.90 GPa) [124], so, as a direct consequence, the stress shielding effect, which occurs in the case of other metallic materials, is not estimated to be present for our Mg-based alloys. A one-way ANOVA analysis considering equal variances based on a confidence level of 95%, in combination with the Tukey and Fisher tests, was performed. These tests suggested that the obtained mean values of the mechanical properties are statistically significant.

Similar compressive strengths were obtained by Prakash et al. [40] for commercially available Mg-Al-Zn-Mn (AZ41) samples. The authors found a value of 337 ± 3 (MPa) for the maximum compressive strength at a strain rate of 2900 s^{-1}. A much more reduced value was found in [125] for cast Mg-0.2Fe-0.8Ca (165.6 MPa). An increase of 209.7 MPa was noticed after a heat treatment at 500 °C for 4 h was applied. This observation was explained by considering the Ca dissolution in the Mg matrix due to high temperature and the grain size reduction. Teslia et al. [126] investigated the compressive strength of Mg-Zn alloys and measured a value of around 250 MPa. In comparison with sintered Mg, which had a compressive strength between 92 and 95 MPa, the increased value obtained after Zn addition was attributed to grain boundary purification processes and crystalized intermetallic presence. Su et al. [127] obtained, for Mg-0.5Ca-x (Sr, Zr, Sn), Young's modulus values of about 45 GPa and different compressive strengths ranging between 150 MPa and 350 MPa as a function of the alloying element. Sumitomo et al. [128] determined for Young's modulus of Mg-Al-Zn alloys a value of 40 GPa and concluded that different factors such as grain size, precipitates, and alloying elements are of utmost importance in elasticity modulus variation.

3.3. In Vitro Corrosion Behavior of the Experimental Samples

Upon immersion of a Mg-based alloy into aqueous solution, a corrosion process of the metallic substrate occurred, and emission of ions and atoms such as Mg^{2+}, OH^-, and H_2 became possible according to Equations (1)–(3) as stated in the Introduction section [129,130]. Further, Mg^{2+} ions react with OH^- and generate a new compound on the substrate surface (Equation (4)). $Mg(OH)_2$ is considered unstable in aqueous media, such as SBF, and according to Equation (5), an exchange chemical reaction occurs.

At the anode, the following Mg alloy dissolution reaction can be observed:

$$Mg \rightarrow Mg^{2+} + 2e^- \tag{1}$$

while at the cathode, the following occurs:

$$2H^+ + 2e^- \rightarrow H_2 \tag{2}$$

$$2H_2O + 2e^- \rightarrow H_2 + 2HO^- \tag{3}$$

Then, the metallic cations react with the hydroxyl anions, which are present in the solution, and generate magnesium hydroxide:

$$Mg^{2+} + 2OH^- \rightarrow Mg(OH)_2 \tag{4}$$

The SBF solution contains chloride ions. In the literature, it is stated that in the cases in which the Cl^- ion concentration is higher than 30 mmol/L, the $Mg(OH)_2$ layer will react, and highly soluble $MgCl_2$ and OH^- result [6,39,131]. For the SBF solution, the chlorine ion concentration is about 131 mmol/L [132]. It can be foreseen that bone substitute samples will be subjected to a severe corrosion process and to a rapid mass loss (Equation (5)).

$$Mg(OH)_2 + 2Cl^- \rightarrow MgCl_2 + 2HO^- \tag{5}$$

The open circuit potential (OCP) and Tafel curves for the investigated alloys are shown in Figure 11, while Table 4 presents the main parameters.

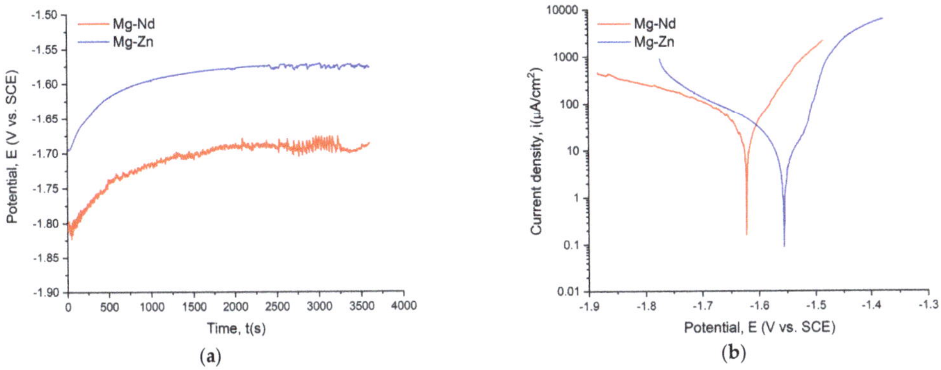

Figure 11. Electrochemical dependencies for Mg-Nd and Mg-Zn alloys: (**a**) OCP curves; (**b**) Tafel curves.

Table 4. The electrochemical parameters measured for experimental samples.

Sample ID	E_{oc} (V)	E_{corr} (mV)	i_{corr} (μA/cm^2)	β_c (mV)	β_a (mV)	Rp (kΩcm^2)	CR (mm/Year)
Mg-Nd	−1.675	−1.620	55.700	268.73	87.28	0.505	1.215
Mg-Zn	−1.567	−1.566	6.859	158.24	381.75	7.110	0.154

By using SBF as an electrolyte, it is observed that all the samples had negative values for E_{oc} and E_{corr}, below −1 V. When a material exhibits an improved corrosion behavior, it presents more electropositive values of E_{oc} and E_{corr}, concomitantly with electronegative values of i_{corr} and higher values of R_p. The ASTM G59-97 standard [133] recommends that researchers compute the CR only for uncoated samples. For excellent corrosion behavior, this parameter must have reduced values.

By considering the E_{oc} value criteria, it can be observed that the Mg-Zn samples had a higher value of −1.567 V than the −1.675 V obtained for the Mg-Nd samples. These findings can be easily verified with the E_{corr} criteria, with higher values of this parameter in the case of Mg-Zn samples: E_{corr} = −1.566 V; and lower values for the Mg-Nd samples: E_{corr} = −1.620. Regarding the corrosion current density i_{corr}, it can be observed that the lowest value, being equivalent to the most electronegative character, was obtained for the Mg-Zn sample (6.859 μA/cm^2). The Tafel curves proved that the corrosion current density shifted in a more noble direction for the Mg-Zn samples, offering increased resistance to the corrosion process and supporting the data presented in Table 4. In [134], the anodic polarization curves are associated with the Mg dissolution process, while the cathodic curves are usually linked to hydrogen emission. Higher values of polarization resistance showed good corrosion resistance. In our case, the highest value was obtained for Mg-Zn (7.102 kΩcm^2). The CR parameter clearly indicates that the Mg-Zn alloy has a significantly reduced value compared to Mg-Nd, which demonstrates better corrosion resistance, as also supported by the other criteria explained in this section.

Many studies in the literature describe the electrochemical corrosion behavior of Mg-based alloys. Zemkova et al. [82] investigated the Y and Nd additions on Mg alloys directly related to their corrosion behavior. It was observed that the Mg-Y secondary phase particles had a detrimental influence on the corrosion resistance [135,136], while in the cases in which only Y was added, it dissolved into the Mg matrix, generating an increase in the corrosion resistance [137]. In addition, the authors noticed that for Mg-4Y-3Nd alloy, the secondary Mg-(Nd, Y) phase did not produce an important galvanic corrosion effect due to a reduced potential difference of 25 mV between α-Mg and the secondary phases. Regarding the Mg-Zn-Mn-Ca alloys, some investigations showed that the secondary phase Mg_2Ca could decrease the corrosion rate in alloys with a Zn/Ca ratio below 1.25. Unfortunately, this ratio is about 2.8 in our case, and the Ca content is lower than 0.5%. It can be concluded that in the absence of the above-mentioned phase, the α-Mg corrodes much faster due to the galvanic coupling phenomena that occurred in the material, while $Ca_2Mg_6Zn_3$ remains in the alloy and generates an increase in the corrosion resistance in the case of an increased Zn content [134].

Electrochemical characterization is an important investigation because it permits researchers to completely understand the kinetic phenomena that usually accompany the Mg degradation process. In this way, parameters such as corrosion rate and polarization resistance can be computed, as they are crucial for analysis dedicated to highly bioactive metals.

3.4. *In Vivo Animal Study*

Mild postoperative complications, such as subcutaneous emphysema, were observed, likely associated with hydrogen gas release from magnesium alloy degradation.

3.4.1. Histopathological Examination

For both materials, at week 1, the histopathological evaluation of skin and subcutaneous lesions displayed minimal necrosis and mild infiltration by polymorphonuclear cells and lymphocytes, with minimal plasma cells, moderate macrophages, and scattered multinucleated giant cells. In the following weeks, polymorphonuclear infiltration and necrosis diminished significantly; however, mild lymphocytes, macrophages, and multinucleated giant cells were still observed in weeks 2, 4, and 8, along with minimal neo-vascularization and fibrosis (Figure 12).

The Lesion Score was higher in week 1, followed by a decreasing trend in both biomaterials; notably, Mg-Zn displayed better compatibility in all time periods (Table 5). At the end of the evaluation, both materials could be considered to have a slight to moderate irritant effect, making them suitable for further testing on the bone defect model.

Table 5. Comparative Lesion Scores in subcutaneous implantation of the experimental alloys (ISO 10993-6).

		Inflammation											
		Acute		**Chronic**									
Samples	**Time**	**Necrosis**	**Polymorphonuclear Neutrophil**	**Lymphocytes**	**Plasma Cells**	**Macrophage**	**Giant Cells**	**Subtotal**	**Neovascularization**	**Fibrosis**	**Fatty Infiltration**	**Subtotal**	**TOTAL**
Mg-Nd	Week 1	1	1.5	1.5	0.5	2	1	**7.5**	1.5	2.5	0.5	**4.5**	**12**
	Week 2	0.5	1	1.5	0	1.5	0.5	**5**	2	2	0	**4**	**9**
	Week 4	1	1	2.25	0.75	2.25	0.75	**8**	1.75	1.75	0	**3.5**	**11.5**
	Week 8	0	0.25	1.5	0	2.5	1.75	**6**	1	1.25	0	**2.25**	**8.25**
Mg-Zn	Week 1	1.5	2.5	1	0	2	1	**8**	1	1	0	**2**	**10**
	Week 2	0	0	2	0.5	1.5	1.5	**5.5**	1	1.5	0	**2.5**	**8**
	Week 4	0	0	0.5	0	0.5	1.5	**2.5**	0	1	0	**1**	**3.5**
	Week 8	0	0	2	0.5	1.5	0.5	**4.5**	0	1	0	**1**	**5.5**

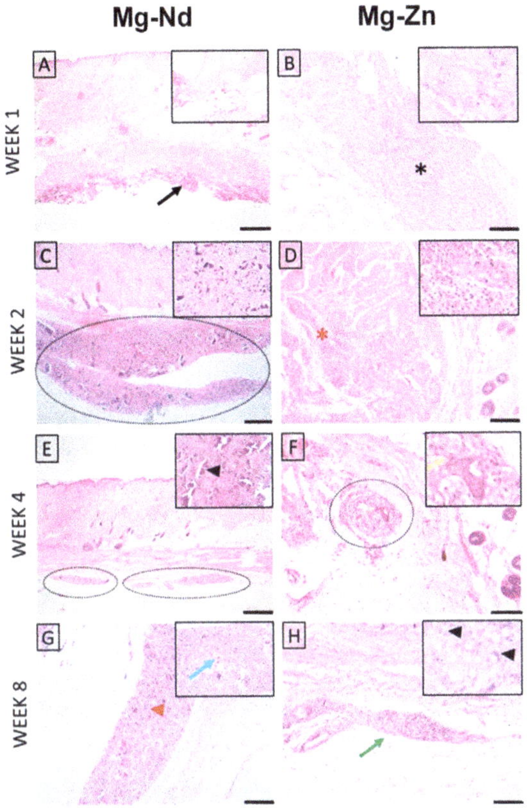

Figure 12. Histopathological evaluation of skin and *subcutis* lesions (Mg-Nd alloy: (**A**,**C**,**E**,**G**), and Mg-Zn alloy: (**B**,**D**,**F**,**H**)): (**A**) Minimal necrosis and mild infiltration by polymorphonuclear cells and lymphocytes, with minimal plasma cells, moderate macrophages, and mild multinucleated giant cells. The lesion is indicated by a black arrow. Inset: higher magnification of the inflammatory infiltrate; (**B**) Moderate necrosis (black asterisk) accompanied by moderate polymorphonuclear cells, minimal lymphocytes, moderate macrophages, and minimal giant cells. Inset: detailed view of necrotic tissue with inflammatory cells; (**C**) Lesion showing minimal necrosis (dotted ellipse) and mild polymorphonuclear cells, mild lymphocytes, moderate macrophages, and minimal giant cell infiltration. Inset: higher magnification highlighting inflammatory cell composition; (**D**) Area with moderate lymphocyte infiltration (red asterisk), minimal plasma cells, mild macrophages, and moderate giant cells. Inset: details of lymphocyte-predominant infiltrate; (**E**) Mild lymphocytes, macrophages, and multinucleated giant cells are observed. Multifocal inflammatory foci are marked (dotted ellipses). Inset: higher magnification of a macrophage-rich focus (black arrowhead); (**F**) Minimal lymphocytes and macrophages with moderate giant cells are seen within a perivascular lesion (dotted circle). Inset: detail showing multinucleated giant cells; (**G**) Mild lymphocytes and moderate macrophages with mild multinucleated giant cells. Minimal neovascularization and fibrosis are present. Lesion core indicated by a red arrowhead; inset shows central necrosis with inflammatory cells (blue arrow); (**H**) Moderate lymphocyte infiltration (green arrow) with minimal macrophages and giant cells. Inset: higher magnification of lymphocytic infiltrate. H&E staining; Ob. 4× (**A**,**C**,**E**), Ob. 10× (**D**,**F**,**H**), Ob. 20× (**B**,**G**); Barr 100 μm (**B**,**G**), 200 μm (**D**,**F**,**H**), 500 μm (**A**,**C**,**E**).

Histological evaluation showed similar early findings in both Mg–Nd and Mg–Zn groups, with the defects visible at week 1 and minimal periosteal and endosteal proliferation at week 2, in the absence of inflammation. By week 4, the Mg–Nd group displayed a pronounced endosteal reaction partially separating the defect from the marrow cavity, while the Mg–Zn group showed incomplete coverage with bone trabeculae. At week 8, the Mg–Zn group achieved complete closure with compact lamellar bone, whereas the Mg–Nd group exhibited extensive endosteal and periosteal proliferation largely covering the defect but without full lamellar organization. These results suggest that Mg–Zn alloy supports earlier maturation of bone tissue, while Mg–Nd alloy triggers a stronger endosteal response that may delay complete structural organization (Figures 13 and 14).

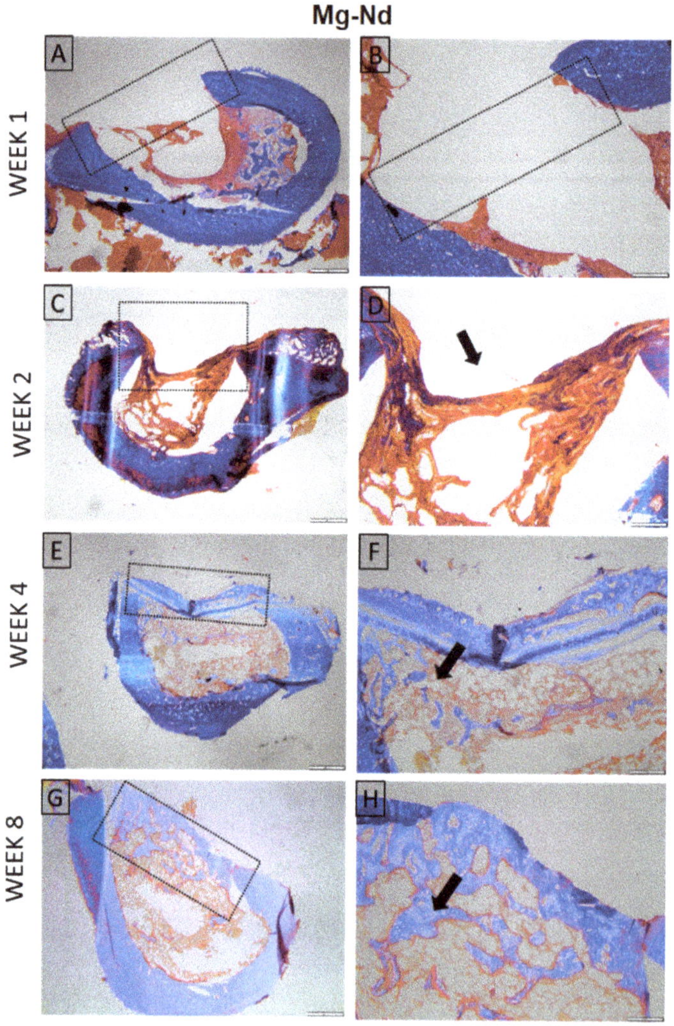

Figure 13. Histological aspect of the Mg-Nd group. Long bone. Week 1 (**A**,**B**) The defect is present, and no signs of inflammation or tissue regeneration are observed, consistent with the early post-injury; Week 2 (**C**,**D**) Proliferation of endosteal and periosteal tissue admixed with collagen that minimally covers the defect, without inflammation (arrow); Week 4 (**E**,**F**) Bone defect incompletely covered by compact bone; severe endosteal reaction partially separating the defect area from the bone marrow (arrow); Week 8 (**G**,**H**) Proliferation of endosteal and periosteal tissue largely covering the defect, without inflammation (arrow). Masson's trichrome stain. Ob. ×4 (**A**,**C**,**E**,**G**) Ob. 10× (**B**,**D**,**F**,**H**). Scale bar 500 µm (**A**,**C**,**E**,**G**) and 200 µm (**B**,**D**,**F**,**H**).

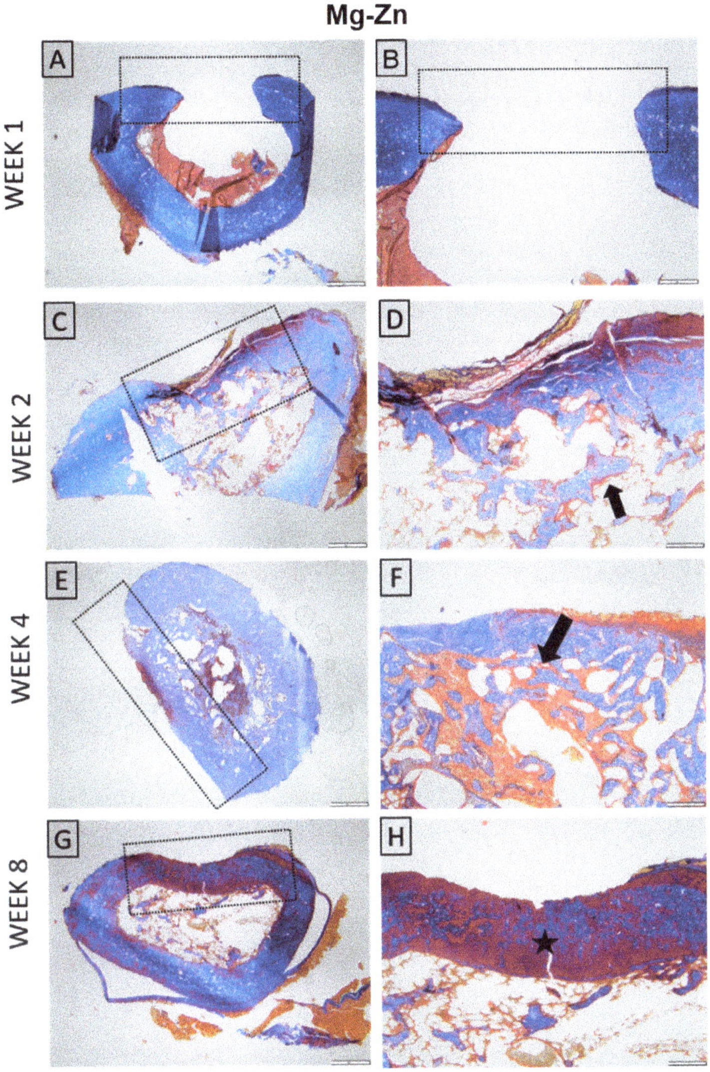

Figure 14. Histological aspect of the Mg-Zn group. Long bone. Week 1 (**A**,**B**) The defect is present, and no signs of inflammation or tissue regeneration are observed, consistent with the early post-injury; Week 2 (**C**,**D**) Proliferation of endosteal and periosteal tissue minimally covers the defect, without inflammation (arrow); Week 4 (**E**,**F**) Bone defect incompletely covered by bone trabecula; (arrow); Week 8 (**G**,**H**) Defect completely covered by bone tissue with a compact lamellar appearance (star). Masson's trichrome stain. Ob. ×4 (**A**,**C**,**E**,**G**) Ob. 10× (**B**,**D**,**F**,**H**). Scale bar 500 μm (**A**,**C**,**E**,**G**) and 200 μm (**B**,**D**,**F**,**H**).

3.4.2. Micro-Computed Tomography Acquisition and Analysis

The comparative μCT analysis of bone morphogenesis between the Mg-Nd and Mg-Zn experimental samples demonstrates distinct patterns in the rate and quality of mineralized tissue growth. Over the time between week 1 and week 8, both groups exhibit a steady advancement in mineral deposition, structural maturation, and host integration, albeit at significantly differing speeds and with distinct morphologies.

In the case of Mg-Nd experimental samples, early time points (weeks 1–2) exhibit a fragmented, low-density mineral phase within the defect area, occupying a relatively restricted volume. The *de novo* tissue exhibits porosity and an irregular morphology, characterized by scattered mineralized nuclei instead of a cohesive mass. The interfacial integration with the femoral cortex is negligible at this time, as indicated by the visible gap between the bone substitute area and the host bone edges, with some exceptions of small anchored formations interfacing the femur. The surface contact is non-continuous, and the biomaterial does not visually match the defect's geometry. By weeks 4 and 8, there is a significant enhancement in mineral volume and density. The bone structure exhibits a more trabecular look, characterized by denser mineral clusters that increasingly resemble physiological bone architecture. The material exhibits a more uniform distribution, hence decreasing porosity. By week 8, the interface with the femoral cortex appears visually continuous, indicating integration via direct bone-to-bone bridging. The defect is filled with tissue that has not only expanded in volume but also in mineral maturity and spatial coherence (Figure 15).

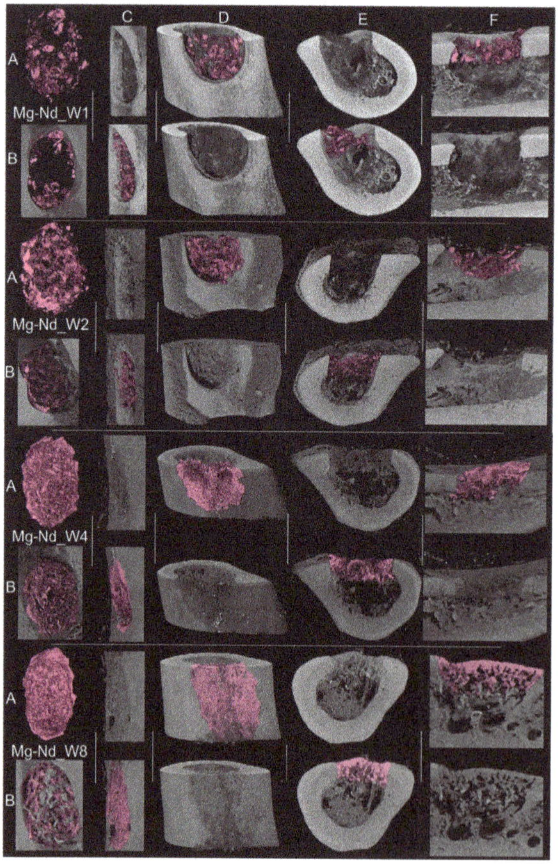

Figure 15. Representative 3D reconstructions and cross-sectional views of mineralized tissue within the femoral defect at multiple time points for the Mg-Nd group: (**A**) segmented bone volume; (**B**) internal view of bone tissue along the medullary canal, within the bone defect; (**C**) side-frontal view; cross-sections of (**D**) COR, (**E**) TRA, and (**F**) SAG planes illustrating bone formation and integration with the host cortex. Each view is shown with and without the segmented bone volume to highlight mineral distribution and interface continuity.

In contrast, the Mg-Zn experimental samples have elevated visible mineral density and a more uniform distribution pattern in early post-implantation images. By week 1, the mineralized phase occupies the defect volume more thoroughly, exhibiting density and a rather smooth interface with the host femur. The biomaterial adheres more closely to the defect borders, and the continuity of cortical alignment is superior. The longitudinal and cross-sectional analyses indicate a reduction in "pores" (probably an efficient natural tissue structuration) and an increase in surface regularity, implying a more efficient initial osteoconductive environment. The advancement from weeks 2 to 8 in the Mg-Zn samples is marked by both augmented mineral volume and a distinct architectural organization. The regenerated tissue exhibits increased density, with elevated greyscale intensity signifying enhanced mineral content. The junction between the newly formed bone and the surrounding femur seems progressively smoother. By week 8, the regenerated bone closely resembles the host in density and shape, indicating both mechanical and biological continuity (Figure 16). The trabecular interweaving between new and old bone is more pronounced than in the Mg-Ng group.

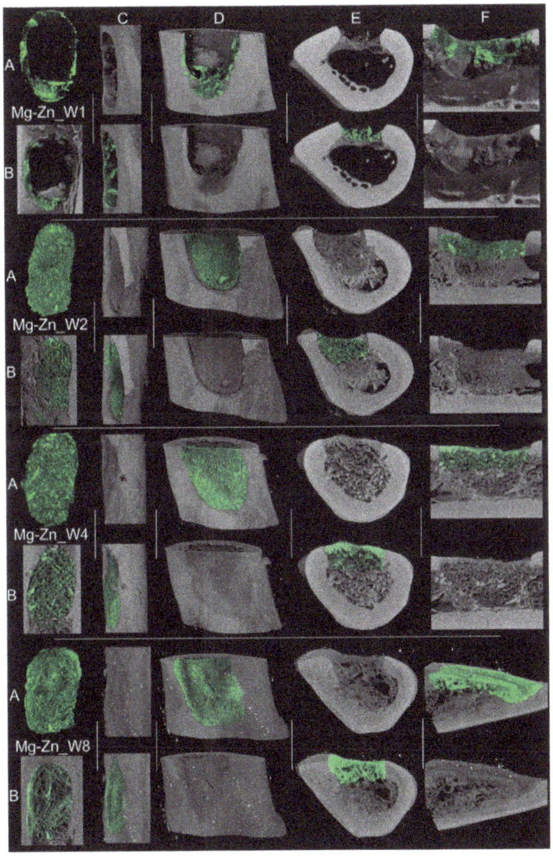

Figure 16. Representative 3D reconstructions and cross-sectional views of mineralized tissue within the femoral defect at multiple time points for the Mg-Zn group: (**A**) segmented bone volume; (**B**) internal view of bone tissue along the medullary canal, within the bone defect; (**C**) side-frontal view; cross-sections of (**D**) COR, (**E**) TRA, and (**F**) SAG planes illustrating bone formation and integration with the host cortex. Each view is shown with and without the segmented bone volume to highlight mineral distribution and interface continuity.

In the comparison of both groups at all time points, Mg-Zn exhibits accelerated and more thorough mineralization, denser and more linked trabeculae, and enhanced defect conformity. The evolution seems more linear and structured, whereas Mg-Nd exhibits delayed mineral deposition, increased porosity, and less consistent interfacial bonding, especially in the initial phases (Figure 17a). The bone surface area-to-volume ratio is elevated in the early Mg-Nd samples, suggesting a more irregular structure, whereas Mg-Zn attains lower ratios more rapidly, indicative of structural consolidation (Figure 17b).

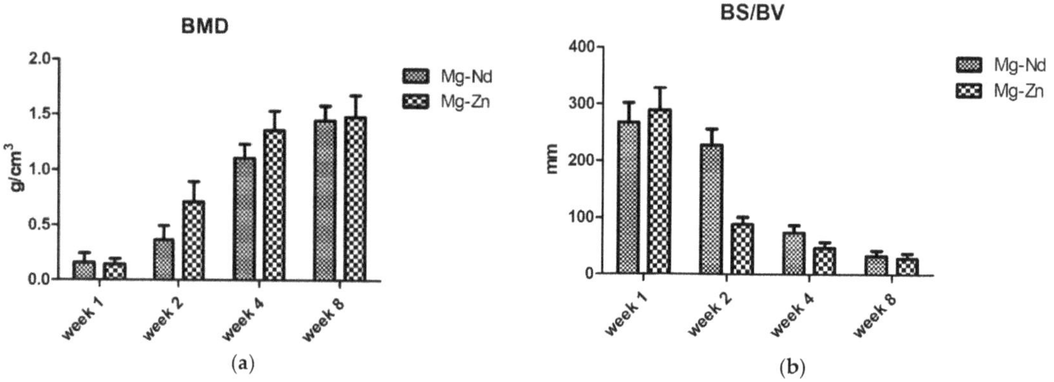

Figure 17. Bone mineral density (**a**) and bone specific surface (BS/BV) (**b**) at different time points.

Microcomputed tomography is considered today an important tool that provides high-resolution three-dimensional images of bone, soft tissue, or other biological structures. Regarding our study, the μCT images were useful in the estimation of the mineralization capacity of the two proposed Mg-based alloys, proving their suitability as possible bone substitute materials.

Collectively, our findings indicate that Mg-Zn facilitates earlier and more cohesive bone regeneration, marked by increased mineral density, enhanced defect filling, smoother cortical transitions, and superior spatial coherence. Mg-Nd, although capable of future integration, adheres to a slower and less efficient trajectory, with mineralization and host interaction trailing behind. These disparities likely indicate changes in material bioactivity, cellular response, and scaffold architecture, eventually affecting regeneration efficacy.

3.4.3. *In Vivo* Study Limitations

It is of utmost importance to emphasize that the results presented in Section 3.4, based on the Sprague Dawley rat model, may not be directly translated to human subjects due to differences in anatomical conditions, such as rat bone density, bone morphology, and metabolic rates, as well as loading conditions when analyzing bone substitutes. Regarding the anatomical differences, the human femur is significantly larger and more complex, exhibiting distinct structural properties when compared to the Sprague Dawley rat femur [138]. In addition, it is well-known that the forces and stresses acting on a bone substitute implanted in a rat's leg cannot be applied to humans because they are significantly reduced, and only extrapolation models can be applied [139,140]. Another aspect worth mentioning is that bone remodeling processes and healing rates differ between species, and the corrosive medium inside living organisms can influence a significant portion of the alloy degradation rate, as well as the bone substitute integration mechanism [141,142]. Consequently, the primary conditions to be considered are the mechanical properties of the bone substitute, such as strength and elastic modulus, which may vary from rat to human, and it is almost impossible to predict their long-term behavior in humans [143]. Regarding biocompatibility and degradation, the breakdown process may differ, involving various responses between species [139].

It is understood that new materials are analyzed to address a clinical need and that animal models must usually be involved; however, scholars should also consider the necessity of investigating many more conditions to better understand the phenomena that occur in the human body.

As a final conclusion, these limitations narrow the application of Mg-Nd and Mg-Zn alloys in near-future clinical trials.

4. Conclusions

The present investigations analyzed two experimental magnesium-based alloys, from the binary systems Mg-Nd and Mg-Zn, designed for the manufacture of biodegradable structures that can be used in the treatment of bone defects. For the analyzed biodegradable structures made with biomaterials that induce osseous regeneration, the mechanical properties are not as critical as in the case of trauma implants. For structures used as bone substitutes, the most important attributes that should be analyzed are biodegradation and interaction with hard tissue, as these prove whether it is feasible for bone defect treatment.

In vitro corrosion tests conducted using an electrochemical approach demonstrated better corrosion resistance for the Mg-Zn alloy due to the Ca and Mn addition in a favorable amount. Also, both experimental magnesium-based alloys exhibited adequate mechanical properties with values close to those of human bone. The properties mentioned above were correlated with the microstructural characterization, directly linked to the addition of the rare earths.

Regarding the mechanical properties, the higher value of Young's modulus obtained for the Mg-Nd alloy compared to the Mg-Zn counterparts could have a potential clinical implication by inducing local stress effects and influencing the bone remodeling process at the defect site. To address this drawback, a careful analysis and design regarding geometry, localization, and degradation rate of Mg-Nd bone substitute must be applied to achieve a mechanical equilibrium state with the natural and incremental load transfer. This fact ensures the correct bone healing mechanism, based on the simultaneous activity of osteoblasts and osteoclasts. It is advisable that patients with Mg-Nd implants should be carefully monitored to analyze the long-term effects of local stress induced in the vicinity of the bone substitute.

Good biocompatibility of both experimental magnesium-based alloys resulted from *in vivo* animal testing, a fact that supports once more the study's initial assumption regarding the suitability of the proposed magnesium-based alloys for this kind of orthopedic need. However, the Mg–Zn alloy exhibited better hard tissue maturation, since the Mg–Nd alloy was characterized by a stronger endosteal response, which is directly linked to a delayed complete bone structural organization. The microcomputed tomography analysis revealed an accelerated mineralization process with dense and linked trabeculae being characterized by a linear evolution, in the case of Mg-Zn samples. On the other hand, for Mg-Nd samples, a delayed mineralization with a reduced interfacial bonding as well as a high porosity grade, combined with a more irregular structure, was noticed.

Based on our experimental results, we can conclude that the magnesium-based alloys, from the binary systems Mg-Nd and Mg-Zn, could be used for designing different biodegradable structures used in the surgical treatment of bone defects. Future studies must be focused on the development of biodegradable structures with patient-adapted geometries and the same long-term studies to evaluate the quality of the regenerated bone after the use of these structures.

To ensure clarity when connecting the alloy microstructure with mechanical properties and corrosion behavior, we will, in future studies, determine the statistical distribution of grain size and analyze how secondary phase distribution and their electrochemical potential modify mechanical properties, such as compressive strength and Young's modulus, due to micro-galvanic phenomena. We want to refine our experimental methods by including a field emission electron microscope and immersion test to be confident in the observed microstructure and the obtained corrosion rates of the alloys.

In order to develop a highly efficient bone substitute, we must consider optimizations for design architecture, application of surface treatments or biocompatible coatings to reduce and control the material degradation. In this way, we will be able to address some of the limitations presented in our paper to ensure a safe clinical translation to humans.

Author Contributions: Conceptualization: I.A., B.S. and V.M.; methodology: A.A., A.I.D., M.I., S.G. and M.M.; software: V.M., M.C.M., S.G. and M.I.; validation: I.A., M.I., B.S. and C.M.C.; formal analysis: A.I.D., R.P., G.M.V., V.M. and M.C.M.; investigation: V.M., M.C.M., C.M.C., R.P., A.A., F.A.T. and G.M.V.; resources: I.A., S.G. and M.M.; data curation: A.A., M.M., R.P., F.A.T., A.I.D. and M.I.; writing—original draft preparation: V.M., M.C.M., A.I.D. and S.G.; writing—review and editing: V.M., A.A., M.M., B.S. and G.M.V.; visualization: I.A., S.G. and B.S.; supervision: I.A., B.S. and C.M.C.; funding acquisition: I.A. All authors have read and agreed to the published version of the manuscript.

Funding: This work was supported by a grant of the Ministry of Research, Innovation, and Digitization, CNCS—UEFISCDI, project number PV-IV-P1-PCE-2023-1452, within PNCDI IV.

Institutional Review Board Statement: The study was conducted in accordance with the Declaration of Helsinki and approved by the Bioethics Committee of the University of Agricultural Sciences and Veterinary Medicine Cluj-Napoca (approval code No. 386 of 15.06.2023), with the approval granted on 16 July 2023. The project was approved by the regional state veterinary authority DSVSA-CN (Project authorization no. 382/25 August 2023).

Informed Consent Statement: Not applicable.

Data Availability Statement: The original contributions presented in the study are included in the article, further inquiries can be directed to the corresponding author.

Conflicts of Interest: The authors declare no conflicts of interest.

References

1. Georgeanu, V.A.; Gingu, O.; Antoniac, I.V.; Manolea, H.O. Current Options and Future Perspectives on Bone Graft and Biomaterials Substitutes for Bone Repair, from Clinical Needs to Advanced Biomaterials Research. *Appl. Sci.* **2023**, *13*, 8471. [CrossRef]
2. Mocanu, A.-C.; Stan, G.E.; Maidaniuc, A.; Miculescu, M.; Antoniac, I.V.; Ciocoiu, R.-C.; Voicu, Ș.I.; Mitran, V.; Cîmpean, A.; Miculescu, F. Naturally-Derived Biphasic Calcium Phosphates through Increased Phosphorus-Based Reagent Amounts for Biomedical Applications. *Materials* **2019**, *12*, 381. [CrossRef]
3. Antoniac, I.; Negrusoiu, M.; Mardare, M.; Socoliuc, C.; Zazgyva, A.; Niculescu, M. Adverse Local Tissue Reaction after 2 Revision Hip Replacements for Ceramic Liner Fracture. *Medicine* **2017**, *96*, e6687. [CrossRef]
4. Botez, P.; Sirbu, P.; Simion, L.; Munteanu, F.I.; Antoniac, I. Application of a Biphasic Macroporous Synthetic Bone Substitutes CERAFORM®: Clinical and Histological Results. *Eur. J. Orthop. Surg. Traumatol.* **2009**, *19*, 387–395. [CrossRef]
5. Allori, A.C.; Sailon, A.M.; Warren, S.M. Biological Basis of Bone Formation, Remodeling, and Repair—Part I: Biochemical Signaling Molecules. *Tissue Eng. Part B Rev.* **2008**, *14*, 259–273. [CrossRef]
6. Antoniac, I.; Manescu, V.; Antoniac, A.; Paltanea, G. Magnesium-Based Alloys with Adapted Interfaces for Bone Implants and Tissue Engineering. *Regen. Biomater.* **2023**, *10*, rbad095. [CrossRef]

7. Quan, P.H.; Antoniac, I.; Miculescu, F.; Antoniac, A.; Manescu, V.; Robu, A.; Bița, A.-I.; Miculescu, M.; Saceleanu, A.; Bodog, A.D.; et al. Fluoride Treatment and *In Vitro* Corrosion Behavior of Mg-Nd-Y-Zn-Zr Alloys Type. *Materials* **2022**, *15*, 566. [CrossRef]
8. Schemitsch, E.H. Size Matters: Defining Critical in Bone Defect Size! *J. Orthop. Trauma* **2017**, *31*, S20–S22. [CrossRef] [PubMed]
9. Xue, N.; Ding, X.; Huang, R.; Jiang, R.; Huang, H.; Pan, X.; Min, W.; Chen, J.; Duan, J.-A.; Liu, P.; et al. Bone Tissue Engineering in the Treatment of Bone Defects. *Pharmaceuticals* **2022**, *15*, 879. [CrossRef] [PubMed]
10. Fernandez-Yague, M.A.; Abbah, S.A.; McNamara, L.; Zeugolis, D.I.; Pandit, A.; Biggs, M.J. Biomimetic Approaches in Bone Tissue Engineering: Integrating Biological and Physicomechanical Strategies. *Adv. Drug Deliv. Rev.* **2015**, *84*, 1–29. [CrossRef] [PubMed]
11. Guazzo, R.; Gardin, C.; Bellin, G.; Sbricoli, L.; Ferroni, L.; Ludovichetti, F.S.; Piattelli, A.; Antoniac, I.; Bressan, E.; Zavan, B. Graphene-Based Nanomaterials for Tissue Engineering in the Dental Field. *Nanomaterials* **2018**, *8*, 349. [CrossRef]
12. Antoniac, I.V.; Burcea, M.; Ionescu, R.D.; Balta, F. IOL's Opacification: A Complex Analysis Based on the Clinical Aspects, Biomaterials Used and Surface Characterization of Explanted IOL's. *Mater. Plast.* **2015**, *52*, 109–112.
13. Şerboiu, C.S.; Aliuș, C.; Dumitru, A.; Țăpoi, D.; Costache, M.; Nica, A.E.; Alexandra-Ana, M.; Antoniac, I.; Grădinaru, S. Gallbladder Pancreatic Heterotopia—The Importance of Diagnostic Imaging in Managing Intraoperative Findings. *Medicina* **2023**, *59*, 1407. [CrossRef]
14. Onisâi, M.; Dumitru, A.; Iordan, I.; Aliuș, C.; Teodor, O.; Alexandru, A.; Gheorghiță, D.; Antoniac, I.; Nica, A.; Mihăilescu, A.-A.; et al. Synchronous Multiple Breast Cancers—Do We Need to Reshape Staging? *Medicina* **2020**, *56*, 230. [CrossRef] [PubMed]
15. Grădinaru, E.S.; Stoicea, M.-C.; Mocanu, L.; Antoniac, I.; Gheorghiță, D.; Grigore, A.G.M. Rare Breast Carcinoma with Paradoxical Plasma Cell Immunoprofile: A Case Report. *Medicina* **2020**, *56*, 62. [CrossRef]
16. Bongio, M.; Van Den Beucken, J.J.J.P.; Leeuwenburgh, S.C.G.; Jansen, J.A. Development of Bone Substitute Materials: From 'Biocompatible' to 'Instructive'. *J. Mater. Chem.* **2010**, *20*, 8747–8759. [CrossRef]
17. Paltanea, G.; Manescu, V.; Antoniac, I.; Antoniac, A.; Nemoianu, I.V.; Robu, A.; Dura, H. A Review of Biomimetic and Biodegradable Magnetic Scaffolds for Bone Tissue Engineering and Oncology. *Int. J. Mol. Sci.* **2023**, *24*, 4312. [CrossRef]
18. Cavalu, S.; Antoniac, I.V.; Mohan, A.; Bodog, F.; Doicin, C.; Mates, I.; Ulmeanu, M.; Murzac, R.; Semenescu, A. Nanoparticles and Nanostructured Surface Fabrication for Innovative Cranial and Maxillofacial Surgery. *Materials* **2020**, *13*, 5391. [CrossRef]
19. Cristescu, I.; Antoniac, I.; Branzei, M.; Ghiban, B.; Ciocoiu, R.; Ciuca, I.; Semenescu, A.; Gradinaru, S.; Pop, D.; Raiciu, A.D. Fractography Study of Explanted Intramedullary Nails. *Mater. Plast.* **2019**, *56*, 759–773. [CrossRef]
20. De Aza, P.N.; Luklinska, Z.B.; Santos, C.; Guitian, F.; De Aza, S. Mechanism of Bone-like Formation on a Bioactive Implant *in Vivo*. *Biomaterials* **2003**, *24*, 1437–1445. [CrossRef]
21. Rahaman, M.N.; Day, D.E.; Bal, B.S.; Fu, Q.; Jung, S.B.; Bonewald, L.F.; Tomsia, A.P. Bioactive Glass in Tissue Engineering. *Acta Biomater.* **2011**, *7*, 2355–2373. [CrossRef]
22. Ali, R.H.; Sghaier, Z.; Ageorges, H.; Ben Salem, E.; Hidouri, M. Magnesium-Substituted Zinc-Calcium Hydroxyfluorapatite Bioceramics for Bone Tissue Engineering. *J. Mech. Behav. Biomed. Mater.* **2025**, *166*, 106933. [CrossRef]
23. Sampaio, R.F.; Carvalho, C.N.; Bradaschia-Correa, V.; Gonçalves, B.L.L.; Arana-Chavez, V.; de Carvalho, A.P.L.; Nogueira, A.P.A.; Grazziotin-Soares, R.; Bauer, J.; Gavini, G.; et al. Apical Sealing and Bioactivity of an Experimental Gutta-Percha Containing Niobium Phosphate Bioglass. *Polymers* **2023**, *15*, 1679. [CrossRef]
24. Kowalska, K.J.; Czechowska, J.P.; Yousef, E.S.; Zima, A. Novel Phosphate Bioglasses and Bioglass-Ceramics for Bone Regeneration. *Ceram. Int.* **2024**, *50*, 45976–45985. [CrossRef]
25. Safarova, Y.; Nessipbekova, A.; Syzdykova, A.; Olzhayev, F.; Umbayev, B.; Kassenova, A.; Fadeeva, I.V.; Askarova, S.; Rau, J.V. Strontium- and Copper-Doped Ceramic Granules in Bone Regeneration-Associated Cellular Processes. *J. Funct. Biomater.* **2024**, *15*, 352. [CrossRef]
26. Dorozhkin, S.V. Calcium Orthophosphate ($CaPO_4$) Containing Composites for Biomedical Applications: Formulations, Properties, and Applications. *J. Compos. Sci.* **2024**, *8*, 218. [CrossRef]
27. Kharissova, O.V.; Méndez, Y.P.; Kharisov, B.I.; Nikolaev, A.L.; Dorozhkin, S.V.; Mena, D.N.; García, B.O. Biomineralization of Calcium Phosphates in Nature. *Nano-Struct. Nano-Objects* **2025**, *41*, 101425. [CrossRef]
28. Colilla, M.; Manzano, M.; Vallet-Regí, M. Recent Advances in Ceramic Implants as Drug Delivery Systems for Biomedical Applications. *Int. J. Nanomed.* **2008**, *3*, 403–414. [CrossRef]
29. Miu, D.-M.; Pavaloiu, R.D.; Sha'at, F.; Vladu, M.-G.; Neagu, G.; Manoiu, V.-S.; Eremia, M.-C. Preparation and Optimization of a Polyhydroxyoctanoate-Hydroxyapatite Composite Available to Scaffolds in Implantable Devices. *Molecules* **2025**, *30*, 730. [CrossRef]
30. Wang, Z.; Shang, J.; Zhang, Z. Composite or Modified Hydroxyapatite Microspheres as Drug Delivery Carrier for Bone and Tooth Tissue Engineering. *Curr. Med. Chem.* **2025**, *32*, 974–981. [CrossRef]
31. Liu, G.; Zhang, B.; Wan, T.; Zhou, C.; Fan, Y.; Tian, W.; Jing, W. A 3D-Printed Biphasic Calcium Phosphate Scaffold Loaded with Platelet Lysate/Gelatin Methacrylate to Promote Vascularization. *J. Mater. Chem. B* **2022**, *10*, 3138–3151. [CrossRef]

32. Sukul, M.; Nguyen, T.B.L.; Min, Y.-K.; Lee, S.-Y.; Lee, B.-T. Effect of Local Sustainable Release of BMP2-VEGF from Nano-Cellulose Loaded in Sponge Biphasic Calcium Phosphate on Bone Regeneration. *Tissue Eng. Part A* **2015**, *21*, 1822–1836. [CrossRef]
33. Tran, N.; Webster, T.J. Nanotechnology for Bone Materials. *Wiley Interdiscip. Rev. Nanomed. Nanobiotechnol.* **2009**, *1*, 336–351. [CrossRef]
34. Chen, D.; Yang, Y.; Li, B.; Yao, Y.; Xu, J.; Liu, R.; Peng, J.; Chang, Z.; Zhao, R.; Hou, R.; et al. Nanocomposite Hydrogels Optimize the Microenvironment by Exterior/Interior Crosstalk for Reprogramming Osteoporotic Homeostasis in Bone Defect Healing. *J. Control. Release* **2025**, *380*, 976–993. [CrossRef]
35. Ibrahim, A.; Hassanein, K.M.A.; Hussein, S.I.Z.; Semieka, M.M.A.; Elshahawy, A.M. Evaluation of a Chitosan/Polyvinyl Alcohol Hydrogel Loaded with Graphene Oxide and Nano TiO_2 for Bone Defect Reconstruction in a Dog Model. *J. Mater. Chem. B* **2025**, *13*, 3581–3592. [CrossRef] [PubMed]
36. Guo, J.; Yao, H.; Chang, L.; Zhu, W.; Zhang, Y.; Li, X.; Yang, B.; Dai, B.; Chen, X.; Lei, L.; et al. Magnesium Nanocomposite Hydrogel Reverses the Pathologies to Enhance Mandible Regeneration. *Adv. Mater.* **2025**, *37*, e2312920. [CrossRef]
37. Xu, Z.; Wang, J.; Gao, L.; Zhang, W. Hydrogels in Alveolar Bone Regeneration. *ACS Biomater. Sci. Eng.* **2024**, *10*, 7337–7351. [CrossRef] [PubMed]
38. He, X.; Li, Y.; Zou, D.; Zu, H.; Li, W.; Zheng, Y. An Overview of Magnesium-Based Implants in Orthopaedics and a Prospect of Its Application in Spine Fusion. *Bioact. Mater.* **2024**, *39*, 456–478. [CrossRef] [PubMed]
39. Antoniac, I.; Miculescu, M.; Mănescu, V.; Stere, A.; Quan, P.H.; Păltânea, G.; Robu, A.; Earar, K. Magnesium-Based Alloys Used in Orthopedic Surgery. *Materials* **2022**, *15*, 1148. [CrossRef]
40. Prakash, G.; Singh, N.K.; Kumar, D.; Chandel, P. Compression and flexure behaviours of magnesium alloy AZ41 at different strain rates. *U.P.B. Sci. Bull. Ser. D* **2020**, *82*, 211–222.
41. Voicu, M.-E.; Golgovici, F. A BIOINTERFACE GROWTH at IMMERSION of a BIODEGRADABLE MAGNESIUM ALLOY in SIMULATED BODY FLUID. *U.P.B. Sci. Bull. Ser. B* **2020**, *82*, 57–68.
42. Zhen, Z.; Liu, X.; Huang, T.; Xi, T.; Zheng, Y. Hemolysis and Cytotoxicity Mechanisms of Biodegradable Magnesium and Its Alloys. *Mater. Sci. Eng. C* **2015**, *46*, 202–206. [CrossRef]
43. Durlach, J.; Durlach, V.; Bac, P.; Bara, M.; Guiet-Bara, A. Magnesium and Therapeutics. *Magnes. Res.* **1994**, *7*, 313–328. [PubMed]
44. Kim, K.-J.; Choi, S.; Sang Cho, Y.; Yang, S.-J.; Cho, Y.-S.; Kim, K.K. Magnesium Ions Enhance Infiltration of Osteoblasts in Scaffolds via Increasing Cell Motility. *J. Mater. Sci. Mater. Med.* **2017**, *28*, 96. [CrossRef]
45. Tian, L.; Sheng, Y.; Huang, L.; Chow, D.H.-K.; Chau, W.H.; Tang, N.; Ngai, T.; Wu, C.; Lu, J.; Qin, L. An Innovative Mg/Ti Hybrid Fixation System Developed for Fracture Fixation and Healing Enhancement at Load-Bearing Skeletal Site. *Biomaterials* **2018**, *180*, 173–183. [CrossRef]
46. Wang, J.; Xu, J.; Wang, X.; Sheng, L.; Zheng, L.; Song, B.; Wu, G.; Zhang, R.; Yao, H.; Zheng, N.; et al. Magnesium-Pretreated Periosteum for Promoting Bone-Tendon Healing after Anterior Cruciate Ligament Reconstruction. *Biomaterials* **2021**, *268*, 120576. [CrossRef]
47. Bita, A.I.; Antoniac, A.; Cotrut, C.; Vasile, E.; Ciuca, I.; Niculescu, M.; Antoniac, I. *In Vitro* Degradation and Corrosion Evaluation of Mg-Ca Alloys for Biomedical Applications. *J. Optoelectron. Adv. Mater.* **2016**, *18*, 394–398.
48. Pogorielov, M.; Husak, E.; Solodivnik, A.; Zhdanov, S. Magnesium-Based Biodegradable Alloys: Degradation, Application, and Alloying Elements. *Interv. Med. Appl. Sci.* **2017**, *9*, 27–38. [CrossRef]
49. Maier, J.A.M.; Bernardini, D.; Rayssiguier, Y.; Mazur, A. High Concentrations of Magnesium Modulate Vascular Endothelial Cell Behaviour *in Vitro*. *Biochim. Biophys. Acta (BBA)—Mol. Basis Dis.* **2004**, *1689*, 6–12. [CrossRef] [PubMed]
50. Bernardini, D.; Nasulewic, A.; Mazur, A.; Maier, J.A.M. Magnesium and Microvascular Endothelial Cells: A Role in Inflammation and Angiogenesis. *Front. Biosci.* **2005**, *10*, 1177–1182. [CrossRef] [PubMed]
51. Rau, J.V.; Antoniac, I.; Fosca, M.; De Bonis, A.; Blajan, A.I.; Cotrut, C.; Graziani, V.; Curcio, M.; Cricenti, A.; Niculescu, M.; et al. Glass-Ceramic Coated Mg-Ca Alloys for Biomedical Implant Applications. *Mater. Sci. Eng. C* **2016**, *64*, 362–369. [CrossRef]
52. Antoniac, I.; Manescu, V.; Paltanea, G.; Antoniac, A.; Nemoianu, I.V.; Petrescu, M.I.; Dura, H.; Bodog, A.D. Additive Manufactured Magnesium-Based Scaffolds for Tissue Engineering. *Materials* **2022**, *15*, 8693. [CrossRef]
53. Manescu, V.; Antoniac, I.; Antoniac, A.; Laptoiu, D.; Paltanea, G.; Ciocoiu, R.; Nemoianu, I.V.; Gruionu, L.G.; Dura, H. Bone Regeneration Induced by Patient-Adapted Mg Alloy-Based Scaffolds for Bone Defects: Present and Future Perspectives. *Biomimetics* **2023**, *8*, 618. [CrossRef]
54. Bita, A.-I.; Antoniac, I.; Ciuca, I. Potential Use of Mg-Ca Alloys for Orthopedic Applications. *U.P.B. Sci. Bull. Ser. B* **2016**, *78*, 173–184.
55. Bița, T.; Corneschi, I.; Ungureanu, E.; Bița, A.-I.; Ignat, N.-D.; Dura, H.; Cârstoc, I.D. INFLUENCE of HEAT TREATMENT on MICROSTRUCTURE and CORROSION BEHAVIOR of BIODEGRADABLE Mg-Ca ALLOY. *U.P.B. Sci. Bull. Ser. B* **2023**, *85*, 247–260.

56. Bița, A.-I.; Antoniac, I.; Miculescu, M.; Stan, G.E.; Leonat, L.; Antoniac, A.; Constantin, B.; Forna, N. Electrochemical and *In Vitro* Biological Evaluation of Bio-Active Coatings Deposited by Magnetron Sputtering onto Biocompatible Mg-0.8Ca Alloy. *Materials* **2022**, *15*, 3100. [CrossRef]
57. Streza, A.; Antoniac, A.; Manescu, V.; Ciocoiu, R.; Cotrut, C.-M.; Miculescu, M.; Miculescu, F.; Antoniac, I.; Fosca, M.; Rau, J.V.; et al. *In Vitro* Studies Regarding the Effect of Cellulose Acetate-Based Composite Coatings on the Functional Properties of the Biodegradable Mg3Nd Alloys. *Biomimetics* **2023**, *8*, 526. [CrossRef] [PubMed]
58. Sun, J.; Du, W.; Fu, J.; Liu, K.; Li, S.; Wang, Z.; Liang, H. A Review on Magnesium Alloys for Application of Degradable Fracturing Tools. *J. Magnes. Alloys* **2022**, *10*, 2649–2672. [CrossRef]
59. Wu, A.; Xia, C. Study of the Microstructure and Mechanical Properties of Mg-Rare Earth Alloys. *Mater. Des.* **2007**, *28*, 1963–1967. [CrossRef]
60. Willbold, E.; Gu, X.; Albert, D.; Kalla, K.; Bobe, K.; Brauneis, M.; Janning, C.; Nellesen, J.; Czayka, W.; Tillmann, W.; et al. Effect of the Addition of Low Rare Earth Elements (Lanthanum, Neodymium, Cerium) on the Biodegradation and Biocompatibility of Magnesium. *Acta Biomater.* **2015**, *11*, 554–562. [CrossRef] [PubMed]
61. Han, L.; Hu, H.; Northwood, D.O.; Li, N. Microstructure and Nano-Scale Mechanical Behavior of Mg–Al and Mg–Al–Ca Alloys. *Mater. Sci. Eng. A* **2008**, *473*, 16–27. [CrossRef]
62. Kirkland, N.T.; Staiger, M.P.; Nisbet, D.; Davies, C.H.J.; Birbilis, N. Performance-Driven Design of Biocompatible Mg Alloys. *JOM* **2011**, *63*, 28–34. [CrossRef]
63. *ASTM E9-09*; Standard Test Methods of Compression Testing of Metallic Materials at Room Temperature. ASTM International: West Conshohocken, PA, USA, 2009.
64. Pietrzyńska, M.; Voelkel, A. Stability of Simulated Body Fluids Such as Blood Plasma, Artificial Urine and Artificial Saliva. *Microchem. J.* **2017**, *134*, 197–201. [CrossRef]
65. Kokubo, T.; Takadama, H. How Useful Is SBF in Predicting *In Vivo* Bone Bioactivity? *Biomaterials* **2006**, *27*, 2907–2915. [CrossRef]
66. *ASTM G5-94*; Standard Reference Test Method for Making Potentiostatic and Potentiodynamic Anodic Polarization Measurements. ASTM International: West Conshohocken, PA, USA, 2011.
67. Dragomir, L.; Antoniac, A.; Manescu, V.; Robu, A.; Dinu, M.; Pana, I.; Cotrut, C.M.; Kamel, E.; Antoniac, I.; Rau, J.V.; et al. Preparation and Characterization of Hydroxyapatite Coating by Magnetron Sputtering on Mg–Zn–Ag Alloys for Orthopaedic Trauma Implants. *Ceram. Int.* **2023**, *49*, 26274–26288. [CrossRef]
68. Antoniac, I.; Miculescu, F.; Cotrut, C.; Ficai, A.; Rau, J.V.; Grosu, E.; Antoniac, A.; Tecu, C.; Cristescu, I. Controlling the Degradation Rate of Biodegradable Mg–Zn-Mn Alloys for Orthopedic Applications by Electrophoretic Deposition of Hydroxyapatite Coating. *Materials* **2020**, *13*, 263. [CrossRef] [PubMed]
69. *ISO 10993-6*; Biological Evaluation of Medical Devices. Part 6: Tests for local effects after implantation. International Organization for Standardization: Geneva, Switzerland, 2021.
70. Destrez, A.; Colin, E.; Testelin, S.; Devauchelle, B.; Dakpé, S.; Naudot, M. Evaluation of a Granular Bone Substitute for Bone Regeneration Using an Optimized *In Vivo* Alveolar Cleft Model. *Bioengineering* **2023**, *10*, 1035. [CrossRef] [PubMed]
71. Verma, N.K.; Kar, A.K.; Singh, A.; Jagdale, P.; Satija, N.K.; Ghosh, D.; Patnaik, S. Control Release of Adenosine Potentiate Osteogenic Differentiation within a Bone Integrative EGCG-g-NOCC/Collagen Composite Scaffold toward Guided Bone Regeneration in a Critical-Sized Calvarial Defect. *Biomacromolecules* **2021**, *22*, 3069–3083. [CrossRef] [PubMed]
72. Park, J.-W.; Bae, S.-R.; Suh, J.-Y.; Lee, D.-H.; Kim, S.H.; Kim, H.; Lee, C.S. Evaluation of Bone Healing with Eggshell-Derived Bone Graft Substitutes in Rat Calvaria: A Pilot Study. *J. Biomed. Mater. Res. Part A* **2008**, *87A*, 203–214. [CrossRef]
73. Nowak, B.; Matuszewska, A.; Nikodem, A.; Filipiak, J.; Landwójtowicz, M.; Sadanowicz, E.; Jędrzejuk, D.; Rzeszutko, M.; Zduniak, K.; Piasecki, T.; et al. Oral Administration of Kaempferol Inhibits Bone Loss in Rat Model of Ovariectomy-Induced Osteopenia. *Pharmacol. Rep.* **2017**, *69*, 1113–1119. [CrossRef]
74. Zhang, Y.; Xu, J.; Ruan, Y.C.; Yu, M.K.; O'Laughlin, M.; Wise, H.; Chen, D.; Tian, L.; Shi, D.; Wang, J.; et al. Implant-Derived Magnesium Induces Local Neuronal Production of CGRP to Improve Bone-Fracture Healing in Rats. *Nat. Med.* **2016**, *22*, 1160–1169. [CrossRef]
75. Song, D.; Fu, Y.; Zhou, Q.; Fu, M.; Wu, X.; Sun, Y.; Bi, W.; Sun, J.; Yang, F.; Guo, H.; et al. Acceleration of Calvarial Bone Regeneration by Stem Cell Recruitment with a Multifunctional Hydrogel. *Adv. Healthc. Mater.* **2025**, *14*, 2501452. [CrossRef] [PubMed]
76. Standard Guide for Metallurgical Characterization of Absorbable Metallic Materials for Medical Implants. Available online: https://store.astm.org/f3160-16.html (accessed on 11 October 2025).
77. Lee, D.-H.; Kim, I.-K.; Cho, H.-Y.; Seo, J.-H.; Jang, J.-M.; Kim, J. Effect of Herbal Extracts on Bone Regeneration in a Rat Calvaria Defect Model and Screening System. *J. Korean Assoc. Oral Maxillofac. Surg.* **2018**, *44*, 79–85. [CrossRef]

78. Zhao, N.; Qin, L.; Liu, Y.; Zhai, M.; Li, D. Improved New Bone Formation Capacity of Hyaluronic Acid-Bone Substitute Compound in Rat Calvarial Critical Size Defect. *BMC Oral Health* **2024**, *24*, 994. [CrossRef] [PubMed]
79. Zhou, H.; Liang, B.; Jiang, H.; Deng, Z.; Yu, K. Magnesium-Based Biomaterials as Emerging Agents for Bone Repair and Regeneration: From Mechanism to Application. *J. Magnes. Alloys* **2021**, *9*, 779–804. [CrossRef]
80. Kaiser, F.; Schröter, L.; Stein, S.; Krüger, B.; Weichhold, J.; Stahlhut, P.; Ignatius, A.; Gbureck, U. Accelerated Bone Regeneration through Rational Design of Magnesium Phosphate Cements. *Acta Biomater.* **2022**, *145*, 358–371. [CrossRef] [PubMed]
81. Bose, S.; Tarafder, S.; Bandyopadhyay, A. Effect of Chemistry on Osteogenesis and Angiogenesis towards Bone Tissue Engineering Using 3D Printed Scaffolds. *Ann. Biomed. Eng.* **2017**, *45*, 261–272. [CrossRef]
82. Zemková, M.; Minárik, P.; Dittrich, J.; Bohlen, J.; Král, R. Individual Effect of Y and Nd on the Microstructure Formation of Mg-Y-Nd Alloys Processed by Severe Plastic Deformation and Their Effect on the Subsequent Mechanical and Corrosion Properties. *J. Magnes. Alloys* **2023**, *11*, 509–521. [CrossRef]
83. Witte, F.; Kaese, V.; Haferkamp, H.; Switzer, E.; Meyer-Lindenberg, A.; Wirth, C.J.; Windhagen, H. *In Vivo* Corrosion of Four Magnesium Alloys and the Associated Bone Response. *Biomaterials* **2005**, *26*, 3557–3563. [CrossRef]
84. Wang, J.; Liu, R.; Dong, X.; Yang, Y. Microstructure and Mechanical Properties of Mg-Zn-Y-Nd-Zr Alloys. *J. Rare Earths* **2013**, *31*, 616–621. [CrossRef]
85. Zhou, H.T.; Zhang, Z.D.; Liu, C.M.; Wang, Q.W. Effect of Nd and Y on the Microstructure and Mechanical Properties of ZK60 Alloy. *Mater. Sci. Eng. A* **2007**, *445–446*, 1–6. [CrossRef]
86. Khan, S.A.; Miyashita, Y.; Mutoh, Y.; Sajuri, Z.B. Influence of Mn Content on Mechanical Properties and Fatigue Behavior of Extruded Mg Alloys. *Mater. Sci. Eng. A* **2006**, *420*, 315–321. [CrossRef]
87. Prasad, S.V.S.; Prasad, S.B.; Verma, K.; Mishra, R.K.; Kumar, V.; Singh, S. The Role and Significance of Magnesium in Modern Day Research-A Review. *J. Magnes. Alloys* **2022**, *10*, 1–61. [CrossRef]
88. Bakhsheshi-Rad, H.R.; Idris, M.H.; Abdul-Kadir, M.R.; Ourdjini, A.; Medraj, M.; Daroonparvar, M.; Hamzah, E. Mechanical and Bio-Corrosion Properties of Quaternary Mg–Ca–Mn–Zn Alloys Compared with Binary Mg–Ca Alloys. *Mater. Des.* **2014**, *53*, 283–292. [CrossRef]
89. Zhang, E.; Yang, L. Microstructure, Mechanical Properties and Bio-Corrosion Properties of Mg–Zn–Mn–Ca Alloy for Biomedical Application. *Mater. Sci. Eng. A* **2008**, *497*, 111–118. [CrossRef]
90. Yang, J.; Wang, J.; Wang, L.; Wu, Y.; Wang, L.; Zhang, H. Microstructure and Mechanical Properties of Mg–4.5Zn–xNd (x = 0, 1 and 2, wt%) Alloys. *Mater. Sci. Eng. A* **2008**, *479*, 339–344. [CrossRef]
91. Li, Q.; Wang, Q.; Wang, Y.; Zeng, X.; Ding, W. Effect of Nd and Y Addition on Microstructure and Mechanical Properties of As-Cast Mg–Zn–Zr Alloy. *J. Alloys Compd.* **2007**, *427*, 115–123. [CrossRef]
92. Yan, K.; Bai, J.; Liu, H.; Jin, Z.-Y. The Precipitation Behavior of $MgZn_2$ and Mg_4Zn_7 Phase in Mg-6Zn (wt.%) Alloy during Equal-Channel Angular Pressing. *J. Magnes. Alloys* **2017**, *5*, 336–339. [CrossRef]
93. Wei, L.Y.; Dunlop, G.L.; Westengen, H. Precipitation Hardening of Mg-Zn and Mg-Zn-RE Alloys. *Metall. Mater. Trans. A* **1995**, *26*, 1705–1716. [CrossRef]
94. Zengin, H.; Turen, Y. Effect of Y Addition on Microstructure and Corrosion Behavior of Extruded Mg–Zn–Nd–Zr Alloy. *J. Magnes. Alloys* **2020**, *8*, 640–653. [CrossRef]
95. Bazhenov, V.E.; Koltygin, A.V.; Sung, M.C.; Park, S.H.; Tselovalnik, Y.V.; Stepashkin, A.A.; Rizhsky, A.A.; Belov, M.V.; Belov, V.D.; Malyutin, K.V. Development of Mg–Zn–Y–Zr Casting Magnesium Alloy with High Thermal Conductivity. *J. Magnes. Alloys* **2021**, *9*, 1567–1577. [CrossRef]
96. Xu, D.K.; Tang, W.N.; Liu, L.; Xu, Y.B.; Han, E.H. Effect of W-Phase on the Mechanical Properties of as-Cast Mg–Zn–Y–Zr Alloys. *J. Alloys Compd.* **2008**, *461*, 248–252. [CrossRef]
97. Luo, Z.P.; Zhang, S.Q. High-Resolution Electron Microscopy on the X-$Mg_{12}ZnY$ Phase in a High Strength Mg-Zn-Zr-Y Magnesium Alloy. *J. Mater. Sci. Lett.* **2000**, *19*, 813–815. [CrossRef]
98. Lee, J.Y.; Kim, D.H.; Lim, H.K.; Kim, D.H. Effects of Zn/Y Ratio on Microstructure and Mechanical Properties of Mg-Zn-Y Alloys. *Mater. Lett.* **2005**, *59*, 3801–3805. [CrossRef]
99. Luo, S.; Tang, A.; Pan, F.; Song, K.; Wang, W. Effect of Mole Ratio of Y to Zn on Phase Constituent of Mg-Zn-Zr-Y Alloys. *Trans. Nonferrous Met. Soc. China* **2011**, *21*, 795–800. [CrossRef]
100. Farahany, S.; Bakhsheshi-Rad, H.R.; Idris, M.H.; Abdul Kadir, M.R.; Lotfabadi, A.F.; Ourdjini, A. In-Situ Thermal Analysis and Macroscopical Characterization of Mg–xCa and Mg–0.5Ca–xZn Alloy Systems. *Thermochim. Acta* **2012**, *527*, 180–189. [CrossRef]
101. Jiang, M.G.; Xu, C.; Nakata, T.; Yan, H.; Chen, R.S.; Kamado, S. Development of Dilute Mg–Zn–Ca–Mn Alloy with High Performance via Extrusion. *J. Alloys Compd.* **2016**, *668*, 13–21. [CrossRef]
102. Schäublin, R.E.; Becker, M.; Cihova, M.; Gerstl, S.S.A.; Deiana, D.; Hébert, C.; Pogatscher, S.; Uggowitzer, P.J.; Löffler, J.F. Precipitation in Lean Mg–Zn–Ca Alloys. *Acta Mater.* **2022**, *239*, 118223. [CrossRef]

103. Du, Y.Z.; Zheng, M.Y.; Qiao, X.G.; Wu, K.; Liu, X.D.; Wang, G.J.; Lv, X.Y. Microstructure and Mechanical Properties of Mg–Zn–Ca–Ce Alloy Processed by Semi-Continuous Casting. *Mater. Sci. Eng. A* **2013**, *582*, 134–139. [CrossRef]
104. Du, Y.Z.; Zheng, M.Y.; Xu, C.; Qiao, X.G.; Wu, K.; Liu, X.D.; Wang, G.J.; Lv, X.Y. Microstructures and Mechanical Properties of As-Cast and as-Extruded Mg-4.50Zn-1.13Ca (wt%) Alloys. *Mater. Sci. Eng. A* **2013**, *576*, 6–13. [CrossRef]
105. Nakata, T.; Xu, C.; Ito, Y.; Kamado, S. Role of Homogenization on Tensile Properties and Microstructures in a Dilute Mg–Zn–Ca–Mn Alloy Sheet. *Mater. Sci. Eng. A* **2022**, *833*, 142541. [CrossRef]
106. Kavyani, M.; Ebrahimi, G.R.; Ezatpour, H.R.; Jahazi, M. Microstructure Refinement, Mechanical and Biocorrosion Properties of Mg–Zn–Ca–Mn Alloy Improved by a New Severe Plastic Deformation Process. *J. Magnes. Alloys* **2022**, *10*, 1640–16652. [CrossRef]
107. Bazhenov, V.E.; Li, A.V.; Komissarov, A.A.; Koltygin, A.V.; Tavolzhanskii, S.A.; Bautin, V.A.; Voropaeva, O.O.; Mukhametshina, A.M.; Tokar, A.A. Microstructure and Mechanical and Corrosion Properties of Hot-Extruded Mg–Zn–Ca–(Mn) Biodegradable Alloys. *J. Magnes. Alloys* **2021**, *9*, 1428–1442. [CrossRef]
108. Cho, D.H.; Nam, J.H.; Lee, B.W.; Cho, K.M.; Park, I.M. Effect of Mn Addition on Grain Refinement of Biodegradable Mg4Zn0.5Ca Alloy. *J. Alloys Compd.* **2016**, *676*, 461–468. [CrossRef]
109. Zhang, Y.; Huang, Y.; Feyerabend, F.; Blawert, C.; Gan, W.; Maawad, E.; You, S.; Gavras, S.; Scharnagl, N.; Bode, J.; et al. Influence of the Amount of Intermetallics on the Degradation of Mg-Nd Alloys under Physiological Conditions. *Acta Biomater.* **2021**, *121*, 695–712. [CrossRef]
110. Gorsse, S.; Hutchinson, C.R.; Chevalier, B.; Nie, J.-F. A Thermodynamic Assessment of the Mg–Nd Binary System Using Random Solution and Associate Models for the Liquid Phase. *J. Alloys Compd.* **2005**, *392*, 253–262. [CrossRef]
111. Thaha, Y.N.; Siswayanti, B.; Elvira, B.R.; Syahid, A.N.; Rachmawati, N.; Erryani, A.; Malau, D.P.; Annur, D.; Lestari, F.P.; Asmaria, T.; et al. Effect of Neodymium on Microstructure, Mechanical Properties, and Corrosion Behavior of Mg-2Si-xNd Alloys Fabricated by Powder Metallurgy. *J. Rare Earths* **2024**, *43*, 2569–2578. [CrossRef]
112. Elkaiam, L.; Hakimi, O.; Goldman, J.; Aghion, E. The Effect of Nd on Mechanical Properties and Corrosion Performance of Biodegradable Mg-5%Zn Alloy. *Metals* **2018**, *8*, 438. [CrossRef]
113. Gong, Z.; Wang, J.; Sun, Y.; Zhu, S.; Wang, L.; Guan, S. Microstructure and Mechanical Properties of the Sub-Rapidly Solidified Mg—Zn—Y—Nd Alloy Prepared by Step-Copper Mold Casting. *Mater. Today Commun.* **2021**, *27*, 102308. [CrossRef]
114. Meng, F.; Lv, S.; Yang, Q.; Qiu, X.; Yan, Z.; Duan, Q.; Meng, J. Multiplex Intermetallic Phases in a Gravity Die-Cast Mg−6.0Zn−1.5Nd−0.5Zr (wt%) Alloy. *J. Magnes. Alloys* **2022**, *10*, 209–223. [CrossRef]
115. Joshi, P.; Rao, R.N.; Golla, C.B.; Özcan, M.; Prasad, P.S. Comprehensive Review of Mg-Based Alloys: Mechanical, Chemical and Biological Properties for Prosthetic and Orthopedic Applications. *J. Mater. Res. Technol.* **2025**, *35*, 2487–2502. [CrossRef]
116. Liu, K.; Zhang, J.; Su, G.; Tang, D.; Rokhlin, L.L.; Elkin, F.M.; Meng, J. Influence of Zn Content on the Microstructure and Mechanical Properties of Extruded Mg–5Y–4Gd–0.4Zr Alloy. *J. Alloys Compd.* **2009**, *481*, 811–818. [CrossRef]
117. Wang, S.; Pan, H.; Xie, D.; Zhang, D.; Li, J.; Xie, H.; Ren, Y.; Qin, G. Grain Refinement and Strength Enhancement in Mg Wrought Alloys: A Review. *J. Magnes. Alloys* **2023**, *11*, 4128–4145. [CrossRef]
118. Ma, C.; Hou, C.; Zhang, X.; Liu, T.; Zhou, N.; Zheng, K. Study on the Microstructure and Mechanical Properties of Mg–Al–Li–Zn–Ti Multi-Component Alloy. *J. Mater. Res. Technol.* **2024**, *29*, 4781–4795. [CrossRef]
119. Azadi, A.; Ebel, T.; Wolff, M.; O'Cearbhaill, E.; Celikin, M. Additive Manufacturing of Magnesium Alloys for Biomedical Applications: Critical Review on Sinterability and Alloy Development. *J. Mater. Res. Technol.* **2025**, *35*, 6986–7007. [CrossRef]
120. Hu, Y.; Guo, X.; Qiao, Y.; Wang, X.; Lin, Q. Preparation of Medical Mg–Zn Alloys and the Effect of Different Zinc Contents on the Alloy. *J. Mater. Sci. Mater. Med.* **2022**, *33*, 9. [CrossRef] [PubMed]
121. *ASTM E111*; Standard Test Method for Young's Modulus, Tangent Modulus, and Chord Modulus. ASTM International: West Conshohocken, PA, USA, 2017.
122. Havaldar, R.; Pilli, S.C.; Putti, B.B. Insights into the Effects of Tensile and Compressive Loadings on Human Femur Bone. *Adv. Biomed. Res.* **2014**, *3*, 101. [CrossRef]
123. Morgan, E.F.; Unnikrisnan, G.U.; Hussein, A.I. Bone Mechanical Properties in Healthy and Diseased States. *Annu. Rev. Biomed. Eng.* **2018**, *20*, 119–143. [CrossRef]
124. Rho, J.Y.; Ashman, R.B.; Turner, C.H. Young's Modulus of Trabecular and Cortical Bone Material: Ultrasonic and Microtensile Measurements. *J. Biomech.* **1993**, *26*, 111–119. [CrossRef]
125. Kurniawan, K.A.; Wicaksono, S.T.; Rasyida, A.; Purniawan, A. Improvement of Compressive Strength of Mg-Fe-Ca Alloy by Heat Treatment as Biodegradable Implant. In Proceedings of the International Conference on Science and Applied Science (ICSAS) 2019, Surakarta, Indonesia, 20 July 2019; AIP Publishing: Melville, NY, USA, 2019.
126. Teslia, S.; Kovalenko, M.; Teslia, M.; Vterkovskiy, M.; Solodkyi, I.; Loboda, P.; Soloviova, T. The Activation of Magnesium Sintering by Zinc Addition. *Alloys* **2024**, *3*, 178–189. [CrossRef]

127. Su, Y.; Lin, J.; Su, Y.; Zai, W.; Li, G.; Wen, C. Investigation on Composition, Mechanical Properties, and Corrosion Resistance of Mg-0.5Ca-X(Sr, Zr, Sn) Biological Alloy. *Scanning* **2018**, *2018*, 6519310. [CrossRef]
128. Sumitomo, T.; Cáceres, C.H.; Veidt, M. The Elastic Modulus of Cast Mg–Al–Zn Alloys. *J. Light Met.* **2002**, *2*, 49–56. [CrossRef]
129. Packham, D.E. Surface Energy, Surface Topography and Adhesion. *Int. J. Adhes. Adhes.* **2003**, *23*, 437–448. [CrossRef]
130. Gambaro, S.; Nascimento, M.L.; Shekargoftar, M.; Ravanbakhsh, S.; Sales, V.; Paternoster, C.; Bartosch, M.; Witte, F.; Mantovani, D. Characterization of a Magnesium Fluoride Conversion Coating on Mg-2Y-1Mn-1Zn Screws for Biomedical Applications. *Materials* **2022**, *15*, 8245. [CrossRef] [PubMed]
131. Li, L.; Huang, Z.; Chen, L.; Zhang, L.; Li, M.; Hou, H.; Zhao, Y. Electrochemical Corrosion Behavior of AZ91D Magnesium Alloy-Graphene Nanoplatelets Composites in Simulated Body Fluids. *J. Mater. Res. Technol.* **2023**, *24*, 449–462. [CrossRef]
132. Sasikumar, Y.; Solomon, M.M.; Olasunkanmi, L.O.; Ebenso, E.E. Effect of Surface Treatment on the Bioactivity and Electrochemical Behavior of Magnesium Alloys in Simulated Body Fluid. *Mater. Corros.* **2017**, *68*, 776–790. [CrossRef]
133. *ASTM G59-97*; Standard Test Method for Conducting Potentiodynamic Polarization Resistance Measurements. ASTM International: West Conshohocken, PA, USA, 2020.
134. Bakhsheshi-Rad, H.R.; Hamzah, E.; Daroonparvar, M.; Kasiri-Asgarani, M.; Medraj, M. Synthesis and Biodegradation Evaluation of Nano-Si and Nano-Si/TiO_2 Coatings on Biodegradable Mg–Ca Alloy in Simulated Body Fluid. *Ceram. Int.* **2014**, *40*, 14009–14018. [CrossRef]
135. Zhang, X.; Li, Y.; Zhang, K.; Wang, C.; Li, H.; Ma, M.; Zhang, B. Corrosion and Electrochemical Behavior of Mg–Y Alloys in 3.5% NaCl Solution. *Trans. Nonferrous Met. Soc. China* **2013**, *23*, 1226–1236. [CrossRef]
136. Sudholz, A.D.; Gusieva, K.; Chen, X.B.; Muddle, B.C.; Gibson, M.A.; Birbilis, N. Electrochemical Behaviour and Corrosion of Mg–Y Alloys. *Corros. Sci.* **2011**, *53*, 2277–2282. [CrossRef]
137. Davenport, A.J.; Padovani, C.; Connolly, B.J.; Stevens, N.P.C.; Beale, T.A.W.; Groso, A.; Stampanoni, M. Synchrotron X-Ray Microtomography Study of the Role of Y in Corrosion of Magnesium Alloy WE43. *Electrochem. Solid-State Lett.* **2006**, *10*, C5. [CrossRef]
138. Rogova, V.-V.; Peev, S.; Yotsova, R.; Gerova-Vatsova, T.; Parushev, I. Histomorphometric Assessment of Non-Decalcified Plastic-Embedded Specimens for Evaluation of Bone Regeneration Using Bone Substitute Materials—A Systematic Review. *Materials* **2024**, *18*, 119. [CrossRef] [PubMed]
139. Pesode, P.; Barve, S. Additive Manufacturing of Magnesium Alloys and Its Biocompatibility. *Bioprinting* **2023**, *36*, e00318. [CrossRef]
140. Wu, M.; Zhao, M.; Cai, Y.; Yao, J.; Wang, P.; Atrens, A. Recent Advances in Bio-Functional Ta-Based Bone Materials: Materials Design and Bioactivity. *Int. J. Extreme Manuf.* **2024**, *6*, 062010. [CrossRef]
141. Taguchi, T.; Lopez, M.J. An Overview of de Novo Bone Generation in Animal Models. *J. Orthop. Res.* **2021**, *39*, 7–21. [CrossRef]
142. Thudium, C.S.; Nielsen, S.H.; Sardar, S.; Mobasheri, A.; van Spil, W.E.; Lories, R.; Henriksen, K.; Bay-Jensen, A.-C.; Karsdal, M.A. Bone Phenotypes in Rheumatology—There Is More to Bone than Just Bone. *BMC Musculoskelet. Disord.* **2020**, *21*, 789. [CrossRef]
143. Niu, Y.; Du, T.; Liu, Y. Biomechanical Characteristics and Analysis Approaches of Bone and Bone Substitute Materials. *J. Funct. Biomater.* **2023**, *14*, 212. [CrossRef] [PubMed]

Disclaimer/Publisher's Note: The statements, opinions and data contained in all publications are solely those of the individual author(s) and contributor(s) and not of MDPI and/or the editor(s). MDPI and/or the editor(s) disclaim responsibility for any injury to people or property resulting from any ideas, methods, instructions or products referred to in the content.

Journal of
Functional Biomaterials

MDPI

Article

Physicochemical and Antioxidant Alterations of Modified and Free Epigallocatechin Gallate Under Thermal Treatment in Air and Vacuum

Lianjing Yu [1], Zi Deng [1], Masato Nakagawa [1], Shanshan Zheng [1], Jun-ichiro Jo [2], Tomonari Tanaka [3] and Yoshitomo Honda [1,*]

[1] Department of Oral Anatomy, Osaka Dental University, 8-1 Kuzuhahanazono-cho, Hirakata 573-1121, Japan; yuyanjings@gmail.com (L.Y.); dengzi1014@gmail.com (Z.D.); m-nakagawa@cc.osaka-dent.ac.jp (M.N.); 33of1213@gmail.com (S.Z.)

[2] Department of Biomaterials, Osaka Dental University, 8-1 Kuzuhahanazono-cho, Hirakata 573-1121, Japan; jo-j@cc.osaka-dent.ac.jp

[3] Department of Biobased Materials Science, Graduate School of Science and Technology, Kyoto Institute of Technology, Matsugasaki, Sakyo-ku, Kyoto 606-8585, Japan; t-tanaka@kit.ac.jp

* Correspondence: honda-y@cc.osaka-dent.ac.jp; Tel.: +81-72-864-3013

Abstract

Epigallocatechin gallate (EGCG)—the most abundant catechin in green tea—is a promising component of advanced composite biomaterials. The pharmacological activity of EGCG is typically attenuated upon thermal processing, although the exact effects of heating free and modified EGCG in air and vacuum are unknown. To bridge this gap, we herein examined the effects of heating free and modified (in gelatin containing beta-tricalcium phosphate granules) EGCG in vacuum and air (100–220 °C, 1–16 h) on its physicochemical and antioxidant properties using water and ethanol solubility measurements, discoloration and antioxidant activity (2,2-diphenyl-1-picrylhydrazyl (DPPH) and 2,2′-azino-bis-3-ethylbenzothiazoline-6-sulfonic acid) assays, ultraviolet–visible spectroscopy, mass spectrometry, nuclear magnetic resonance spectroscopy, and attenuated total reflectance Fourier transform infrared spectroscopy. The antioxidant activity of EGCG-modified gelatin sponges was assessed in vitro using the DPPH assay and in vivo using a calvarial bone defect model in eight-week-old male Sprague–Dawley rats. Free and modified EGCG showed antioxidant activities, which were largely retained after heating in vacuum at 150 °C. These findings show that appropriate heating procedures preserve the antioxidant activity of EGCG and provide insights for the development of EGCG-based biomaterials.

Keywords: epigallocatechin gallate; antioxidant; thermal treatment; vacuum heating; DPPH assay; ABTS assay; gelatin; β-TCP

1. Introduction

Tea prepared from the leaves of *Camellia sinensis* contains various polyphenols, such as catechins and their derivatives, which exhibit diverse pharmacological effects and health benefits. Hence, tea is widely consumed worldwide as a beverage and health food, especially in Asia [1,2], where green tea sales are expected to exceed USD 3.14 billion in 2024 [3]. Epigallocatechin gallate (EGCG, a catechin) is the most abundant [1] and pharmacologically active [4] component of green tea, exhibiting antioxidant [4,5], anti-inflammatory [6], antibacterial [7], and anticancer [8] activities, promoting osteoblastic differentiation [9] and growth factor release for dental diseases [10], and inhibiting osteoclastogenesis [11].

EGCG is a polyphenol containing four substituted rings (A–D). Rings B and D exhibit a higher physiological activity because of their multiple hydroxyl groups [12], which play an important role in protecting cells and tissues from oxidative stress [5]. However, these pharmacological effects are readily attenuated by various stimuli, including physicochemical (heating in water [13] and irradiation [14]) and chemical [15] factors. Although the pharmacological effects of EGCG have been extensively characterized, most studies have focused on its consumption in the form of tea [16]. The scope of EGCG applications in medicine and dentistry has been broadened through its hybridization with various materials, including polymers (gelatin, collagen, hyaluronic acid, chitosan, and polycaprolactone) [17–22], metals (gold nanoparticles) [23,24], and inorganic nanomaterials (hydroxyapatite, Fe_2O_3, and TiO_2) [25–27]. Furthermore, extensive efforts in nanoparticle research have expanded EGCG applications as a functional drug in drug delivery systems [28–30]. In addition to the effects of EGCG mentioned above, these composites have a broad application scope and can exhibit anticancer [23,24,28] and antibacterial [21] properties, regulate senescence-associated secretory phenotype factors from senescent cells [31], act as stem cell seeding carriers [31], and promote bone regeneration [17,19,22].

Our group has developed EGCG-conjugated materials for bone regeneration, including EGCG-conjugated gelatin with or without β-tricalcium phosphate (β-TCP) [20,31,32]. Gelatin was subjected to dehydrothermal treatment (vacuum heating) in the presence of EGCG to enhance crosslinking and obtain an optimal scaffold for cells. The resulting materials showed a very robust crosslinked network and exhibited a remarkably augmented bone-forming ability [20,31] and antisenescence activity even after heating [31]. Although studies dealing with the vacuum heating of EGCG-containing biomaterials are few, hydrothermal treatment has been applied to EGCG-capped α-Fe_2O_3 nanoparticles [26] and EGCG/Zn-coated Ti [33].

The thermal and digestion treatments of free EGCG cause its degradation through epimerization and autoxidation and afford diverse products, such as monomeric catechins, quinone derivatives, and dimerized oxidation products (theasinensin A, D, and P-2) [34,35]. Specifically, when heated in water, EGCG undergoes accelerated decomposition [12,34,36,37] and loses its antioxidant capacity [13]. For example, 50% of EGCG in aqueous solution undergoes notable decomposition after 40 min of heating at 120 °C [37]. After 45 min of heating at 180 °C in water (using oil bath), total phenolic content decreases by >50%, and the rate of 2,2-diphenyl-1-picrylhydrazyl (DPPH) radical scavenging decreases by 90% [13]. The stability of catechins in solution generally depends on pH and temperature [37]. Oxygen concentration, which relates to oxidation, is an important factor promoting instability [36], and the oxidation of EGCG is closely attributed to peroxyl radicals generated from atmospheric oxygen [38].

The application of heating for EGCG and EGCG-based biomaterials in air or under vacuum possibly increase, as with water-based heating. However, comprehensive insights into the changes in the physicochemical properties and antioxidant ability of EGCG upon heating in a vacuum are currently scarce. The aim of this study is to evaluate the effects of vacuum and atmospheric heating at different durations and temperatures on the physicochemical and antioxidant properties of free EGCG. Furthermore, to determine whether modified EGCG on biomaterials exhibits similar effects, the antioxidant properties of EGCG-conjugated gelatin sponges containing β-TCP were evaluated both in vivo and in vitro after vacuum heating as a model experiment.

https://doi.org/10.3390/jfb17010018

2. Materials and Methods

2.1. VHT and AHT

EGCG powder (Taiyo Kagaku Co., Ltd., Tokyo, Japan) was provided by Toagosei Co., Ltd. (Tokyo, Japan). Samples with non-heating treatment were denoted as NT. For each thermal treatment, the EGCG powder was placed into four separate amber glass vials, which were heat-treated independently (n = 4). Vacuum heating treatment (VHT) was performed using a vacuum dryer (AVO-250NS, AS ONE Co., Osaka, Japan), and atmospheric heating treatment (AHT) was performed using a convection oven (LC-122, TABAI ESPEC Co., Osaka, Japan). To investigate the effect of temperature, samples were heated for 4 h at 100–220 °C (VHT) and 100–200 °C (AHT). To evaluate the effect of treatment duration, samples were subjected to VHT or AHT at 150 °C for 1–16 h. Hereafter, vacuum and atmospheric heating treatment are designated as VHT or AHT (temperature, time), e.g., vacuum heating treatment at 150 °C for 16 h: VHT (150 °C, 16 h). All samples were sealed and stored at −20 °C after heating and dissolved as needed for assays. A stereomicroscope (SZ61; Olympus Corporation, Tokyo, Japan) and digital camera (GR III, RICOH Company, Ltd., Tokyo, Japan) were used for macro photography. Color differences (ΔE, CIEDE2000) were measured using a spectrophotometer (Spectro 1 Pro, Variable, Chattanooga, TN, USA) and black art paper as a reference. Relative ΔE (%) was calculated as 100% × (ΔE of experimental group/ΔE of EGCG with NT).

2.2. Antioxidant Activity Assay

Antioxidant activity was measured using 2,2′-azino-bis-3-ethylbenzothiazoline-6-sulfonic acid (ABTS, a methemoglobin oxidase inhibitor) and 2,2-diphenyl-1-picrylhydrazyl (DPPH). The ABTS assay was performed following manufacturer's instructions (709001; Cayman Chem. Co., Ann Arbor, MI, USA): EGCG powder was dissolved in antioxidant assay buffer (1×) to a concentration of 50 μg/mL. Absorbance (Abs) at 405 nm was measured using a microplate reader (SpectraMax M5, Molecular Devices, San Jose, CA, USA). ΔAbs values were calculated as the difference in absorbance between each sample and the blank. The rate of antioxidant activity retention was calculated as 100% × (ΔAbs of experimental group/ΔAbs of EGCG with NT).

The DPPH assay was performed using the corresponding kit (D678, Dojindo Lab., Kumamoto, Japan) according to manufacturer's protocol. EGCG powder was dissolved in 80% aqueous ethanol (EtOH) to a concentration of 5 μg/mL, and the solution was mixed with the DPPH working solution in a 96-well plate and incubated at 25 °C in the dark. After 30 min, absorbance at 517 nm was measured using a microplate spectrophotometer (SpectraMax M5). The rate of antioxidant activity retention was calculated as 100% × (ΔAbs of experimental group/ΔAbs of EGCG with NT).

For sponges (see Section 2.4), 80% EtOH was premixed with the DPPH working solution. One-quarter of the sponge (~9 mg) was added to 400 μL of the premixed solution and ground at 30 times/s for 2 min with 0.8 mm–diameter glass beads (Toshinriko, Tokyo, Japan) using a mixer mill (MM301, Retsch, Haan, Germany) (n = 4). The mixture was centrifuged, the supernatant transferred to a 96-well plate, and absorbance at 517 nm measured in the dark using a microplate spectrophotometer (SpectraMax M5).

2.3. Physicochemical Characterization

Ultraviolet–visible (UV–vis) spectra were obtained using a Nanodrop spectrophotometer (ND1000, Thermo Fisher Scientific, Waltham, MA, USA). Mass spectrometry analysis (JMS-700k, JEOL Ltd., Tokyo, Japan) was conducted using m-nitrobenzyl alcohol as the matrix. ^{1}H nuclear magnetic resonance (NMR) spectra (BioSpin AV-600, Bruker Co., Billerica,

https://doi.org/10.3390/jfb17010018

MA, USA) were recorded in deuterium oxide. Attenuated total reflection Fourier transform infrared (ATR-FTIR) spectra (FT/IR-4600, JASCO Co., Tokyo, Japan) were recorded in the wavenumber range of 400–4000 cm^{-1}. Powder X-ray diffraction (XRD) patterns (XRD-6000, Shimadzu Co., Kyoto, Japan) were collected within the range of 2θ = 20–40°.

2.4. Preparation of Gelatin Sponges

Vacuum-heated gelatin sponges containing 80 wt% porous β-tricalcium phosphate (β-TCP) particles (500–1000 μm) (vh-GSβ), vacuum-heated chemically synthesized gelatin sponges containing 80 wt% porous β-TCP particles (vhc-GSβ), and vacuum-heated chemically synthesized gelatin sponges containing 0.28 wt% intact EGCG and 80 wt% porous β-TCP particles (vhEc-GSβ) were prepared as described elsewhere [32] and provided by Toagosei Co., Ltd. The specimen size was fixed to 9 mm × 1 mm, and VHT was conducted at 150 °C for 16 h.

2.5. Sponge Morphology Observation

A stereomicroscope (SZ61; Olympus Corporation) was used for macro photography. Surface microstructure was observed using field-emission scanning electron microscopy (FE-SEM; S-4800; Hitachi, Ltd., Tokyo, Japan) at an acceleration voltage of 5 kV. Prior to FE-SEM imaging, all samples were coated with O_sO_4 using an osmium coating device (HPC-20 Device; Vacuum Device Co., Ltd., Ibaraki, Japan). Pore size was quantified as the maximum Feret diameter extracted from binary FE-SEM images using the ImageJ software (v. 1.54g, National Institutes of Health, Bethesda, MD, USA). Pores with a minimum Feret diameter of <10 μm were excluded because of their insufficiency for cellular infiltration [39].

2.6. Animal Experiments

Animal experiments were conducted with the approval of the Osaka Dental University Ethics Committee (Approval No. 25-06005). Sixteen eight-week-old male Sprague–Dawley rats purchased from Shimizu Laboratory Supplies Co., Ltd. (Kyoto, Japan) were used to construct a 9 mm-diameter critical-sized bone defect at calvaria under deep anesthesia. All animals were housed in a well-maintained animal facility under standard experimental conditions throughout the study period. No specific inclusion or exclusion criteria were established, and all animals used in the experiment were included in the analysis. The animals were randomly divided into four groups: control, vh-GSβ, vhc-GSβ, and vhEc-GSβ (n = 4 per group), with the experimental unit defined as a single animal. In the experimental groups, a single hydrogel sheet (vh-GSβ, vhc-GSβ, or vhEc-GSβ) prepared using Ringer's solution (90 μL) was implanted into the defect site, while the control group received no material. Humane endpoints were defined as feeding difficulty, abnormal posture, or >20% weight loss; however, no animals reached these endpoints. All rats underwent cardiac perfusion fixation with a 4% polyformaldehyde solution (No. 15291; Muto Pure Chemicals Co., Ltd., Tokyo, Japan) on the third postoperative day. Samples were immersed in Morse solution (135-17071; FUJIFILM Wako Pure Chemical Corporation, Osaka, Japan) for three days for decalcification and then subjected to frozen-tissue sectioning using the Kawamoto method [40]. Sections were cut at a thickness of 8 μm (Leica CM1950, Leica Biosystems, Wetzlar, Germany).

2.7. Immunofluorescence Staining

Frozen sections of animal tissues were blocked and permeabilized in phosphate-buffered saline (164-28713; FUJIFILM Wako Pure Chemical Corporation, Osaka, Japan) containing 5% goat serum (S-1000, Vector Laboratories, Inc., Newark, CA, USA) and 0.3% Triton X-100 (Code 12967-32, Nacalai, Kyoto, Japan) for 30 min. Immunostaining was performed using an anti-4-hydroxynonenal (4-HNE) polyclonal antibody (bs-6313R-Cy5, Bioss Antibodies, Woburn, MA,

USA; 1:100) in the dark at 4 °C overnight. After washing three times with phosphate-buffered saline, the cell nuclei were stained with 4′,6-diamidino-2-phenylindole (DAPI) Fluoromount-G (Southern Biotech, Birmingham, AL, USA). Images were captured using a digital microscope (BZ-800; Keyence, Osaka, Japan). Each experimental group contained four animals, and three random sections per animal were analyzed for statistical evaluation. The fraction of 4-HNE-positive staining was determined using immunofluorescence images acquired at a magnification of 10×. Regions of interest (ROIs) were defined within the calvarial defect area in each section. For groups with material implantation, ROIs were restricted to material-filled defect regions. Quantitative analysis was performed based on these predefined ROIs using the ImageJ software (v. 1.54g), with the quantitative metric calculated as area of 4-HNE fluorescence/area of DAPI fluorescence (n = 4).

2.8. Statistical Analysis

Data processing and analysis were performed using the Prism 10.2.1 software (GraphPad Software, San Diego, CA, USA). Data normality was assessed using the Shapiro–Wilk test, and the homogeneity of variances was evaluated using the Brown–Forsythe test. For comparisons involving more than two groups, one-way analysis of variance (ANOVA) followed by Tukey's multiple comparisons test was used when data met the assumptions of normality and homogeneity of variances. When normality was satisfied, but variances were unequal, Welch's ANOVA followed by Dunnett's T3 multiple comparisons test was applied. When data were not normally distributed, the Kruskal–Wallis test followed by Dunn's multiple comparisons test was used. For comparisons between two groups, Welch's t test was applied when normality was satisfied but variances were unequal. All results were presented as means ± standard deviations. Differences were considered statistically significant at $p < 0.05$.

2.9. Use of Artificial Intelligence

During the preparation of this manuscript, the authors used Paperpal (v. 4.24.0, Cactus Communications Pvt, Ltd., Mumbai, Maharashtra, India) for English language polishing. After using this tool, the authors reviewed and edited the content as required and take full responsibility for the writing of this article and its publication.

3. Results

3.1. Color Change of Free EGCG After Heat Treatment

EGCG without heating (NT) exhibited a pale pinkish-white color (Figure 1A). The increase in temperature augmented the discoloration for both EGCG with VHT and AHT. EGCG with VHT (200 °C, 4 h) appeared brown, while EGCG with AHT (200 °C, 4 h) was black and lost its powdery appearance, exhibiting a plate-like particle morphology (Figure 1A). EGCG with VHT (220 °C, 4 h) was nearly black yet maintained a powdery state (Figure 1A). At 150 °C, the pink color of EGCG with VHT minimally intensified with an increase in heating time from 1 to 16 h (Figure 1B, upper part). AHT caused a more pronounced color change than VHT when heating was performed for ≥4 h; in particular, EGCG with AHT (150 °C, 16 h) was brown and completely different in color from EGCG with VHT (150 °C, 16 h) (Figure 1B). These observations were supported by the quantitative color differences (Figure 1C).

3.2. Solubility Behavior of Free EGCG in Ultrapure Water (UPW) and EtOH

The solubility of EGCG in the solvent appears to affect the process of combining this molecule with biomaterials. Figure 2 shows images of the aqueous and ethanolic solutions of heat-treated EGCG. EGCG with NT and EGCG subjected to VHT at 100–120 °C for 4 h and 150 °C for 1 h completely dissolved in UPW and EtOH to form transparent solutions

 https://doi.org/10.3390/jfb17010018

(Figure 2A). This conclusion was supported by UV–vis spectroscopic data (Figure 2B). EGCG subjected to VHT above 200 °C partially precipitated in UPW but fully dissolved in EtOH to yield a brown solution (Figure 2A), although quantitative analysis did not show significant differences (Figure 2B). EGCG with VHT (220 °C, 4 h) was poorly soluble in UPW, causing its slight discoloration, yet dissolved in EtOH to yield a black solution (Figure 2A). EGCG with AHT (150 °C, 4 h and 8 h) dissolved in UPW and EtOH to afford light pink solutions, whereas a notably deeper color was obtained when heating was performed for 16 h (Figure 2A). EGCG with AHT (200 °C, 4 h) was insoluble in UPW and almost insoluble in EtOH, affording an orange liquid (Figure 2A,B).

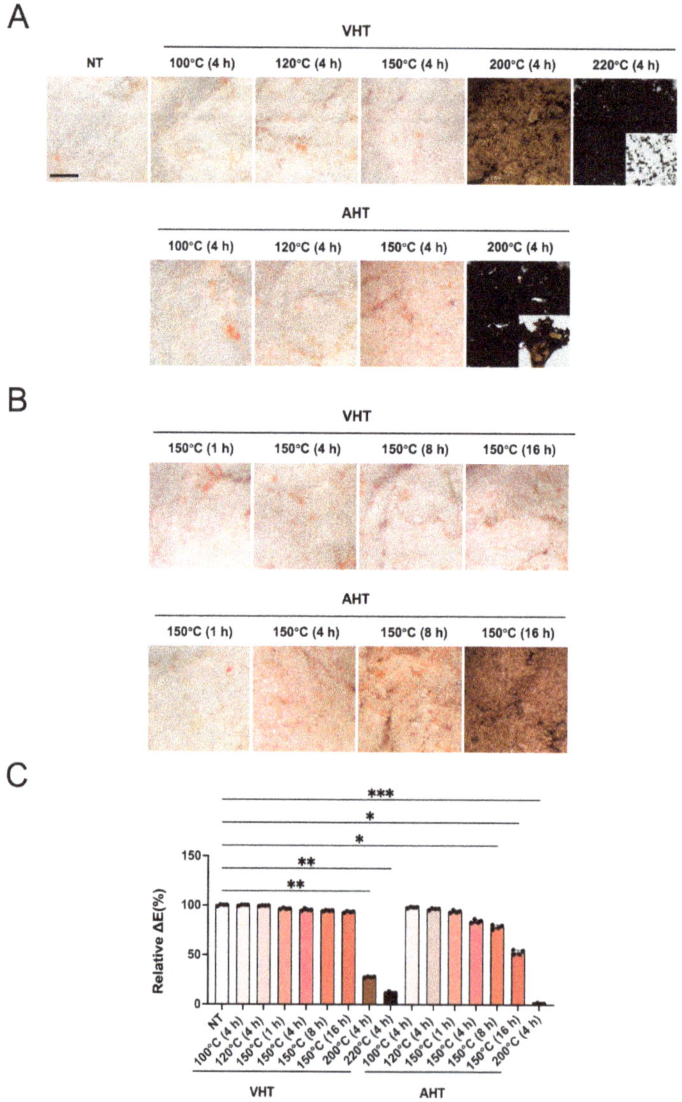

Figure 1. Stereomicroscopic and quantitative analysis of epigallocatechin gallate (EGCG) powder treated with or without heating at different temperatures and durations. (**A**) Stereomicroscopic images

https://doi.org/10.3390/jfb17010018

of EGCG treated at different heating temperatures for 4 h. NT: non-heating treatment; VHT: vacuum heating treatment; AHT: atmospheric heating treatment. Insets: the dispersions of corresponding samples. Scale bar = 500 µm. (**B**) Stereomicroscopic images of EGCG heated at 150 h for different durations. (**C**) Quantitative analysis of color differences. $n = 4$; Welch's ANOVA with Dunnett's T3 multiple comparisons test for comparisons with the control group; * $p < 0.05$, ** $p < 0.01$, *** $p < 0.001$.

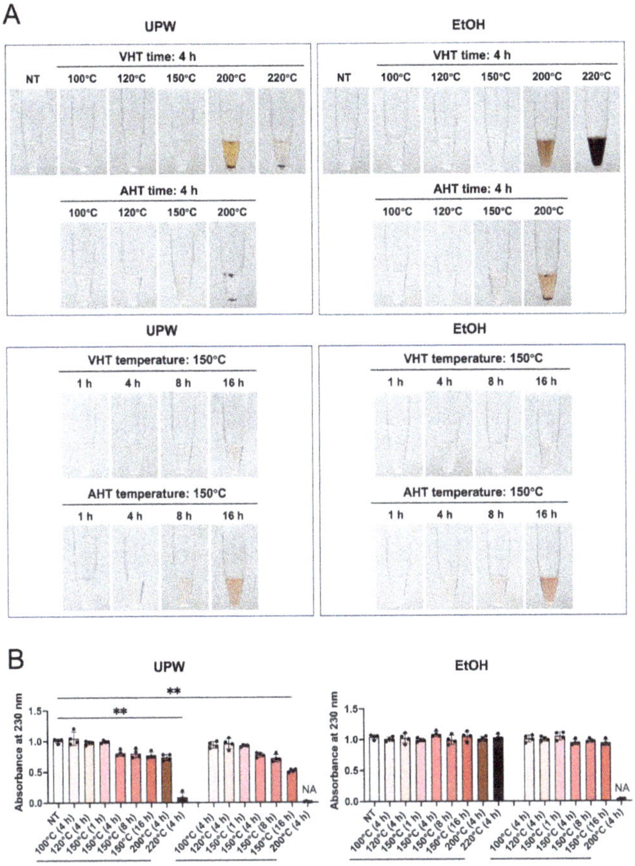

Figure 2. Solubility of EGCG powder treated with or without heating at different temperatures and durations. UPW: ultrapure water; EtOH: 80% ethanol. NT: non-heating treatment; VHT: vacuum heating treatment; AHT: atmospheric heating treatment. (**A**) Representative images showing the solubility of EGCG after different treatments. (**B**) Quantitative analysis of Ultraviolet-visible (UV–vis) spectra absorbance at 230 nm for EGCG-dissolved solution. NA = not available. **Left** (UPW): Kruskal–Wallis test with Dunn's multiple comparisons test; **Right** (EtOH): Welch's ANOVA with Dunnett's T3 multiple comparisons test. $n = 4$; ** $p < 0.01$. Statistical analyses were calculated for comparisons with the control group.

3.3. Antioxidant Activity of Free EGCG After Heat Treatment

The effects of heating on the antioxidant activity of EGCG are presented in Figure 3. Four-hour VHT had negligible effects when performed at ≤200 °C but significantly decreased antioxidant activity when performed at 220 °C (Figure 3A). EGCG retained its antioxidant activity after up to 16 h of VHT at 150 °C (Figure 3B), whereas AHT at 150 °C significantly reduced the antioxidant activity of EGCG after only 4 h (Figure 3C,D).

 https://doi.org/10.3390/jfb17010018

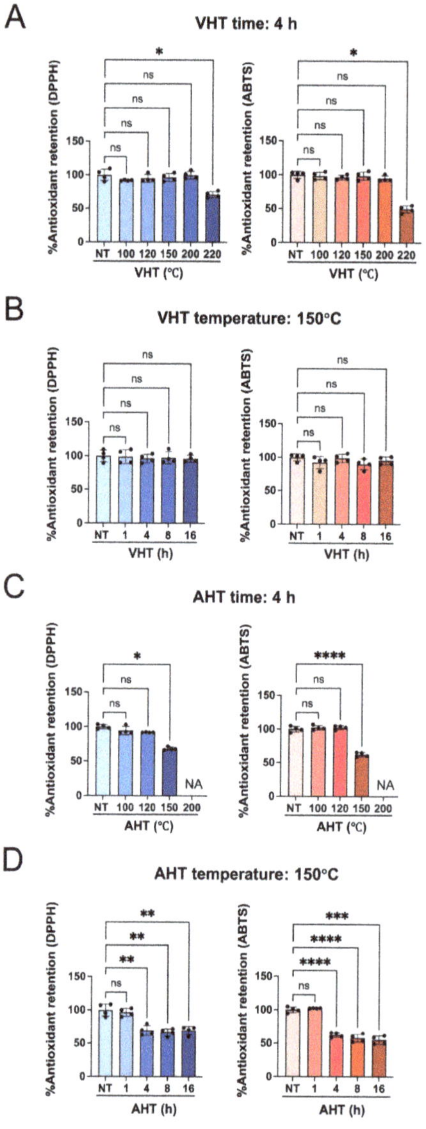

Figure 3. Antioxidant activity of EGCG subjected to different treatments. NT: non-heating treatment; VHT: vacuum heating treatment; AHT: atmospheric heating treatment. (**A–D**) **Left**: 2,2-diphenyl-1-picrylhydrazyl (DPPH) assay; **right**: 2,2′-azino-bis-3-ethylbenzothiazoline-6-sulfonic acid (ABTS) assay. NA = not available. $n = 4$; ns: not significant; * $p < 0.05$; ** $p < 0.01$; *** $p < 0.001$; **** $p < 0.0001$. (**A**) EGCG treated with VHT for 4 h at different temperatures; Kruskal–Wallis test with Dunn's multiple comparisons test. (**B**) EGCG treated with VHT at 150 °C for different heating durations; Welch's ANOVA with Dunnett's T3 multiple comparisons test (**left**) and Kruskal–Wallis test with Dunn's multiple comparisons test (**right**). (**C**) EGCG treated with AHT for 4 h at different temperatures; Kruskal–Wallis test with Dunn's multiple comparisons test (**left**) and Welch's ANOVA followed by Dunnett's T3 multiple comparisons test (**right**). (**D**) EGCG treated with AHT at 150 °C for different heating durations; Welch's ANOVA followed by Dunnett's T3 multiple comparisons test. All statistical analyses were calculated for comparisons with the control group.

 https://doi.org/10.3390/jfb17010018

3.4. Physicochemical Properties of Free EGCG After Heat Treatment

To explore the mechanisms underlying the heating-induced decrease in antioxidant activity, we evaluated the UV–vis spectra of heated EGCG and quantified absorbance at 276 nm. The absorption spectra of VHT samples almost matched those of EGCG without treatment in all cases except EGCG with VHT (220 °C, 4 h); at that high temperature, absorbance at 276 nm was markedly decreased (Figure 4A). In the case of AHT, absorbance at 276 nm significantly decreased upon heating at 150 °C for 4 h and 16 h. We could not obtain data for EGCG with AHT (200 °C, 4 h), possibly because of its low solubility (Figure 4B).

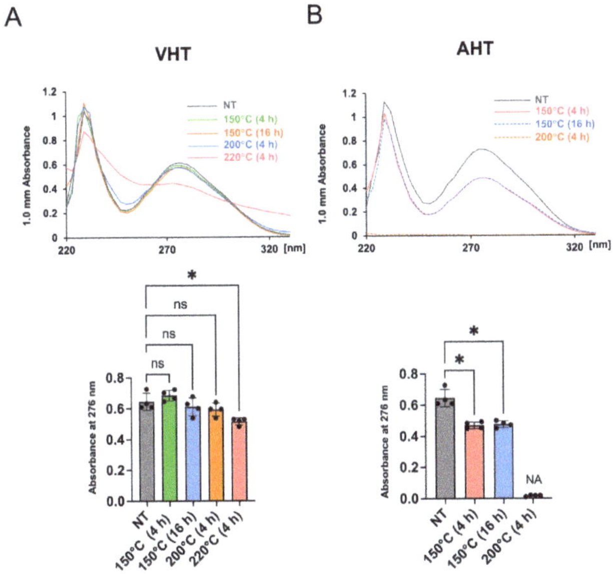

Figure 4. The UV–vis analysis and its quantitative data of EGCG treated with or without heating at different temperatures and durations. NT: non-heating treatment; VHT: vacuum heating treatment; AHT: atmospheric heating treatment. (**A**) Absorbance of EGCG after different treatments under VHT (**top**) and quantification of absorbance at 276 nm (**bottom**). (**B**) Absorbance of EGCG after different treatments under AHT (**top**) and quantification of absorbance at 276 nm (**bottom**). NA = not available. $n = 4$; Welch's ANOVA with Dunnett's T3 multiple comparisons test for comparisons with the control group; ns: not significant, * $p < 0.05$.

Subsequently, we performed mass spectrometric, NMR spectroscopic, and ATR-FTIR spectroscopic analyses to probe the physicochemical properties of EGCG treated at 150 °C for 16 h with vacuum heating as well as to determine instability of EGCG with VHT (220 °C, 4 h) and AHT (200 °C, 4 h) (Figures 5–7). This is because the former sample showed high stability, whereas the latter differed remarkably from intact EGCG. Mass spectrometry, NMR spectroscopy, and ATR-FTIR spectroscopy revealed no notable differences between EGCG with NT and with VHT (150 °C, 16 h) (Figures 5A, 6A and 7). However, EGCG with VHT (220 °C, 4 h) and AHT (200 °C, 4 h) exhibited distinct decomposition patterns, producing more low-m/z fragments than EGCG with NT and with VHT (150 °C, 16 h) (Figure 5B). The ^{1}H NMR spectra of EGCG with NT and with VHT (150 °C, 16 h) showed signals at 2.7–3.0 ppm (C-ring CH_2), 4.9 and 5.5 ppm (C-ring CH), 6.0 ppm (A-ring CH), 6.5 ppm (B-ring CH), 6.9 ppm (D-ring CH) in agreement with a previous study [41]. Unlike the spectrum of EGCG with VHT (150 °C, 16 h), that of EGCG with VHT (220 °C, 4 h) had

 https://doi.org/10.3390/jfb17010018

very weak A-ring signals, and no C-ring and D-ring signals were observed, whereas the B-ring signals were detectable. Two new signals appeared at 6.7 and 7.0 ppm (Figure 6B, green arrows). The specific peak of EGCG, as mentioned above, was undetectable in the spectrum of EGCG with AHT (200 °C, 4 h), possibly because this sample was insoluble in D_2O (Figure 6B). Compared with those of EGCG with NT and with VHT (150 °C, 16 h), the ATR-FTIR spectra of EGCG with VHT (220 °C, 4 h) and with AHT (200 °C, 4 h) featured broader peaks in the range of 400–1700 cm^{-1}, where the peaks of EGCG with NT were located [29]. The peak at 1689 cm^{-1} tentatively assigned to C=O bonds (Figure 7A, green arrow) remarkably weakened upon VHT at 220 °C for 4 h. The ratio of absorbance at 1689 cm^{-1} to that at 1613 cm^{-1} was not significantly different between EGCG with NT and EGCG with VHT (150 °C, 16 h) (Figure 7B).

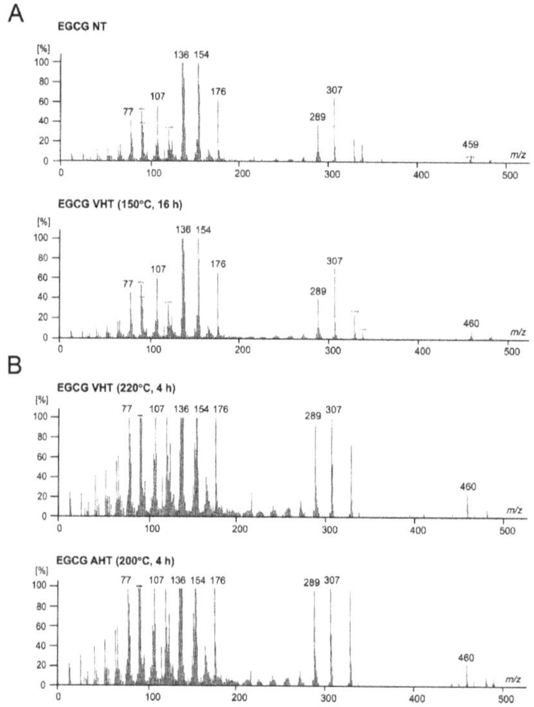

Figure 5. Mass spectrometric analysis of EGCG treated with or without heating at different temperatures and times. NT: non-heating treatment; VHT: vacuum heating treatment; AHT: atmospheric heating treatment. (**A**) **Top**: EGCG with NT; **bottom**: EGCG with VHT (150 °C, 16 h). (**B**) **Top**: EGCG with VHT (220 °C, 4 h); **bottom**: EGCG with AHT (200 °C, 4 h).

3.5. *In Vivo and In Vitro Antioxidant Effects of EGCG-Modified Gelatin*

Depending on the modified state, the physicochemical properties of EGCG can change upon heat treatment. To confirm whether polymer (gelatin)-modified EGCG retains its antioxidant properties after VHT at 150 °C for 16 h, we prepared gelatin-based materials with (vhEc-GSβ) and without (vhc-GSβ and vh-GSβ) EGCG and compared their antioxidant properties in vitro and in vivo. All three types of sponges showed porous structures (Figure 8A) and contained β-TCP (Figure 8C) [42] and gelatin (Figure 8D) [43]. Quantitative analysis of sponge pore size based on FE-SEM images indicated only minor variations among the samples (Figure 8B).

https://doi.org/10.3390/jfb17010018

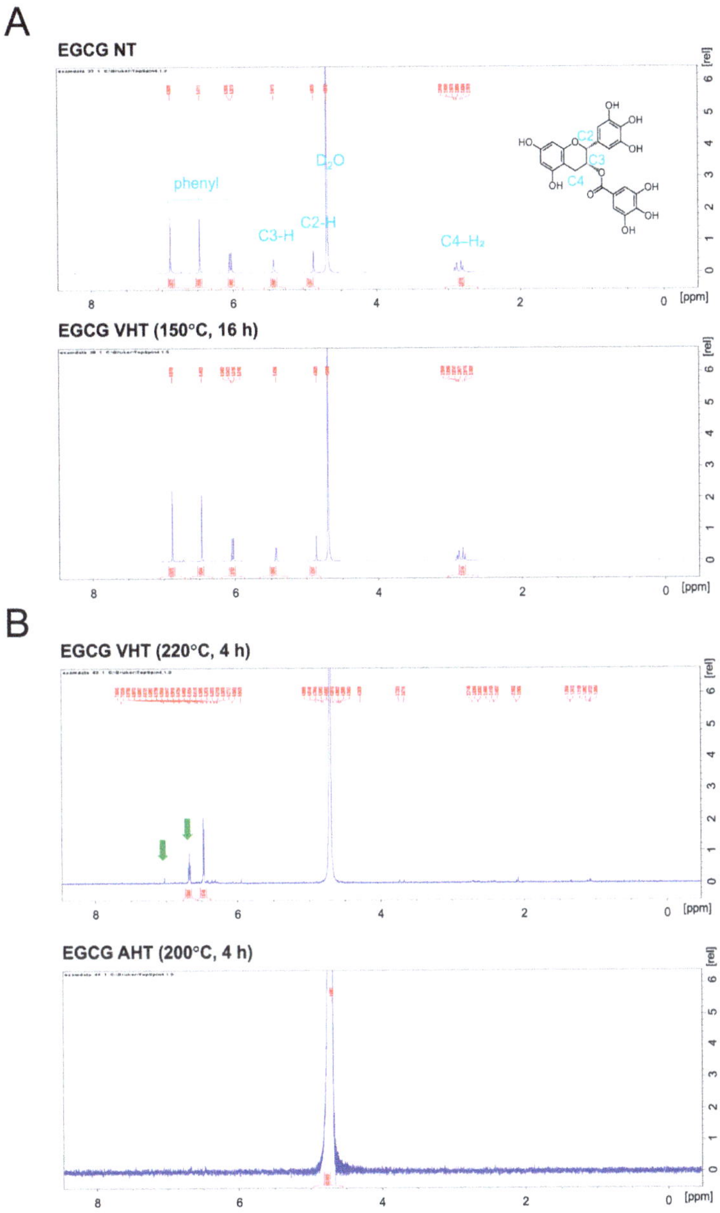

Figure 6. [1]H Nuclear Magnetic Resonance analysis of EGCG treated with or without heating at different temperatures and times. NT: non-heating treatment; VHT: vacuum heating treatment; AHT: atmospheric heating treatment. (**A**) **Top**: EGCG with NT; **bottom**: EGCG with VHT (150 °C, 16 h). (**B**) **Top**: EGCG with VHT (220 °C, 4 h), green arrows: peaks at 6.7 and 7.0 ppm; **bottom**: EGCG with AHT (200 °C, 4 h), not detectable due to poor solubility.

In the DPPH radical scavenging assay, the reaction solutions of vh-GSβ and vhc-GSβ were deep purple, whereas that of vhEc-GSβ showed a significantly lighter pale-yellow color (Figure 8E). In agreement with these color changes, the DPPH absorbance (517 nm)

https://doi.org/10.3390/jfb17010018

was significantly lower for vhEc-GSβ than for vh-GSβ and vhc-GSβ, which indicates that EGCG retained its high DPPH radical scavenging activity upon modification with gelatin (Figure 8F).

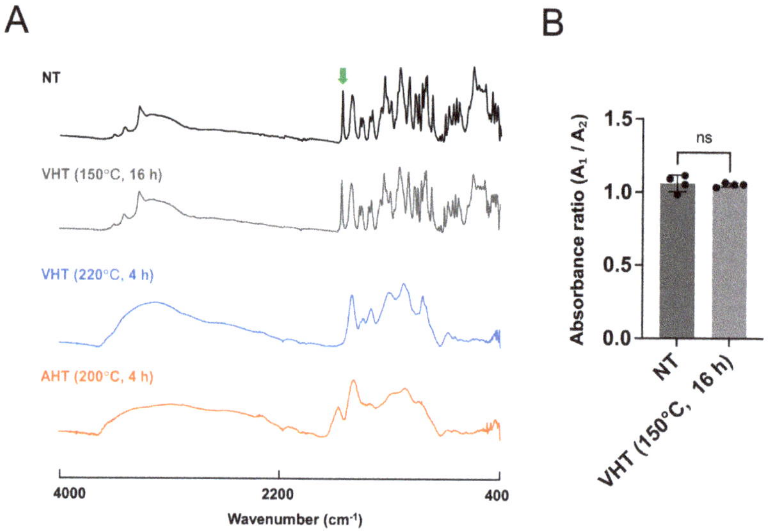

Figure 7. Attenuated total reflection-Fourier transform infrared (ATR–FTIR) spectroscopic analysis of EGCG powder treated with or without heating at different temperatures and times. NT: non-heating treatment; VHT: vacuum heating treatment; AHT: atmospheric heating treatment. (**A**) ATR–FTIR spectra of EGCG powder under the indicated treatment conditions. From top to bottom: EGCG with NT; EGCG with VHT (150 °C, 16 h); EGCG with VHT (220 °C, 4 h); EGCG with AHT (200 °C, 4 h). Green arrow: peak at 1689 cm^{-1} tentatively assigned to C=O bond. (**B**) Ratio of absorbance at 1689 cm^{-1} (A_1, peak of interest) to that at 1613 cm^{-1} (A_2, aromatic C=C stretching vibration used as an internal reference). $n = 4$; Welch's t test; ns: not significant.

β-TCP, a well-known bone substitute, promotes the generation of oxidative stress in vivo and in vitro [44–46]. To evaluate the in vivo antioxidant properties of the three types of sponges, we implanted them in a critical bone defect model at the rat calvaria (9 mm) and subjected them to immunofluorescence staining for a reactive oxygen species (ROS) marker (4-hydroxynonenal) at three days post-surgery. Strong staining was observed for the sponges without EGCG (vh-GSβ and vhc-GSβ), possibly at the inner part of β-TCP. vhEc-GSβ showed negligible staining despite having the same amount of β-TCP, which demonstrates that modified EGCG subjected to VHT at 150 °C for 16 h suppressed β-TCP-induced oxidative stress in vivo and retained its antioxidant properties (Figure 8G,H).

https://doi.org/10.3390/jfb17010018

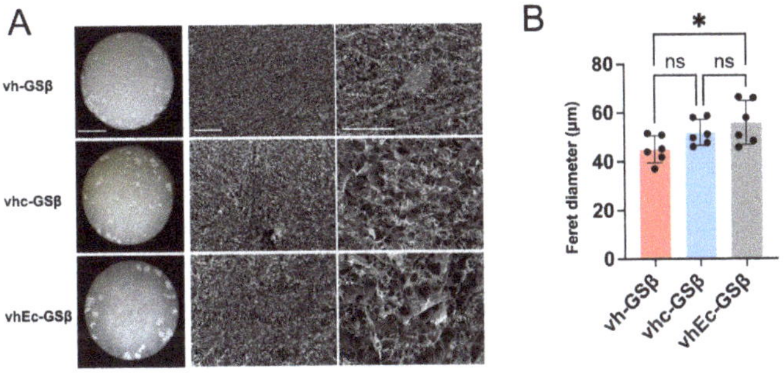

Figure 8. *Cont.*

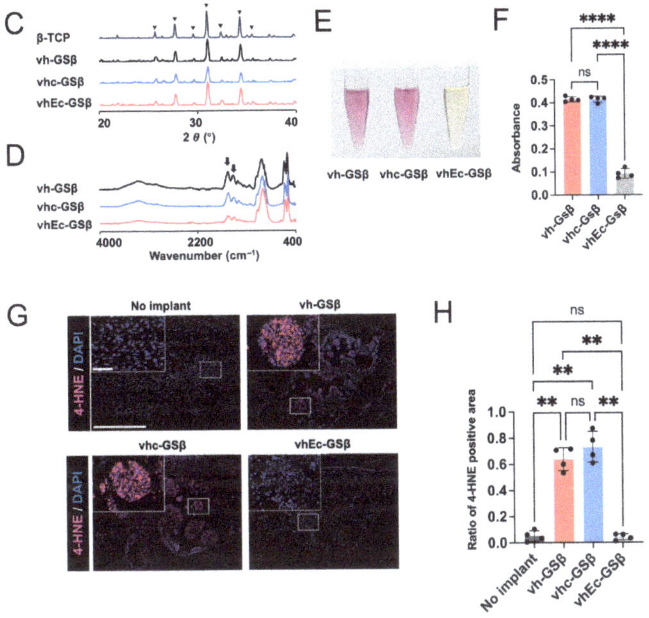

Figure 8. Characterization and antioxidant activity of samples. (**A**) Three types of sponges: vacuum-heated (150 °C, 16 h) gelatin sponges containing β-tricalcium phosphate (β-TCP) granules (vh-GSβ), vacuum-heated chemically synthesized gelatin sponges containing β-TCP (vhc-GSβ), and vacuum-heated chemically synthesized gelatin sponges containing EGCG and β-TCP (vhEc-GSβ). Macroscopic appearances (**left**; scale bar = 5 mm) and field-emission scanning electron microscopic appearances (**middle**; scale bar = 500 μm, **right**; scale bar = 200 μm). (**B**) Quantitative analysis of pore size based on the average of the maximum Feret diameter. $n = 6$; one-way ANOVA with Tukey's multiple comparisons test; ns: not significant; * $p < 0.05$. (**C**) X-ray powder diffraction patterns. Arrow heads: the characteristic diffraction peaks of β-TCP. (**D**) ATR-FTIR spectra of the samples. Arrows: the characteristic peaks of amide bands. (**E**) DPPH assay. (**F**) Quantitative analysis of DPPH scavenging activity using Welch's ANOVA followed by Dunnett's T3 multiple comparisons test. $n = 4$, ns: not significant, **** $p < 0.0001$. (**G**) Immunofluorescence staining for 4-Hydroxynonenal (4-HNE) in rat calvarial bone defects implanted with or without samples. Scale bars: **top** = 50 μm; **bottom** = 500 μm. (**H**) Quantitative analysis of the ratio of 4-HNE to 4′,6-diamidino-2-phenylindole (DAPI) fluorescence area. $n = 4$; Welch's ANOVA followed by Dunnett's T3 multiple comparisons test; ns: not significant; ** $p < 0.01$.

https://doi.org/10.3390/jfb17010018

4. Discussion

Given the worldwide consumption of tea as beverages and foods, various studies have evaluated changes in EGCG properties upon heating, especially in aqueous media [13]. Recent advances in regenerative therapies have widened the application scope of EGCG to biomaterial use [17–27]. However, the changes in the properties of modified and free EGCG upon thermal treatment in vacuum remain underexplored. Herein, we evaluated the changes in the physicochemical properties and antioxidant activity of free EGCG upon VHT and AHT. Additionally, the antioxidant ability of modified EGCG in EGCG-conjugated gelatin containing β-TCP was evaluated in vivo and in vitro after vacuum heating. VHT caused negligible changes in the physicochemical and antioxidant properties of free EGCG when performed at 200 °C for 4 h and 150 °C for 16 h. EGCG modified with gelatin also retained its antioxidant capacity even after 16 h of VHT at 150 °C.

Color is an intuitive indicator of EGCG stability [34]. Upon VHT, the color of EGCG changed more gradually than upon AHT (Figure 1). Previous studies have reported similar color changes to pinkish-brown for EGCG powder stored for long periods under atmospheric conditions [34]. This color change may stem from colored compounds formed by oxidation, such as catechin diquinone or other catechin derivatives [34]. The enzyme-catalyzed oxidation of EGCG is known to yield colored products such as dehydrotheasinene and its quinoxaline derivatives [47]. Samples stored under low-humidity and high-oxygen conditions have been reported to exhibit increased theasinensin formation and color development compared with those stored under high-humidity conditions [34]. Although we could not isolate the components responsible for color changes after heating, AHT may cause the coloration of EGCG by promoting oxidation affording colored compounds. Discoloration change was observed even upon VHT, becoming more pronounced with increasing temperature and time. The low-humidity, low-oxygen conditions of VHT may promote the dehydrative condensation of EGCG and, hence, the formation of polymeric species.

Previous studies have shown that the solubility of EGCG and other catechins partially depends on solvent polarity [48]. Herein, EGCG treated at higher temperatures exhibited reduced solubility. Moreover, the solubility of EGCG in EtOH exceeded that in UPW (Figure 2). Mass spectrometric, NMR spectroscopic, and ATR-FTIR spectroscopic analyses revealed differences in EGCG subjected to VHT (220 °C, 4 h) and AHT (200 °C, 4 h) (Figures 6 and 7). EGCG subjected to VHT or AHT (particularly VHT above 220 °C or AHT above 200 °C) may undergo structural changes, such as oxidation and decomposition, which increase the amount of low-polarity small molecules. Those changes might elicit the alteration of solubility.

The antioxidant activity of free EGCG showed little change upon VHT (even at 200 °C for 4 h and 150 °C for 16 h) but remarkably decreased after AHT at 150 °C for 4 h (Figure 3C). The absorption peak of EGCG in the 270–276 nm region ($\lambda_{max} \approx 275$ nm) is mainly due to the formation of a conjugated π electron system from multiple phenolic rings (especially the polyphenol and gallate group in ring D) [49,50]. With the decomposition of EGCG, this peak loses intensity [51]. Furthermore, epigallocatechin (EGC), which does not have a gallate group (D-ring) and has a lower antioxidant activity than EGCG, features a notably weaker absorbance at ~270 nm at the same concentration [38,52,53]. Herein, the absorbance at 276 nm was significantly attenuated upon VHT (220 °C, 4 h) and AHT (150° C, 4 h and 16 h) (Figure 4). ATR-FTIR spectroscopic analysis showed that the C=O peak (1689 cm^{-1}) lost intensity after VHT (220 °C, 4 h), possibly because of the degradation of the ester bond between the C- and D-rings and galloyl group detachment (Figure 7). Taken together,

 https://doi.org/10.3390/jfb17010018

these results suggested that high-temperature VHT and AHT induced changes in the galloyl group.

The minimization of oxygen exposure through VHT resulted in antioxidant activity retention significantly exceeding that observed upon AHT (Figure 3). The effects of heating EGCG in water or under humid conditions on its antioxidant properties have been actively studied because of the widespread consumption of tea, the general consensus being that EGCG is heat-sensitive and loses its antioxidant activity upon heating [13]. Murakami et al. reported that the radical scavenging activity of EGCG almost vanished after 120 min of heating at 180 °C (oil bath) under aqueous conditions. The degradation of EGCG produces gallic acid and gallocatechin gallate (GCG, epimer of EGCG) [13]; during this degradation process, GCG is thought to be produced by the epimerization of EGCG, while gallic acid and EGC are produced by degalloylation [34]. Unlike that of EGC, the antioxidant activity of GCG is not remarkably lower than that of EGCG [54]. In our study, although EGCG decomposed at temperatures exceeding 220 °C, its antioxidant ability was maintained even after prolonged VHT at 150 °C for 16 h or 200 °C for 4 h. Although we could not reveal whether EGCG is converted into GCG after VHT at 150 °C, given the results of antioxidant activity and UV–vis analyses, VHT (at least up to 150 °C) might not cause degalloylation owing to the absence of oxygen, preserving antioxidant activity.

Given the context of our study, the identification of compounds produced from modified EGCG in gelatin after VHT can prove the ability of VHT to prevent the alteration of EGCG. However, this identification faces technical challenges, such as the difficulty in isolating modified EGCG in the pure state. Moreover, heating is expected to promote the denaturation of both base materials (gelatin in our experiments) and modified EGCG, which complicates result interpretation. Heating is known to change the antioxidant activity of gelatin [55]. VHT induces dehydrative condensation and promotes the formation of crosslinked structures in various polymers, such as gelatin [20,56] and collagen [57,58]. Crosslinking modifies the polymers used in biomaterials, altering their stability, solubility, and controlled-drug-release ability [58,59]. Given that heating is expected to affect gelatin properties, we determined the antioxidant activities of sponges with or without EGCG subjected to the same VHT at 150 °C, under conditions that did not affect free EGCG. The radical scavenging ability of EGCG was reported to decrease upon heating under aqueous conditions [13], whereas in our work, modified EGCG retained its antioxidant activity even after VHT at 150 °C for 16 h.

Generally, unheated EGCG has been reported to directly scavenge ROS [38,52] and suppress ROS-producing cells [5]. Previous studies showed that implanted β-TCP is phagocytosed by macrophage-like cells, which then generate ROS [44,60]. Macrophage ROS production is suppressed by unheated free EGCG in vivo [61] and in vitro [62]. The results of our in vitro evaluation indicate that modified EGCG (conjugated to gelatin) maintains its direct ROS-scavenging effect after VHT (Figure 8E,F). Histological examination revealed significantly reduced staining for 4-HNE in β-TCP granules, although the question of whether modified EGCG after VHT removed ROS directly or indirectly via inhibiting ROS-producing cells remains unanswered (Figure 8G,H). However, our results suggest that adequate conditions of VHT preserve the antioxidant ability of modified EGCG.

Our study demonstrates that free and modified EGCG exhibit excellent antioxidant activity after VHT at ≤150 °C. Further research should evaluate physicochemical property changes under prolonged heating and identify the products of decomposition upon VHT at 220 °C and AHT at 200 °C. Moreover, although we used free and modified EGCG (in gelatin), other base materials may have different effects on EGCG stability. Thus,

 https://doi.org/10.3390/jfb17010018

careful consideration is necessary to elucidate the mechanisms of VHT-induced changes in EGCG performance.

5. Conclusions

This study evaluated the effects of VHT and AHT on the physicochemical properties and antioxidant activity of free and modified (gelatin-hybridized) EGCG. VHT caused the discoloration and decreased the solubility of free EGCG, with these effects becoming more pronounced with increasing temperature. However, physicochemical and antioxidant property analyses showed no significant changes upon VHT at 150 °C up to 16 h, with marked changes appearing from 220 °C onwards. AHT caused greater changes than VHT, including discoloration and solubility loss, even when performed at 150 °C for 4 h, which indicated that degradation started at lower temperatures under ambient conditions. The AHT-induced decrease in antioxidant activity was partly attributed to decomposition-induced degalloylation. Similar to free EGCG, modified EGCG maintained its antioxidant activity in vivo and in vitro after VHT at 150 °C. Although EGCG is considered a thermally unstable material, our results suggest that free EGCG and EGCG-containing biomaterials subjected to VHT at up to 150 °C may retain their antioxidant capacity and exert pharmacological effects. These insights facilitate the development of novel biomaterials for biomedicine and materials engineering.

Author Contributions: Conceptualization, L.Y. and Y.H.; methodology, L.Y., Y.H. and T.T.; formal analysis, L.Y.; investigation, L.Y.; validation, L.Y.; writing—original draft preparation, L.Y. and Y.H.; writing—review and editing, L.Y., Y.H., Z.D., M.N., S.Z., J.-i.J. and T.T.; data curation, L.Y., Y.H. and Z.D.; visualization, L.Y., Y.H., Z.D. and M.N.; supervision, Y.H.; project administration, Y.H.; funding acquisition, Y.H. All authors have read and agreed to the published version of the manuscript.

Funding: This work was partially funded by a joint research project with Toagosei Co., Ltd.

Institutional Review Board Statement: The animal study protocol was approved by Osaka Dental University Institutional Animal Care and Use Committee (approval no. 25-06005, approval date: 22 July 2025).

Informed Consent Statement: Not applicable.

Data Availability Statement: The original contributions presented in the study are included in the article, further inquiries can be directed to the corresponding author.

Acknowledgments: We thank Xinfei Du (Department of Oral Anatomy, Osaka Dental University) and Yojiro Maeda (Department of Oral Implantology, Osaka Dental University) for support with animal experiments, Jiaru Lyu (Department of Oral Anatomy, Osaka Dental University) for support with SEM image analysis and Hiroya Fujishima (Kyoto Institute of Technology) for support with ATR-FTIR spectroscopy.

Conflicts of Interest: This study received funding from Toagosei Co., Ltd. The funder was involved in the preparation of sponges but not in the study design, data collection, analysis, and interpretation, the writing of this article, or the decision to submit it for publication.

Abbreviations

The following abbreviations are used in this manuscript:

https://doi.org/10.3390/jfb17010018

EGCG	Epigallocatechin gallate
VHT	Vacuum heating treatment
AHT	Atmospheric heating treatment
DPPH	2,2-Diphenyl-1-picrylhydrazyl
ABTS	2,2′-Azino-bis-3-ethylbenzothiazoline-6-sulfonic acid
UV–vis	Ultraviolet–visible
XRD	X-ray powder diffraction
ATR	Attenuated total reflection
FTIR	Fourier transform infrared
FE-SEM	Field-emission scanning electron microscopy
β-TCP	β-Tricalcium phosphate
NT	Non-heating treatment
EtOH	Ethanol
UPW	Ultrapure water
NMR	Nuclear magnetic resonance
4-HNE	4-Hydroxynonenal
ROI	Region of interest
ANOVA	Analysis of variance
ns	Not significant
GCG	Gallocatechin gallate
EGC	Epigallocatechin
NA	Not available

References

1. Graham, H.N. Green tea composition, consumption, and polyphenol chemistry. *Prev. Med.* **1992**, *21*, 334–350. [CrossRef]
2. Chow, H.-H.S.; Cai, Y.; Hakim, I.A.; Crowell, J.A.; Shahi, F.; Brooks, C.A.; Dorr, R.T.; Hara, Y.; Alberts, D.S. Pharmacokinetics and Safety of Green Tea Polyphenols after Multiple-Dose Administration of Epigallocatechin Gallate and Polyphenon E in Healthy Individuals. *Clin. Cancer Res.* **2003**, *9*, 3312–3319.
3. Asia Pacific Green Tea Market by Container Type (Plastic Bottles, Cans, and Others), Product Type (Flavoured and Unflavoured), and Region-Industry Analysis, Size, Share, Growth, Trends, and Forecasts (2025 to 2033). Available online: https://www.marketdataforecast.com/market-reports/asia-pacific-green-tea-market (accessed on 28 October 2025).
4. Nanjo, F.; Mori, M.; Goto, K.; Hara, Y. Radical Scavenging Activity of Tea Catechins and Their Related Compounds. *Biosci. Biotechnol. Biochem.* **1999**, *63*, 1621–1623. [CrossRef]
5. Xie, L.-W.; Cai, S.; Zhao, T.-S.; Li, M.; Tian, Y. Green tea derivative (−)-epigallocatechin-3-gallate (EGCG) confers protection against ionizing radiation-induced intestinal epithelial cell death both in vitro and in vivo. *Free Radic. Biol. Med.* **2020**, *161*, 175–186. [CrossRef] [PubMed]
6. Liu, Q.; Qian, Y.; Chen, F.; Chen, X.; Chen, Z.; Zheng, M. EGCG attenuates pro-inflammatory cytokines and chemokines production in LPS-stimulated L02 hepatocyte. *Acta Biochim. Biophys. Sin.* **2014**, *46*, 31–39. [CrossRef] [PubMed]
7. Yang, Y.; Han, X.; Chen, Y.; Wu, J.; Li, M.; Yang, H.; Xu, W.; Wei, L. EGCG Induces Pro-inflammatory Response in Macrophages to Prevent Bacterial Infection through the 67LR/p38/JNK Signaling Pathway. *J. Agric. Food Chem.* **2021**, *69*, 5638–5651. [CrossRef] [PubMed]
8. Higdon, J.V.; Frei, B. Tea Catechins and Polyphenols: Health Effects, Metabolism, and Antioxidant Functions. *Crit. Rev. Food Sci. Nutr.* **2003**, *43*, 89–143. [CrossRef]
9. Kaida, K.; Honda, Y.; Hashimoto, Y.; Tanaka, M.; Baba, S. Application of Green Tea Catechin for Inducing the Osteogenic Differentiation of Human Dedifferentiated Fat Cells in Vitro. *Int. J. Mol. Sci.* **2015**, *16*, 27988–28000. [CrossRef]
10. Li, L.; Gao, F.; Zhang, H.; Yu, Y.; Ma, B.; Ge, S. A protective growth factor delivery strategy based on polyphenol-protein self-assembly to promote inflammatory bone regeneration. *Biomaterials* **2025**, *320*, 123272. [CrossRef]
11. Kanzaki, H.; Shinohara, F.; Itohiya-Kasuya, K.; Ishikawa, M.; Nakamura, Y. Nrf2 Activation Attenuates Both Orthodontic Tooth Movement and Relapse. *J. Dent. Res.* **2015**, *94*, 787–794. [CrossRef]
12. Wang, R.; Zhou, W.; Jiang, X. Reaction Kinetics of Degradation and Epimerization of Epigallocatechin Gallate (EGCG) in Aqueous System over a Wide Temperature Range. *J. Agric. Food Chem.* **2008**, *56*, 2694–2701. [CrossRef]
13. Murakami, M.; Yamaguchi, T.; Takamura, H.; Matoba, T. Change in the Radical-Scavenging Activity of Quercetin and Epigallocatechin Gallate during Heat Treatment. *J. Home Econ.* **2004**, *55*, 213–217. [CrossRef]

 https://doi.org/10.3390/jfb17010018

14. Bianchi, A.; Marchetti, N.; Scalia, S. Photodegradation of (−)-epigallocatechin-3-gallate in topical cream formulations and its photostabilization. *J. Pharm. Biomed. Anal.* **2011**, *56*, 692–697. [CrossRef] [PubMed]
15. Mochizuki, M.; Yamazaki, S.-I.; Kano, K.; Ikeda, T. Kinetic analysis and mechanistic aspects of autoxidation of catechins. *Biochim. Biophys. Acta (BBA)-Gen. Subj.* **2002**, *1569*, 35–44. [CrossRef]
16. Chen, Z.-Y.; Zhu, Q.Y.; Tsang, D.; Huang, Y. Degradation of Green Tea Catechins in Tea Drinks. *J. Agric. Food Chem.* **2001**, *49*, 477–482. [CrossRef] [PubMed]
17. Chu, C.; Deng, J.; Xiang, L.; Wu, Y.; Wei, X.; Qu, Y.; Man, Y. Evaluation of epigallocatechin-3-gallate (EGCG) cross-linked collagen membranes and concerns on osteoblasts. *Mater. Sci. Eng. C* **2016**, *67*, 386–394. [CrossRef]
18. Elder, S.; Clune, J.; Walker, J.; Gloth, P. Suitability of EGCG as a Means of Stabilizing a Porcine Osteochondral Xenograft. *J. Funct. Biomater.* **2017**, *8*, 43. [CrossRef]
19. Lee, J.S.; Lee, J.S.; Lee, M.S.; An, S.; Yang, K.; Lee, K.; Yang, H.S.; Lee, H.; Cho, S.-W. Plant Flavonoid-Mediated Multifunctional Surface Modification Chemistry: Catechin Coating for Enhanced Osteogenesis of Human Stem Cells. *Chem. Mater.* **2017**, *29*, 4375–4384. [CrossRef]
20. Honda, Y.; Takeda, Y.; Li, P.; Huang, A.; Sasayama, S.; Hara, E.; Uemura, N.; Ueda, M.; Hashimoto, M.; Arita, K.; et al. Epigallocatechin Gallate-Modified Gelatin Sponges Treated by Vacuum Heating as a Novel Scaffold for Bone Tissue Engineering. *Molecules* **2018**, *23*, 876. [CrossRef]
21. Moreno-Vásquez, M.J.; Plascencia-Jatomea, M.; Sánchez-Valdes, S.; Tanori-Córdova, J.C.; Castillo-Yañez, F.J.; Quintero-Reyes, I.E.; Graciano-Verdugo, A.Z. Characterization of Epigallocatechin-Gallate-Grafted Chitosan Nanoparticles and Evaluation of Their Antibacterial and Antioxidant Potential. *Polymers* **2021**, *13*, 1375. [CrossRef]
22. Xiao, S.; Wei, J.; Jin, S.; Xia, X.; Yuan, L.; Zou, Q.; Zuo, Y.; Li, J.; Li, Y. A Multifunctional Coating Strategy for Promotion of Immunomodulatory and Osteo/Angio-Genic Activity. *Adv. Funct. Mater.* **2023**, *33*, 2208968. [CrossRef]
23. Safwat, M.A.; Kandil, B.A.; Elblbesy, M.A.; Soliman, G.M.; Eleraky, N.E. Epigallocatechin-3-Gallate-Loaded Gold Nanoparticles: Preparation and Evaluation of Anticancer Efficacy in Ehrlich Tumor-Bearing Mice. *Pharmaceuticals* **2020**, *13*, 254. [CrossRef] [PubMed]
24. Wakayama, C.; Inubushi, S.; Kunihisa, T.; Mizumoto, S.; Baba, M.; Tanino, H.; Cho, I.S.; Ooya, T. Protein corona formation on epigallocatechin gallate-Au nanoparticles suppressed tumor accumulation. *JCIS Open* **2023**, *9*, 100074. [CrossRef]
25. Ren, J.; Sun, L.; Xiao, C.; Zhou, S.; Liang, Q.; Sun, S.; Deng, C. Chemical bonding of Epigallocatechin-3-gallate to the surface of nano-hydroxyapatite to enhance its biological activity for anti-osteosarcoma. *Smart Mater. Med.* **2023**, *4*, 396–406. [CrossRef]
26. Ali, K.; Rakesh, S.; Khalid, S.; Khan, A.U. Moist Heat Synthesis of Magnetic EGCG-Cappedα-Fe2O3 Nanoparticles and Their In Vitro and In Silico Interactions with Pristine HSA- and NDM-1-Producing Bacteria. *ACS Omega* **2023**, *8*, 48775–48786. [CrossRef]
27. Miu, B.A.; Stan, M.S.; Mernea, M.; Dinischiotu, A.; Voinea, I.C. Pure Epigallocatechin-3-gallate-Assisted Green Synthesis of Highly Stable Titanium Dioxide Nanoparticles. *Materials* **2024**, *17*, 275. [CrossRef]
28. Ke, F.; Zhang, M.; Qin, N.; Zhao, G.; Chu, J.; Wan, X. Synergistic antioxidant activity and anticancer effect of green tea catechin stabilized on nanoscale cyclodextrin-based metal–organic frameworks. *J. Mater. Sci.* **2019**, *54*, 10420–10429. [CrossRef]
29. Avadhani, K.S.; Manikkath, J.; Tiwari, M.; Chandrasekhar, M.; Godavarthi, A.; Vidya, S.M.; Hariharapura, R.C.; Kalthur, G.; Udupa, N.; Mutalik, S. Skin delivery of epigallocatechin-3-gallate (EGCG) and hyaluronic acid loaded nano-transfersomes for antioxidant and anti-aging effects in UV radiation induced skin damage. *Drug Deliv.* **2017**, *24*, 61–74. [CrossRef]
30. Bi, D.; Qu, F.; Xiao, W.; Wu, J.; Liu, P.; Du, H.; Xie, Y.; Liu, H.; Zhang, L.; Tao, J.; et al. Reactive Oxygen Species-Responsive Gel-Based Microneedle Patches for Prolonged and Intelligent Psoriasis Management. *ACS Nano* **2023**, *17*, 4346–4357. [CrossRef]
31. Deng, W.; Jo, J.-i.; Tanaka, T.; Morikuni, H.; Hashimoto, Y.; Matsumoto, N.; Honda, Y. A Senomorphic-Conjugated Scaffold for Application of Senescent Cells in Regenerative Medicine. *Adv. Ther.* **2023**, *6*, 2200276. [CrossRef]
32. Gao, B.; Honda, Y.; Yamada, Y.; Tanaka, T.; Takeda, Y.; Nambu, T.; Baba, S. Utility of Thermal Cross-Linking in Stabilizing Hydrogels with Beta-Tricalcium Phosphate and/or Epigallocatechin Gallate for Use in Bone Regeneration Therapy. *Polymers* **2022**, *14*, 40. [CrossRef]
33. Xie, M.; Zhong, Y.; Wang, S.; Tian, L.; Wang, X.; Hu, Y.; Zeng, R. EGCG/Zn coating on titanium implants by one-step hydrothermal method for improving anticorrosion, antibacterial and osteogenesis properties. *Mater. Chem. Phys.* **2022**, *292*, 126872. [CrossRef]
34. Li, N.; Taylor, L.S.; Ferruzzi, M.G.; Mauer, L.J. Color and chemical stability of tea polyphenol (−)-epigallocatechin-3-gallate in solution and solid states. *Food Res. Int.* **2013**, *53*, 909–921. [CrossRef]
35. Neilson, A.P.; Hopf, A.S.; Cooper, B.R.; Pereira, M.A.; Bomser, J.A.; Ferruzzi, M.G. Catechin Degradation with Concurrent Formation of Homo- and Heterocatechin Dimers during in Vitro Digestion. *J. Agric. Food Chem.* **2007**, *55*, 8941–8949. [CrossRef]
36. Zimeri, J.; Tong, C.H. Degradation Kinetics of (−)-Epigallocatechin Gallate as a Function of pH and Dissolved Oxygen in a Liquid Model System. *J. Food Sci.* **1999**, *64*, 753–758. [CrossRef]

https://doi.org/10.3390/jfb17010018

37. Li, N.; Taylor, L.S.; Ferruzzi, M.G.; Mauer, L.J. Kinetic Study of Catechin Stability: Effects of pH, Concentration, and Temperature. *J. Agric. Food Chem.* **2012**, *60*, 12531–12539. [CrossRef] [PubMed]
38. Valcic, S.; Burr, J.A.; Timmermann, B.N.; Liebler, D.C. Antioxidant Chemistry of Green Tea Catechins. New Oxidation Products of (−)-Epigallocatechin Gallate and (−)-Epigallocatechin from Their Reactions with Peroxyl Radicals. *Chem. Res. Toxicol.* **2000**, *13*, 801–810. [CrossRef]
39. Markel, D.C.; Bou-Akl, T.; Wu, B.; Pawlitz, P.; Yu, X.; Chen, L.; Shi, T.; Ren, W. In Vivo Cell Migration and Growth Within Electrospun Porous Nanofibrous Scaffolds with Different Pore Sizes in a Mouse Pouch Model. *J. Funct. Biomater.* **2025**, *16*, 181. [CrossRef] [PubMed]
40. Kawamoto, T. Use of a new adhesive film for the preparation of multi-purpose fresh-frozen sections from hard tissues, whole-animals, insects and plants. *Arch. Histol. Cytol.* **2003**, *66*, 123–143. [CrossRef] [PubMed]
41. Fu, N.; Zhou, Z.; Jones, T.B.; Tan, T.T.Y.; Wu, W.D.; Lin, S.X.; Chen, X.D.; Chan, P.P.Y. Production of monodisperse epigallocatechin gallate (EGCG) microparticles by spray drying for high antioxidant activity retention. *Int. J. Pharm.* **2011**, *413*, 155–166. [CrossRef]
42. Raynaud, S.; Champion, E.; Bernache-Assollant, D.; Thomas, P. Calcium phosphate apatites with variable Ca/P atomic ratio I. Synthesis, characterisation and thermal stability of powders. *Biomaterials* **2002**, *23*, 1065–1072. [CrossRef]
43. Muyonga, J.H.; Cole, C.G.B.; Duodu, K.G. Fourier transform infrared (FTIR) spectroscopic study of acid soluble collagen and gelatin from skins and bones of young and adult Nile perch (*Lates niloticus*). *Food Chem.* **2004**, *86*, 325–332. [CrossRef]
44. Maruyama, K.; Cheng, J.-Y.; Ishii, H.; Takahashi, Y.; Zangiacomi, V.; Satoh, T.; Hosono, T.; Yamaguchi, K. Activation of NLRP3 Inflammasome Complexes by Beta-Tricalcium Phosphate Particles and Stimulation of Immune Cell Migration in vivo. *J. Innate Immun.* **2021**, *14*, 207–217. [CrossRef] [PubMed]
45. Liu, L.; Dai, H.; Wu, Y.; Li, B.; Yi, J.; Xu, C.; Wu, X. In vitro and in vivo mechanism of hepatocellular carcinoma inhibition by β-TCP nanoparticles. *Int. J. Nanomed.* **2019**, *14*, 3491–3502. [CrossRef]
46. Wang, X.; Honda, Y.; Zhao, J.; Morikuni, H.; Nishiura, A.; Hashimoto, Y.; Matsumoto, N. Enhancement of Bone-Forming Ability on Beta-Tricalcium Phosphate by Modulating Cellular Senescence Mechanisms Using Senolytics. *Int. J. Mol. Sci.* **2021**, *22*, 12415. [CrossRef]
47. Tanaka, T.; Matsuo, Y.; Kouno, I. A Novel Black Tea Pigment and Two New Oxidation Products of Epigallocatechin-3-O-gallate. *J. Agric. Food. Chem.* **2005**, *53*, 7571–7578. [CrossRef]
48. Sevillano, D.M.; van der Wielen, L.A.M.; Trifunovic, O.; Ottens, M. Model Comparison for the Prediction of the Solubility of Green Tea Catechins in Ethanol/Water Mixtures. *Ind. Eng. Chem. Res.* **2013**, *52*, 6039–6048. [CrossRef]
49. Pires, F.; Tzeli, D.; Jones, N.C.; Hoffmann, S.V.; Raposo, M. Electronic States of Epigallocatechin-3-Gallate in Water and in 1,2-dipalmitoyl-sn-glycero-3-phospho-(1′-rac-glycerol) (Sodium Salt) Liposomes. *Int. J. Mol. Sci.* **2025**, *26*, 1084. [CrossRef]
50. Tang, J. Using UV–Vis Titration to Elucidate Novel Epigallocatechin Gallate (EGCG)-Induced Binding of the c-MYC G-Quadruplex. *Pharmaceuticals* **2025**, *18*, 719. [CrossRef] [PubMed]
51. Liu, K.; Shi, Z.; Zhou, S. Reduction of hexavalent chromium using epigallocatechin gallate in aqueous solutions: Kinetics and mechanism. *RSC Adv.* **2016**, *6*, 67196–67203. [CrossRef]
52. He, J.; Xu, L.; Yang, L.; Wang, X. Epigallocatechin Gallate Is the Most Effective Catechin Against Antioxidant Stress via Hydrogen Peroxide and Radical Scavenging Activity. *Med. Sci. Monit.* **2018**, *24*, 8198–8206. [CrossRef] [PubMed]
53. Honda, Y.; Han, X.; Gong, Y.; Miyake, A.; Matsumoto, T.; Zhao, J.; Kashiwagi, T.; Tanaka, T.; Yoshikawa, K.; Matsumoto, N.; et al. Inhibitory effect of galloyl moiety in epigallocatechin gallate for gelation of gelatin. *J. Osaka Dent. Univ.* **2020**, *54*, 251–255. [CrossRef]
54. Wang, W.; Le, T.; Wang, W.-W.; Yin, J.-F.; Jiang, H.-Y. The Effects of Structure and Oxidative Polymerization on Antioxidant Activity of Catechins and Polymers. *Foods* **2023**, *12*, 4207. [CrossRef] [PubMed]
55. Kchaou, H.; Benbettaieb, N.; Jridi, M.; Nasri, M.; Debeaufort, F. Influence of Maillard reaction and temperature on functional, structure and bioactive properties of fish gelatin films. *Food Hydrocoll.* **2019**, *97*, 105196. [CrossRef]
56. Watanabe, R.; Hayashi, R.; Kimura, Y.; Tanaka, Y.; Kageyama, T.; Hara, S.; Tabata, Y.; Nishida, K. A Novel Gelatin Hydrogel Carrier Sheet for Corneal Endothelial Transplantation. *Tissue Eng. Part A* **2011**, *17*, 2213–2219. [CrossRef]
57. Sionkowska, A.; Skopinska-Wisniewska, J.; Gawron, M.; Kozlowska, J.; Planecka, A. Chemical and thermal cross-linking of collagen and elastin hydrolysates. *Int. J. Biol. Macromol.* **2010**, *47*, 570–577. [CrossRef]
58. Drexler, J.W.; Powell, H.M. Dehydrothermal Crosslinking of Electrospun Collagen. *Tissue Eng. Part C Methods* **2010**, *17*, 9–17. [CrossRef] [PubMed]
59. Hu, W.; Wang, Z.; Xiao, Y.; Zhang, S.; Wang, J. Advances in crosslinking strategies of biomedical hydrogels. *Biomater. Sci.* **2019**, *7*, 843–855. [CrossRef]
60. Arbez, B.; Manero, F.; Mabilleau, G.; Libouban, H.; Chappard, D. Human macrophages and osteoclasts resorb β-tricalcium phosphate in vitro but not mouse macrophages. *Micron* **2019**, *125*, 102730. [CrossRef]

61. Álvarez, E.; Leiro, J.; Orallo, F. Effect of (−)-epigallocatechin-3-gallate on respiratory burst of rat macrophages. *Int. Immunopharmacol.* **2002**, *2*, 849–855. [CrossRef]
62. Hossen, I.; Kaiqi, Z.; Hua, W.; Junsong, X.; Mingquan, H.; Yanping, C. Epigallocatechin gallate (EGCG) inhibits lipopolysaccharide-induced inflammation in RAW 264.7 macrophage cells via modulating nuclear factor kappa-light-chain enhancer of activated B cells (NF-κB) signaling pathway. *Food Sci. Nutr.* **2023**, *11*, 4634–4650. [CrossRef] [PubMed]

Disclaimer/Publisher's Note: The statements, opinions and data contained in all publications are solely those of the individual author(s) and contributor(s) and not of MDPI and/or the editor(s). MDPI and/or the editor(s) disclaim responsibility for any injury to people or property resulting from any ideas, methods, instructions or products referred to in the content.

https://doi.org/10.3390/jfb17010018

Journal of
Functional Biomaterials

MDPI

Article

Electrospinning of Bovine Split Hide Collagen and Collagen/Glycosaminoglycan for a Study of Stem Cell Adhesion and Proliferation on the Mats: Influence of Composition and Structural Morphology

Todorka G. Vladkova [1,*], Dilyana N. Gospodinova [2], Peter D. Dineff [2], Milena Keremidarska-Markova [3,4], Kamelia Hristova-Panusheva [3] and Natalia Krasteva [3]

1 Department of Polymer Engineering, Faculty of Chemical Technologies, University of Chemical Technology and Metallurgy, 1756 Sofia, Bulgaria
2 Department of Electrical Apparatus, Electro-Technical Faculty, Technical University of Sofia, 1797 Sofia, Bulgaria; dilianang@tu-sofia.bg (D.N.G.); dineff_pd@abv.bg (P.D.D.)
3 Institute of Biophysics and Biomedical Engineering, Bulgarian Academy of Sciences, BAS, 1113 Sofia, Bulgaria; m_keremidarska@uni-sofia.bg (M.K.-M.); kamelia.t.hristova@gmail.com (K.H.-P.); nataly@bio21.bas.bg (N.K.)
4 Department of Animal and Human Physiology, Faculty of Biology, Sofia University 'St. Kliment Ohridski', 1164 Sofia, Bulgaria
* Correspondence: tgv@uctm.edu; Tel.: +359-887839374

Abstract: Electrospun collagen-based fibrous mats are of increasing interest for cell culture, regenerative medicine, and tissue engineering. The focus of this investigation is on the assessment of the electrospinning ability of bovine split hide collagen (BSHC), the effect of glycosaminoglycan (GAG) incorporation on the mats' structural morphology, and the impact on the adhesion and proliferation of human adipose-derived mesenchymal stem cells (hAD-MSCs). Electrospun mats were prepared using benign and fluoroalcohol solutions of BSHC and BSHC/GAGs under varied operation conditions. SEM observations and analysis were employed to characterize the structural morphology of the mats. Several parameters were used to evaluate the hAD-MSC behavior: cytotoxicity, cell morphology, cell number and spreading area, cytoskeleton, focal adhesion contacts, and cell proliferation. Electrospinning using benign solvents was impossible. However, fiber mats were successfully prepared from hexafluoropropanol (HFP) solutions. Different structural morphologies and fiber diameters of the electrospun mats were observed depending on the composition and concentration of the electrospinning solutions. Both BSHC and BSHC/GAG mats supported the in vitro adhesion, growth, and differentiation of hAD-MSCs, with some variations based on their composition and structural morphology. The absence of cytotoxicity and the good hAD-MSC adhesiveness make them promising substrates for cell adhesion, proliferation, and further stem cell differentiation.

Keywords: bovine split hide collagen; glycosaminoglycan; electrospun fiber mats; benign solvents; human adipose-derived mesenchymal stem cells

1. Introduction

The rapid evolution of electrospinning technology has enabled the preparation of nanofiber mats with three-dimensional porous structures, which closely imitate the microstructure of the natural extracellular matrix (ECM). These materials have attracted increased attention as scaffolds for tissue engineering, drug delivery, wound dressing, periodontal regeneration, vascular reconstruction, etc. [1,2]

To date, a wide range of natural polymers (e.g., chitosan, glycosaminoglycans, and collagen) and synthetic polymers (e.g., polyglycolic acid (PGA), polylactic acid (PLA), and poly (lactic-co-glycolic acid) (PLGA)), as well as composite materials, have been successfully electrospun [3–8]. A critical factor in their effectiveness is their ability to closely mimic the topography and function of the natural extracellular matrix (ECM). Key parameters, such as the fiber diameter, diameter distribution, and alignment, significantly influence the properties of electrospun scaffolds [3]. Collagen, a major structural component of the ECM, is particularly advantageous due to its excellent biocompatibility, biodegradability, low immunogenicity, and ability to support cell and tissue growth. Electrospun collagen matrices are highly tunable, making them ideal for tailored tissue engineering [1–5]. Consequently, collagen has become a major component in the development of in vitro cell culture substrates, tissue engineering scaffolds, and drug delivery platforms. In these applications, the collagen used is derived from animal (bovine, porcine, equine, and chicken) and marine sources (jellyfish, sea cucumber, squids, starfish, sponges, crustaceans, and mussels), each offering specific properties. For instance, fish collagen demonstrates inherent antimicrobial activity, broadening its potential applications [6].

Compared to traditional collagen sponges and other porous materials, electrospun collagen and collagen composite matrices provide superior biomimicry, closely replicating the structural and mechanical properties of the native ECM [4,7–9]. The structural and functional properties of the electrospun collagen matrices depend on the electrospinning parameters as well as the collagen origin and type [10]. By optimizing the processing conditions, one can precisely control the fiber morphology to match specific tissue requirements [11].

Natural collagen fibrils serve as the primary structural element in load-bearing tissues such as tendons, ligaments, skin, cornea, and bone. A variety of electrospun collagen scaffolds can be engineered with a fiber diameter and porosity matching those of the native ECM, enhancing their regenerative potential [1,2,4]. Collagen is typically electrospun using fluoroalcohol solutions, but the degradation observed during electrospinning has led some researchers to question whether collagen electrospinning is merely an expensive way of producing gelatin [12]. Furthermore, concerns about potential toxicity from residual organic solvents in the electrospinning processes have driven interest in the development of alternative approaches using benign and aqueous-based systems. The first successful electrospinning of collagen nanofibers using phosphate-buffered solutions (a benign solvent system) was reported in 2009–2010 [13,14]. However, subsequent reports about electrospun collagen mats using similar mild solvents remain notably absent in the literature. The recent work of Visser et al. (2023) [15] has provided new insights through the enzymatic and spectroscopic analysis of electrospun rat tail collagen. Their study demonstrated that the triple-helical structure of collagen was disrupted when it was electrospun using acetic acid/ethanol mixtures at various temperatures, revealing that structural damage occurs independently of fluorinated solvents. This finding challenges the previous assumption that fluorinated solvents were solely responsible for collagen denaturation during electrospinning. The authors emphasize that the impact of electrospinning parameters on collagen's biochemical properties must be carefully considered when fabricating ECM-mimicking fibrous constructs. While numerous electrospun porous mats fabricated from different collagen types and sources have been reported in the literature [1,2,4,6], no report was found about mats electrospun specifically from bovine split hide collagen (BSHC).

The development of composite collagen scaffolds that better mimic the natural extracellular matrix (ECM) has been carried out [7], including those biofunctionalized with molecules such as recombinant ICOS-Fc to enhance bone remodeling [16]. Among these, collagen/glycosaminoglycan mixtures are particularly promising for fabricating biomimetic scaffolds that provide not only mechanical support but also deliver biochemical signals to facilitate cell adhesion, proliferation, and differentiation [17,18]. However, challenges remain in ensuring these scaffolds accurately replicate the tissue-specific GAG expression profiles found in native human tissues, a critical consideration for their eventual commercialization [19]. The physical–chemical, mechanical, and biological properties of the collagen/GAG scaffolds can be precisely controlled by varying their composition (collagen origin and type; GAG type and amount) and the manufacturing conditions [20–22].

Collagen/GAG scaffolds have been extensively investigated as ECM analogues for regenerating a variety of tissues, including skin [23–25], peripheral nerves [26], muscles [27], cartilage [28], tendons [17,29], bone tissue [30,31], etc. Studies have also examined the behavior of different eukaryotic cells on collagen and collagen/GAG scaffolds, including the maturation of osteoblast cells [32,33]; osteoblast differentiation and matrix mineralization [34]; tendon cell recruitment, alignment, and metabolic activity [29]; human trabecular meshwork cell behavior [35]; different cell adhesion types [36], etc.

Stem cells possess two defining characteristics that distinguish them from other cell types: the ability to self-renew through prolonged proliferation and the capacity to differentiate into specialized cell lineages (e.g., myocytes, hematopoietic cells, and neurons). Among these, mesenchymal stem cells (MSCs) have emerged as particularly valuable for regenerative medicine due to their multipotent differentiation potential, immunomodulatory properties, and trophic factor secretion [37]. MSCs demonstrate remarkable plasticity in their biological responses, with their phenotype being highly sensitive to micro-environmental cues. This adaptability, combined with their multiline differentiation capacity (osteogenic, chondrogenic, adipogenic, etc.), makes them exceptionally versatile for therapeutic applications [38]. Adipose-derived mesenchymal stromal cells attract significant attention in both pre-clinical and clinical research due to their potential in treating a wide array of conditions, ranging from various autoimmune diseases such as inflammatory bowel disease, systemic sclerosis, type 1 diabetes, and rheumatoid arthritis, to neurological disorders including amyotrophic lateral sclerosis, ischemic stroke, and pinal cord injuries [39]. Notably, recent research has shown that MSCs exhibit significant antimicrobial activity through both indirect and direct mechanisms, partially mediated by the secretion of antimicrobial peptides and proteins [40]. These multifunctional properties position MSCs as promising candidates for addressing complex clinical challenges in regenerative medicine, antimicrobial therapy, and inflammation management.

Adipose-derived MSCs (AD-MSCs) were chosen for this study due to their high yield (100–500× more cells/gram than bone marrow), minimally invasive harvesting, and superior proliferation and differentiation potential [41,42]. Their strong adhesion to collagen-based scaffolds and robust immunomodulatory properties make them ideal for evaluating electrospun matrices. Compared to other MSC sources, AD-MSCs demonstrate better angiogenic potential, lower donor variability, and excellent cryopreservation stability, ensuring experimental reproducibility [43,44]. These advantages make AD-MSCs optimal candidates for tissue engineering applications requiring ECM interaction and regenerative capacity. Although electrospinning enables the fabrication of fibrous mats with properties closely resembling the natural ECM, which is ideal for seeding and growing different cell types, including MSCs [4], no previous studies have reported on the electrospinning of bovine split hide collagen (BSHC) or BSHC/glycosaminoglycan (BSHC/GAGs) composites.

This study aims to fabricate and characterize electrospun BSHC and BSHC/GAGs fibrous mats from benign and conventional solvents and to evaluate hAD-MSCs' behavior on them. The structural morphology of the electrospun mats was optimized by systematically varying the concentration and composition of the BSHC and BSHC/GAGs solutions, applied voltage, flow rate, needle diameter, and tip-to-collector distance. Two distinct approaches, direct contact and indirect exposure to mats, were used to evaluate hAD-MSCs' responses.

2. Experiment

2.1. Materials

Bovine split hide type I collagen, BSHC (Collapro Bovine Premium, Hulsh Protein Technologies, Oost Gelre, The Netherlands) was used in this investigation; Glutar aldehyde (GA, 50% water solution, Alfa Aesar, Karlsrue, Germany) was used as a cross-linking agent. Modifying agents were the glycosaminoglycans (GAGs) chondroitin sulfate (CS (Alfa Aesar, Heysham, UK) and hyaluronic acid, HA (Alfa Aesar, Heysham, UK), both as 10% water solutions.

The solvents 1,1,1,3,3,3-hexafluoro-2-propanol (HFP, 99.0%, Alfa Aesar, Heysham, UK) and ethyl alcohol (96.0%, p.a.), as well as a standard phosphate buffer (PBS, pH = 7.0 at 20 °C, CPA Chem, Bogomilovo, Bulgaria), were used in their delivered form.

All salts used for preparation of a concentrated phosphate buffer (PBS×20): NaCl, KCl, KH_2PO_4, and $Na_2HPO_4 \cdot 12H_2O$ were p.a. Concentrated phosphate buffer, PBS×20, was prepared as follows: 80 g of NaCl, 2 g of KCl, 2 g of KH_2PO_4, and 28.8 g of $Na_2HPO_4 \cdot 12H_2O$ was dissolved in 500 mL of distilled water, and pH was fixed at seven via addition of HCl.

2.2. Electrospinning Solutions

Based on reported advantages of the collagen electrospinning from benign solvents [13,14] and established methods using acetic acid/ethanol solutions [15], we prepared electrospinning solutions of BSHC in standard phosphate buffer (PBS), concentrated phosphate buffer (PBS20x), and mixed solvents: PBS/ethanol, PBS20x/ethanol, and acetic acid/ethanol. HFP electrospinning solutions of BSHC and BSHC/GAG were also prepared for this investigation.

2.2.1. BSHC Solutions

- PBS and PBS20x solutions of BSHC

PBS and PBS20x solutions of BSHC with a step-varied concentration of 5–20 wt.% (step of 1 wt.%) were prepared via 3 h of intense stirring of BSHC preliminarily swelled in the corresponding solvent for 24 h.

- PBS/ethanol and PBS20x/ethanol solutions of BSHC

Solutions of BSHC with a step-varied concentration of 5–20 wt.% (step of 1 wt.%) in both PBS/ethanol (volume ratios, *v/v* 1:4; 1:2; and 1:1) and PBS×20/ethanol (volume ratio, *v/v* 1:1) were prepared via 3 h of intense stirring of BSHC preliminarily swelled in the corresponding mixed solvent for 24 h.

- Acetic acid/ethanol solutions of BSHC

Acetic acid/ethanol (*v*/*v*, 1:1) solutions of BSHC with concentrations of 6 wt.%, 8 wt.%, and 10 wt.% were prepared via 3 h of intense stirring of BSHC, preliminarily swelled in the mixture solvent for 24 h.

- HFP solutions BSHC

HFP solutions with step-varied concentration of 3–12 wt.% (step of 1 wt.%) were prepared by dissolving BSHC preliminarily swelled in HFP for 24 h.

2.2.2. BSHC/GAGs Solutions

HFP solutions of BSHC/GAGs were prepared by premixing the corresponding GAG with BSHC, followed by adding HFP, 24 h of swelling at room conditions, and stirring for 3 h. The GAGs used in this investigation were hyaluronic acid (HA), chondroitin sulfate (CS), and combinations of them (HA/CS). Solutions with different amounts of GAGs were prepared as follows:

- Hyaluronic acid:
  BSHC/HA5 (5 wt.% HA), BSHC/HA10 (10 wt.%), BSHC/HA15 (15 wt.%).
- Chondroitin sulfate:
  BSHC/CS5 (5 wt.%), BSHC/CS10 (10 wt.%).
- Both hyaluronic acid and chondroitin sulfate:
  BSHC/HA5/CS5 (5 wt.% HA/ 5 wt.% CS);
  BSHC/HA10/CS5 (10 wt.% HA/ 5 wt. CS%);
  BSHC/HA10/CS10 (10 wt.% HA/10 wt.% CS).

2.3. Electrospinning

Numerous parameters are known to influence the electrospinning process, including solution properties (solvent type and solution concentration); process variables (flow rate, applied voltage, tip-to-collector distance, and needle diameter); as well as environmental characteristics, like temperature, humidity, etc. [1,2,4,45,46]. For this study, fibrous mats were electrospun under ambient laboratory conditions using the following operation conditions: electrospinning voltage, 10–30 kV; needle diameter, 0.4–0.9 mm; flow rate, 1–100 μLmin^{-1}; and tip-to-collector distance, 10–30 cm for every solution. All experiments were conducted using a custom-built electrospinning apparatus (Figure 1), with parameters systematically varied for each solution formulation.

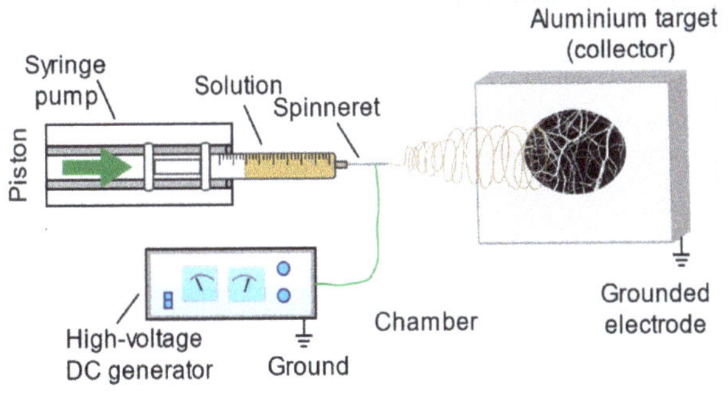

Figure 1. Schematic presentation of the electrospinning equipment employed for this investigation.

All electrospun samples were cross-linked via 24 h exposure in vapors of glutar aldehyde, which is one of the most commonly used cross-linking agents for collagen, among other others [47].

2.4. Scanning Electron Microscopy

The morphology of the electrospun BSHC and BSHC/GAG mats was observed using SEM (SEM/FIB Lyra I XMU, Tescan, Brno, Czech Republic). ImageJ, version 1.54g, NIH, USA, software was employed to accurately measure fiber diameter from the SEM images, analyzing 300 randomly selected areas.

2.5. Cell Experiments

- Cell culture conditions

Human adipose-derived mesenchymal stem cells (hAD-MSCs) were obtained from Lonza (Spain) at passage 2 and cultured in Dulbecco's Modified Eagle Medium (DMEM) supplemented with 10% fetal bovine serum (FBS), 1% antibiotic/antimycotic solution (Sigma, Baden-Württemberg, Germany), and 2 mM.

L-glutamine (Roche Diagnostics, Mannheim, Germany) and 1 mM sodium pyruvate (Gibco BRL, Waltham, MA, USA) were used. The cells were incubated at 37 °C in a humidified atmosphere with 5% CO_2. Media changes were performed twice a week. For experimental procedures, cells were detached using Trypsin-EDTA (Lonza, Verviers, Belgium), counted using a Neubauer haemocytometer, and seeded in 24-well plates. Depending on the experimental conditions, the cells were subsequently assayed according to the protocols outlined below.

- Indirect Cytotoxicity Evaluation

Cells were seeded onto plain cover glasses (CG) at a concentration of 2×10^4 cells/mL and incubated in DMEM supplemented with 10% FBS for 24 h. After this incubation, the culture medium was replaced with medium pre-incubated for 4 days with different mats (NFs) to assess whether the mats release any toxic substances that could affect cell viability. After 2 and 24 h of incubation, the medium was removed, and the cells were stained with 0.001% fluorescein diacetate (FDA) (Sigma, Munich, Germany), dissolved in acetone, for 2 min. The cells were then washed several times with PBS, and representative images were captured using a fluorescent microscope (Zeiss, Axiovert 25, Oberkochen, Germany) equipped with a digital camera.

- Overall cell morphology, cell number, and spreading area

Human AD-MSCs were seeded onto studied mats at concentration of 2×10^4 cells/mL and incubated in serum-free DMEM for 2 h. At the end of the incubation, the medium was removed, and the cells were processed for FDA staining, as described above. Representative images were captured using a fluorescent microscope. Plain (uncoated) glass coverslips (CG) and fibronectin (FN)-coated CG (20 $\mu g \cdot mL^{-1}$, Roche Diagnostics, Monza, Italy) served as negative and positive controls, respectively. Cell number and average cell spreading area were quantified using ImageJ software based on FDA-stained micrographs captured at 10x magnification. Data shown are representative of three separate experiments, and values are given as mean ± SD. Differences to control (FN-coated CG) with (p) < 0.05 and (p) < 0.01 were considered statistically insignificant and marked with one and two asterisks: * $p < 0.05$; ** $p < 0.01$.

- Immunostaining of actin cytoskeleton and focal adhesion contacts

Human AD-MSCs were seeded onto nanofibers, as previously described. After 2 h of incubation in serum-free medium, the cells were fixed with 3% paraformaldehyde, permeabilized with 0.5% Triton-X100 (v/v), and stained to visualize focal adhesion complexes and the actin cytoskeleton. Specifically, cells were stained with anti-vinculin antibody to visualize focal adhesion complexes or FITC-conjugated phalloidin to stain the actin cytoskeleton. After three washes with PBS, vinculin samples were incubated with a secondary antibody, followed by additional washing to remove unbound dye. Then, the cells were mounted and examined using a fluorescent microscope.

- Cell proliferation assay

Cells with concentration of 2×10^4 cells/mL were seeded in 24-well plates containing collagen mats in DMEM supplemented with 10% FBS. At various time points, cells were transferred to a new plate, washed with PBS, and assessed for proliferation using the CCK-8 assay, according to the manufacturer's instructions. The newly synthesized yellow formazan dye was quantified spectrophotometrically using a standard microplate reader (BioRad) at a wavelength of 450 nm. Values are expressed as means ± SEM (n = 3).

3. Results and Discussion

Electrospinning trials were performed for BSHC and BHSC/GAG using both benign solvents and HFP under a wide range of solution concentrations and electrospinning operation conditions. The structural morphology of the resulting electrospun BSHC and BSHC/GAGs mats was analyzed using SEM images, while the behavior of hAD-MSCs was evaluated using direct and indirect methods.

3.1. *Electrospinning of BSHC from Benign Solvents*

The electrospinning of collagen from fluoroalcohol solutions presents several challenges, including toxicity, high cost, and significant destruction of the triple-helical structure of the collagen [4]. Given previous reports on the electrospinning of collagen from benign solvents, such as phosphate buffers and their ethanol mixtures [13,14], as well as the successful use of acetic acid/ethanol for rat tail collagen [15], we initiated our study by exploring the electrospinning of BSHC from similar solvents. A wide range of operation conditions were tested for each BSHC solution, including voltages of 10 kV to 30 kV, flow rate from 1 μL/min to 100 μL/min, tip-to-collector distance of 10 cm to 30 cm, and needle diameters ranging from 0.4 mm to 0.9 mm.

Electrospinning of both PBS and PBS×20 solutions of BSHC across a broad concentration range (4–20 wt.% (step of 1 wt.%) was unsuccessful. Phase separation occurred in the mixed solvents, including PBS/ethanol (1:4, 1:2, 1:1, *v/v*) and PBSx20/ethanol (1:1, *v/v*), with ethanol forming the upper phase. As a result, electrospinning was not feasible under any tested conditions.

Similarly, attempts to electrospinning BSHC solutions prepared in acetic acid/ethanol (1:1, *v/v*) at concentrations of 4, 6, 8, 10, and 12 wt.% were unsuccessful under all operation conditions. Notably, the formation of Taylor cones was not observed, further indicating the unsuitability of these solvent systems for the electrospinning of BSHC. The extensive number of trials conducted across a wide range of solution concentrations and electrospinning conditions demonstrated that the fabrication of electrospun BSHC mats from benign solvents (phosphate buffers and their ethanol mixtures or acetic acid/ethanol) was not feasible. Therefore, subsequent experiments focused on the electrospinning of BSHC and BSHC/GAG mats using HFP solutions.

3.2. *Electrospun BSHC Mats*

The electrospinning of BSHC from HFP solutions (3–12 wt.%) exhibited spray formation at 3 wt.% and stopped the needle at 10 wt.% and 12 wt.% solutions under varied operation conditions, including voltage ranging from 10 kV to 30 kV, needle diameter of 0.6–0.9 mm, flow rates of 1–100 μL/min, and tip-to-collector distance of 10–30 cm.

Fibrous mats with different fiber diameters were successfully electrospun from 6–9 wt.% BSHC solutions in HFP at an applied voltage of 25 kV, flow rate of 50–60 μL/min, needle diameters of 0.7 and 0.9 mm, and tip-to-collector distance of 15 cm.

Figure 2 presents the morphology (a) and fiber diameter distribution (b) of a fibrous mat electrospun from 6 wt.% BSHC solution in HFP, cross-linked via 24 h of exposure to glutaraldehyde vapors.

It is evident that the electrospun BSHC mat consists of defect-free fibers (Figure 2a) with micron and submicron diameters, and a relatively narrow fiber diameter distribution (Figure 2b). The mean diameter is 1.165 µm, with a standard deviation of 0.454 µm, a minimum value of 0.353 µm, and a maximum value of 3.003 µm.

Submicron-diameter fibers are observed in the mats electrospun from 7 wt.% BSHC solutions (Figure 3b) and higher concentrations in HFP. However, these fibers exhibit defects in the form of balls, as shown in Figure 3a.

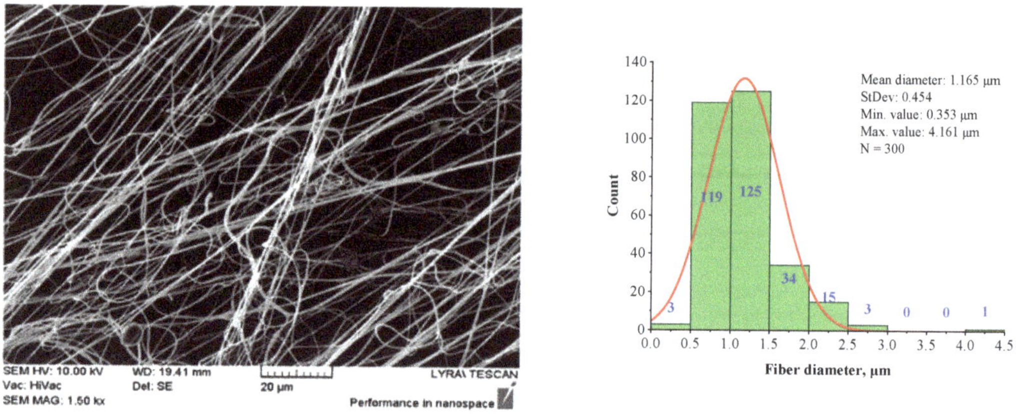

(a) (b)

Figure 2. SEM image (**a**) and histogram (**b**) of BSHC fibrous mat electrospun from 6 wt.% HFP solution after cross-linking via 24 h of exposure in glutaraldehyde vapors.

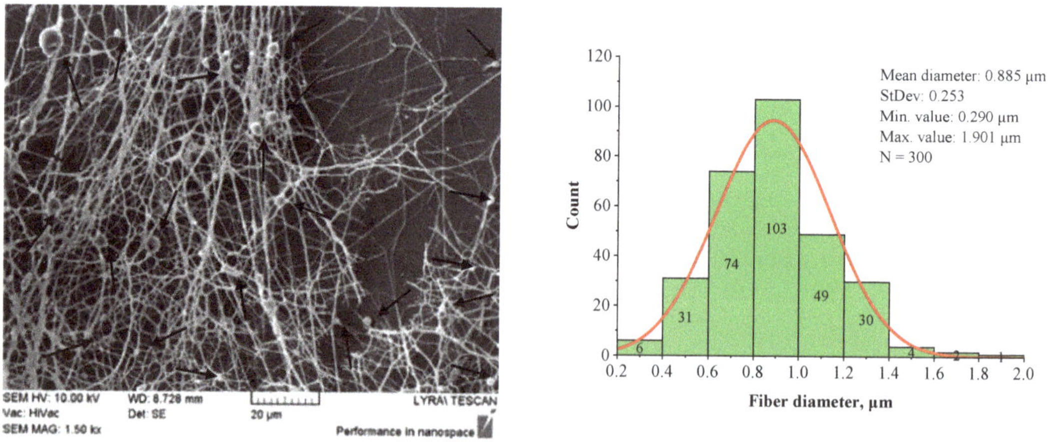

(a) (b)

Figure 3. SEM image (**a**) and histogram (**b**) of BSHC mat electrospun from 9 wt.% HFP solution.

The SEM image (Figure 3a) reveals defects in the BSHC mat, whereas the corresponding histogram (Figure 3b) shows the formation of fibers with a mean diameter of 0.885 µm, a standard deviation of 0.253 µm, a minimum value of 0.290 µm, and a maximum value of 1.901 µm. Comparing this to the BSHC mat, electrospun from 6 wt.% HFP solution (Figure 2), it is evident that the electrospinning of BSHC from more concentrated HFP solutions (7 wt.%

and above) results in finer fibers, but with noticeable defects. These trials demonstrate that electrospinning of BSHC from HFP solutions enables the fabrication of fibrous BSHC mats with fibers in the micron and submicron scale. However, to obtain defect-free fibers, the optimal concentration for electrospinning is between 5 wt.% and 6 wt.%.

3.3. Electrospun BSHC/GAG Mats

Glycosaminoglycans (GAGs) are a main component of the natural ECM of different tissues, playing an active role in regulating cell proliferation, differentiation, and tissue repair [48]. To create biomimetic scaffolds that better replicate the natural ECM, GAGs are often incorporated into porous collagen mats [22,34]. Collagen is most commonly combined with hyaluronic acid (HA) and chondroitin sulfate (CS), both of which are crucial for maintaining the ECM microenvironment and facilitating signaling molecules in living tissues [19,49,50]. Since the cells' activity, including viability, proliferation, and distribution within the scaffold, depends on the composition and concentration of the GAGs [33,35], electrospun BSHC/GAG mats were fabricated for this study with varying concentrations of HA and CS, and combinations of both HA and CS.

The electrospinning conditions used to fabricate the BSHC/GAGs mats were optimized to produce defect-free fibers: 6 wt.% HFP solutions, an electrospinning voltage of 25 kV, a flow rate of 60 µL/min, a needle diameter of 0.7 mm, and a tip-to-collector distance of 15 cm. All samples were cross-linked via exposure to glutaraldehyde vapors for 24 h. The concentrations of HA, CS, and HA/CS were systematically varied, and the resulting mat samples were labeled as described in the Experiment section.

It was observed that when the total GAG concentration in the HFP solutions reached 15 wt.% or higher (samples BSHC/HA20, BSHC/HA10/CS5, and BSHC/HA10/CS10), electrospinning became impossible due to clogging of the needle. Fibrous mats were successfully electrospun from 6 wt.% HFP solutions containing lower GAG concentrations, including BSHC/HA5, BSHC/HA10, BSHC/CS5, BSHC/CS10, and BSHC/HA5/CS5.

The structural morphology (a) and fiber diameter distribution (b) of BSHC/HA5 fibrous mat, electrospun from 6 wt.% HFP solution containing 5 wt.% HA, are presented in Figure 4.

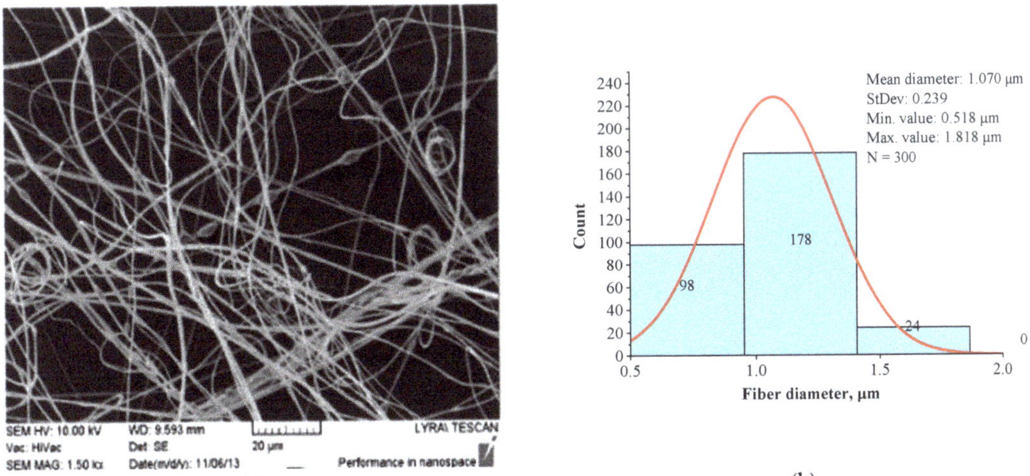

(a) (b)

Figure 4. SEM image (**a**) and histogram (**b**) of BSHC/HA5 mat electrospun from 6 wt.% HFP solution of BSHC in presence of 5 wt.% HA.

A comparison of Figures 3 and 4 demonstrates that the presence of HA results in slightly finer, defect-free fibers with a relatively narrow fiber diameter distribution. The histogram in Figure 4b indicates that the average diameter of BSHC/HA5 fiber is 1.070 μm, with a standard deviation of 0.518 μm), whereas BSHC fibers have a slightly larger average diameter of 1.165 μm, with a standard deviation of 0.459 μm, as shown in Figure 3b.

The presence of CS in the BSHC/CS5 solution significantly alters the structural morphology of the electrospun mat, as shown in Figure 5.

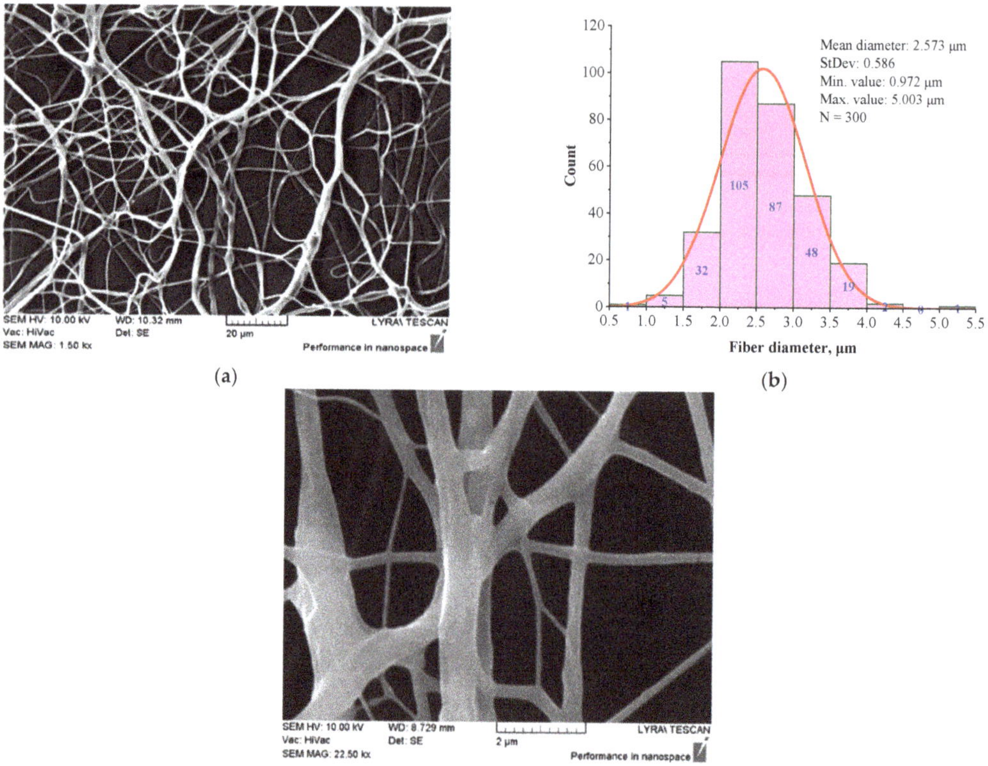

(**a**) (**b**)

(**c**)

Figure 5. Morphology (**a**), (**c**) and histogram (**b**) of BSHC/CS5 mat, electrospun from 6 wt.% HFP solution of BSHC in presence of 5 wt.% CS, cross-linked in glutaraldehyde vapors for 24 h.

The fibers appear highly branched, as seen in Figure 5a, with greater clarity at high magnification, as shown in Figure 5c. The histogram in Figure 5b shows that their mean diameter is 2.573 μm (standard deviation of 0.972 μm), which is significantly larger than that of BSHC mat fibers (mean diameter of 1.165 μm, standard deviation of 0.459 μm, Figure 3b) and BSHC/HA5 mat fibers (mean diameter of 1.070 μm, standard deviation of 0.239 μm, Figure 4b). This indicates that the presence of CS in the BSHC electrospinning solution results in the formation of thicker fibers, whereas the presence of HA leads to the formation of finer fibers. The reason for the altered structural morphology of the BSHC/CS5 mat could be the possible hydrogen bonding of the CS to the collagen that has already been observed by other researchers [51]. The small amounts of the GAG used by us, both CS and HA (5 or 10 wt.% as 10% water solutions), make the detection of their hydrogen bonding to the collagen molecules difficult, and therefore, such experimental data are not presented here, the details can be found in the Supplementary Materials.

Figure 6 presents the structural morphology (a) and fiber diameter distribution (b) of the BSHC/HA5/CS5 mat electrospun from the 6 wt.% HFP solution of BSHC containing both 5 wt.% HA and 5 wt.% CS.

In contrast to the highly branched fibers of the BSHC/CS5 mat containing CS (Figure 5a,c), the BSHC/HA5/CS5 mat, which incorporates both HA and CS, primarily consists of linear fibers. Their mean diameter of 0.960 μm (standard deviation: 0.233 μm, Figure 6b) is lower than that of the defect-free linear fibers of both BSHC (2.573 μm, standard deviation: 0.233 μm, Figure 2b), and BSHC/HA mats (1.070 μm, standard deviation: 0.518 μm, Figure 4b). Notably, the fibers in the BSHC/HA5/CS5 mat containing both HA and CS exhibit a high degree of entanglement.

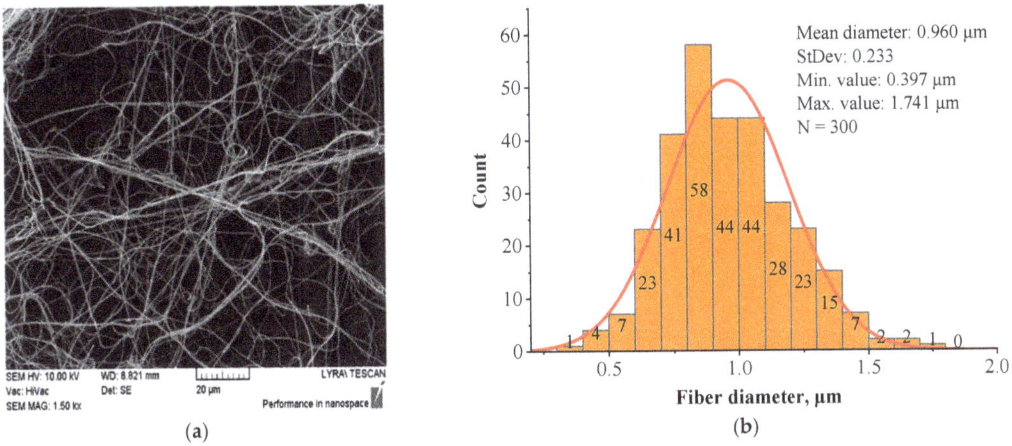

Figure 6. Morphology (**a**) and fiber diameter distribution (**b**) of the BSHC/HA5/CS5 mat electrospun from a 6 wt.% HFP solution of BSHC incorporating 5 wt.% HA and 5 wt.% CS cross-linked in glutaraldehyde vapors for 24 h.

It could be concluded that the structural morphology of the studied BSHC/GAGs fibrous mats, electrospun under identical operation conditions, is specific and dependent on their composition, as previously reported [48]. The fibers within these mats can be either linear or branched, exhibiting variations in mean diameter and diameter distribution, yet consistently falling within the micron and submicron scale. The presence of HA in BSHC/HA5 and BSHC/HA5/CS5 results in a lower mean diameter compared to the BSHC mat, which lacks HA. In contrast, the inclusion of CS in the BSHC/CS5 solution leads to the formation of mats with highly branched and relatively thick fibers. Notably, while the fibers in the BSHC/HA5/CS/5 mats are finer, they exhibit a greater degree of entanglement compared to other defect-free fibers.

3.4. Behavior of Human Adipose-Derived Mesenchymal Stem Cells (hAD-MSCs)

It is known that the behavior of human cells on porous collagen materials is influenced by their composition, structural morphology, and the collagen origin. GAGs are often incorporated to improve the biological activity of these materials. The type and concentration of GAGs significantly affect cell activity, including viability, proliferation, and spatial distribution within the scaffold [33,35,36,41]. In this study, we investigated electrospun fibrous mats based on BSHC and BSHC/GAGs with varying composition and structural morphology, which had not been previously studied. Table 1 demonstrates the selected electrospun mats for this investigation, with different compositions, structural morphologies, and fiber diameters.

Two distinct approaches, direct contact and indirect exposure, were employed to evaluate comprehensively the behavior of the hAD-MSCs [51]. The direct contact approach involves seeding cells directly onto the mats, providing insights into how their physical and bioactive properties influence cellular responses, including changes in cell attachment, membrane integrity, and morphology [52,53]. In contrast, the indirect exposure method assesses the potential release of toxic substances from the mats over time and allows for the evaluation of the impact of leached compounds on cell viability and morphology without direct physical interaction [54,55]. By combining these two approaches, a more comprehensive understanding can be gained regarding both the benefits and potential risks of the electrospun mats for biomedical applications.

Table 1. Structural morphology and composition of mats used to study hAD-MSCs' behavior: those not containing GAGs (BHSC), those containing 5 wt.% or 10 wt.% HA (BSHC/HA5 and BSHC/HA10, respectively), those containing 5 wt.% or 10 wt.% CS (BSHC/CS5 and BSHC/CS10, respectively), and those containing 5 wt.% HA and 5 wt.% CS simultaneously (BSHC/HA5/CS5).

Sample	Structure	Mean Fiber Diameter (μm)	Min Diameter (μm)	Max Diameter (μm)
BHSC	Linear	1.165 ± 0.454	0.353	3.003
BSHC/HA5	Linear	1.070 ± 0.518	0.518	1.818
BSHC/HA10	Linear	0.832 ± 0.270	0.129	1.707
BSHC/CS5	Branched	2.573 ± 0.586	0.972	5.003
BSHC/CS10	Branched	2.196 ±0.838	0.705	5.193
BSHC/HA5/CS5	High degree of entanglement	0.960 ± 0.233	0.397	1.741

3.4.1. Indirect Method

The indirect cytotoxicity of the studied BSHC and BSHC/GAGs fibrous mats was evaluated using fluorescein diacetate (FDA) staining to observe cell viability and overall cell health after 2 and 24 h exposures to DMEM pre-incubated with the mats [56]. The fluorescence intensity directly correlates with the number of viable cells. Morphology changes such as cell rounding, shrinkage, or detachment were considered as indicators of a toxic effect.

The indirect cytotoxicity of the studied mats toward the used test hAD-MSCs is presented in Figure 7.

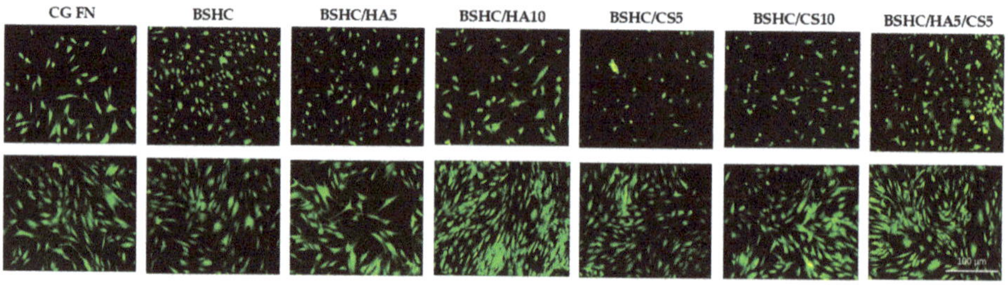

Figure 7. Indirect cytotoxicity toward hAD-MSCs cultured for 2 h (upper panel) and 24 h (lower panel) on positive control (CG-FN); BSHC mat without GAGs; and BSHC mats containing HA (BSHC/HA5 and BSHC/HA10), CS (BSHC/CS5 and BSHC/CS10), or both HA and CS (BSHC/HA5/CS5).

The hAD-MSCs seeded on CG-FN (positive control) exhibit strong green fluorescence due to the enzymatic conversion of FDA to fluorescein, indicating healthy and viable cells. Some differences in cell viability were observed after 2 h of exposure to DMEM on all tested BSHC and BSHC/GAG mats compared to the CG-FN control (Figure 7, upper panel). After 24 h of exposure to DMEM, FDA fluorescence intensity increased in both control and all electrospun mats, suggesting cell proliferation and growth, with confluence being reached (Figure 7, lower panel). These findings indicate that any potential toxic substances leaching from the studied mats are negligible after 2 h and are over longer exposure periods (24 h and beyond). This highlights the biocompatibility of the electrospun pure BSHC mats, as well as those containing different amounts of HA, CS, or their combinations: BSHC, BSHC/HA5, BSHC/CS5, BSHC/CS10, BSHC/HA5/CS5, and BSHC/HA10/CS5.

3.4.2. Direct Approach

In the direct approach, hAD-MSCs were seeded directly onto the electrospun mats with different compositions and structural morphologies, including BSHC, BSHC/HA5, BSHC/CS5, BSHC/CS10, BSHC/HA5/CS5, and BSHC/HA10/CS5. This allowed for a detailed assessment of hAD-MSCs' viability, morphology, attachment, spreading area, and cytoskeletal organization.

- Overall cell morphology, cell number, and spreading area

The overall morphology of viable hAD-MSC cells after 2 h of attachment to the electrospun mats, BSHC, BSHC/HA5, BSHC/HA10, BSHC/CS5, BSHC/CS10, and BSHC/HA5/CS5, is shown in Figure 8. For comparison, a glass coverslip (CG) and fibronectin-coated glass coverslip (CG-FN) were used as negative and positive controls, respectively. Fibronectin, a highly adhesive extracellular matrix (ECM) protein known to promote robust cell adhesion and spreading [57–59], was chosen as a positive control.

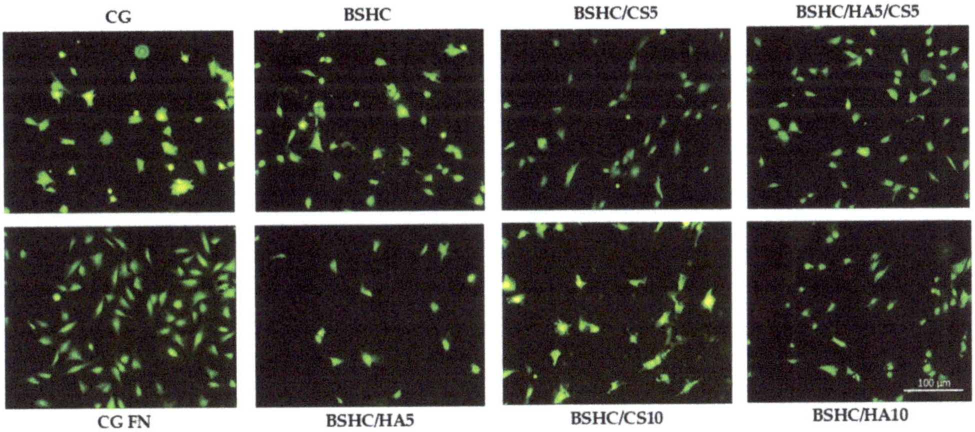

Figure 8. Fluorescence images of attached viable hAD-MSCs (stained with fluorescein diacetate, FDA) on negative control (CG), positive control (CG-FN), BSHC mat without GAGs, and BSHC mats containing HA (BSHC/HA5 and BSHC/HA10), CS (BSHC/CS5 and BSHC/CS10), or both HA and CS (BSHC/HA5/CS5).

Figure 8 illustrates that hAD-MSCs cultured on CG-FN (positive control) exhibit strong adhesion and spreading, displaying a characteristic fibroblast-like, spindle-shaped, and elongated morphology [60]. These cells' features, a centrally located nucleus, thin cytoplasmic extensions, prominent filopodia (finger-like projections), and lamella podia (sheet-like extensions) at their edges, are indicative of strong adhesion and active cytoskeletal organization [60]. In contrast, hAD-MSCs on CG (negative control) appear more compact, shrunken, less spread, and display irregular protrusions. As a result, the hAD-MSCs exhibit weak adhesion, failing to develop leading and trailing edges typically observed in the FN-coated positive control, CG-FN. This weaker attachment also results in the absence of directed projections such as lamella podia and filopodia, which are hallmarks of cells interacting with more adhesive surfaces [61].

Favorable spreading of hAD-MSCs on the studied electrospun BSHC and BSHC/GAGs mats was expected, as their major component, collagen fibers, is well known to promote cell adhesion and growth. Moreover, the presence of HA, CS, or their combination was expected to further enhance cell attachment and proliferation, contributing to a more supportive microenvironment [62–65].

Significant morphological differences were observed among hAD-MSCs seeded on the studied electrospun mats, BSHC mat, and BSHC/GAGs mats (BSHC/HA5, BHSC/HA10, BSHC/CS5, BSHC/CS10, and BSHC/HA5/CS5) and compared to the negative (CG) and positive (CG-FN) controls (Figure 8). Cells on the BSHC mat, which lacks GAGs, exhibited the most spreading. In contrast, on the BSHC/CS10 mat, some cells appeared shrunken, while others displayed a stellate-like morphology, indicating poor adhesion. Notably, the BSHC mat consists of micron-diameter linear fibers, whereas the second one, BSHC/CS10 containing CS, consists of micron-diameter, highly branched fibers. Our study confirms that the fiber alignment has a significant effect on cellular behavior, inducing cell adhesion and proliferation [66].

Although fewer in number, hAD-MSCs on the electrospun BSHC/HA5 and BSHC/HA10 mats, both composed of micron-diameter linear fibers, also exhibited an elongated morphology, suggesting that HA may provide some degree of contact guidance influencing their morphology, as presented in [67].

The results for the number of attached cells (a) and the spreading area (b) are shown in Figure 9.

As expected, the number of adhered hAD-MSCs on the negative control, CG (Figure 9a, light gray column), is significantly lower than on the positive control, CG-FN (Figure 9b, the dark gray column). The number of adhered hAD-MSCs on the studied electrospun mats (Figure 9a, the green, orange, blue and rose columns) is also lower than on the positive control, CG-FN (Figure 9a, dark gray column) suggesting that the adhesiveness of BSHC and BSHC/GAGs mats for hAD-MSCs may be lower that of fibronectin.

The number of adhered hAD-MSCs varies among the studied electrospun mats, both those without GAGs (Figure 9a) and those containing GAGs (Figure 9a, the green, orange, blue, and rose columns). These differences likely reflect the influence of the variations in composition and structural morphology, which is a known effect [25].

The number of attached hAD-MSCs on the BSHC/HA10 mat with a lower fiber diameter is higher than that on the BSHC/HA5 mat, for which the fiber diameter is higher. The number of attached hAD-MSCs on the BSHC/HA10 mat with a lower branched fiber diameter is also higher than that on the BSHC/HA5 mat, for which the branched fiber diameter is higher (see Table 1). This confirms the known influence of the fiber's diameter on cell adhesion [67].

Among the tested mats, BSHC/HA10 (dark blue column) exhibited the highest number of attached hAD-MSCs, followed by the BSHC mat without GAGs (the green column). Both are composed of non-branched fibers with micron-scale diameters. In contrast, BSHC/HA5 (light blue column) shows the lowest number of attached cells. The difference in cell attachment between BSHC, BSHCHA5, and BSHC/HA10 mats may be attributed to variations in the fiber thickness and the fiber diameter distribution, despite the three having fiber diameters within the micron-scale range (see Figures 2 and 4 and Table 1). This is one more confirmation of the known fiber diameter's influence on the cells' attachment to electrospun collagen matrices [67].

Figure 9b reveals an intriguing observation: despite having the lowest number of attached cells, hAD-MSCs on BSHC/HA5 (light blue column) exhibit the largest spreading area, approaching that of cells on fibronectin-coated glass (Figure 9, dark gray column). This suggests that while certain electrospun mats may not support optimal initial cell attachment, they may still promote significant cell spreading when adhesion is weaker. These findings highlight a potential compromise between cell attachment and spreading on fibrous surfaces, emphasizing that the composition and structural morphology of the mats play distinct roles in regulating hAD-MSCs' behavior [67].

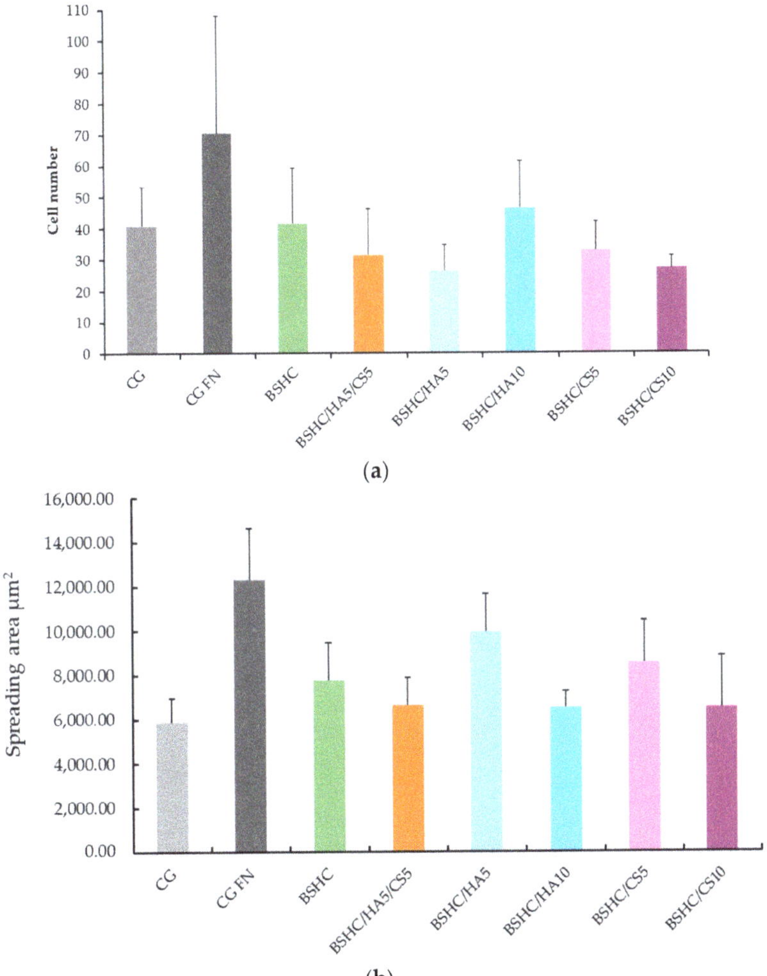

Figure 9. Number (**a**) and spreading area (**b**) of attached hAD-MSCs on negative control (CG), positive control (CG-FN); BSHC mat without GAGs; and BSHC mats containing HA (BSHC/HA5 and BSHC/HA10), CS (BSHC/CS5 and BSHC/CS10), or both HA and CS (BSHC/HA5/CS5).

- Organization of actin cytoskeleton

By visualizing the actin distribution, the effects of mat fibers on cell adhesion, spreading, and cytoskeletal dynamics can be assessed, providing valuable insight into how hAD-MSCs attach to both the mat fibers and the control surfaces. To further investigate cytoskeletal integrity, immunostaining was performed for the cytoskeletal protein F-actin. The arrangement of actin filaments plays a crucial role in determining cell morphology, influencing whether cells appear round, spread, or elongated. Disruptions of actin filament organization can serve as an indicator for cellular stress or cytotoxicity [67,68]. Specifically, fragmented or disorganized actin structures may suggest that mat fibers interfere with normal cell function or trigger apoptotic changes [65]. Additionally, actin filaments are essential for focal adhesion contact formation, where cells anchor to the extracellular matrix and surfaces [66,67].

Figure 10 presents the organization of the actin cytoskeleton in hAD-MSCs cultured on control surfaces and the studied electrospun BSHC and BSHC/GAGs mats.

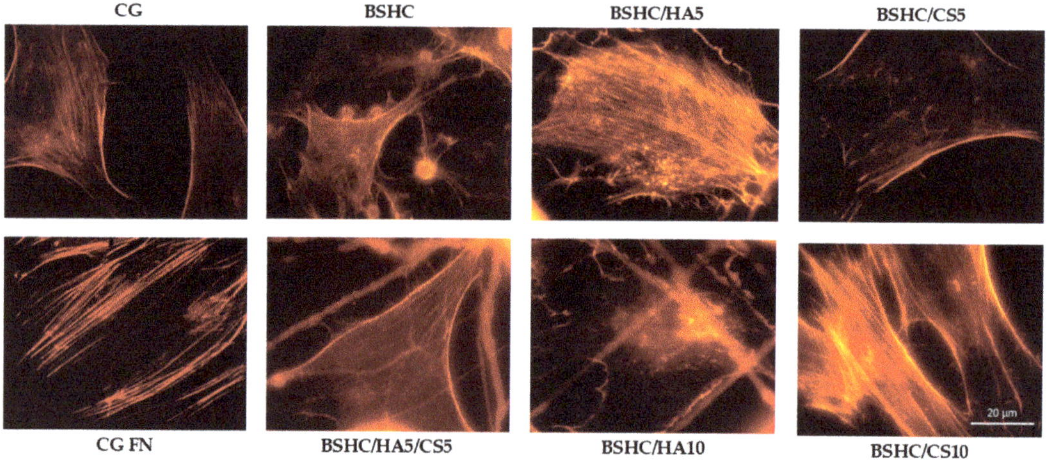

Figure 10. Organization of actin cytoskeleton in hAD-MSCs cultured in serum-free medium for 2 h on negative control (CG); positive control (CG-FN); BSHC mat without GAGs; and BSHC mats containing HA (BSHC/HA5 and BSHC/HA10), CS (BSHC/CS5 and BSHC/CS10), or both HA and CS (BSHC/HA5/CS5).

Actin is well-expressed across all studied electrospun mats, both those without GAG (BSHC) and those containing GAGs (HA, BSHC/HA5, and BSHC/HA10; CS, BSHC/CS5, and BSHC/CS10 or a combination of both HA and CS, BSHC/HA5/CS5). This indicates that these mats provide a suitable substrate for cell attachment and growth. Actin filaments are observed not only at the cell periphery but also throughout the cell body, supporting cell spreading and maintaining structural integrity. This suggests that the studied mats effectively promote cytoskeletal organization and facilitate cellular interactions, reinforcing their potential as biocompatible materials for tissue engineering applications.

- Vinculin Expression and Focal Adhesion Formation on Electrospun Fibrous Mats

Focal adhesion contacts are particularly important in the context of biomaterials, as they reflect how well cells interact with a surface. Well-formed focal adhesion contacts indicate strong cell-substrate adhesion and proper cytoskeletal organization, which are key markers of biocompatibility. Controversially, poorly formed or absent focal adhesion contacts may suggest weak adhesion, cellular stress, or cytotoxic effects, often resulting from suboptimal surface properties or material toxicity [66,69].

To assess these interactions, immunostaining for vinculin, a key structural and regulatory protein in focal adhesion complexes, was performed. Vinculin plays a pivotal role in the cell attachment to the extracellular matrix (ECM) or synthetic substrates by linking actin filaments to adhesion plaque proteins, such as talin and paxillin. This molecular bridging is essential for transmitting mechanical forces; regulating cell signaling pathways; and influencing cell behavior, spreading, migration, and differentiation [66,68]. Figure 11 depicts the distribution and morphology of focal adhesion contacts in hAD-MSCs cultured on negative and positive controls, CG and CG-FN, respectively, and on electrospun BSHC and BSHC/GAGs mats.

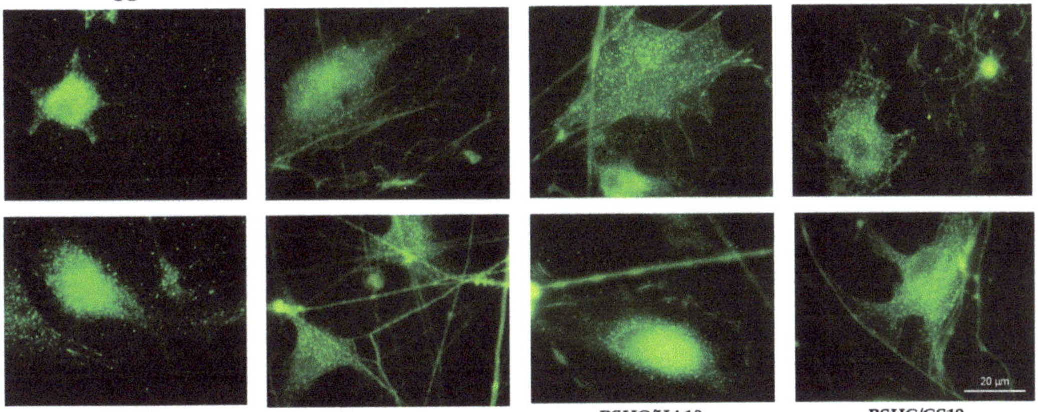

Figure 11. Organization of focal adhesion contacts in hAD-MSCs cultured for 2 h in a serum-free medium on negative control (CG); positive control (CG-FN); BSHC mat without GAGs; and BSHC mats containing HA (BSHC/HA5 and BSHC/HA10), CS (BSHC/CS5 and BSHC/CS10), or both HA and CS (BSHC/HA5/CS5).

Immunostaining analysis revealed that vinculin is well expressed in the hAD-MSCs adhered to all studied fibrous mats (BSHC and BSHC/GAG), confirming the active formation of focal adhesion complexes. Notably, well-defined focal adhesions were predominantly observed on BSHC, BSHC/HA5, and BSHC/HA10, composed of linear micron-diameter fibers. This suggests that electrospun mats with linear fiber architectures provide a more favorable substrate for focal adhesion assembly, probably due to their optimized structural and compositional properties. The presence of vinculin-rich focal adhesions on these substrates (BSHC, BSHC/HA5, and BSHC/HA10) underscores their ability to support stable cell adhesion and proper cytoskeletal organization, both critical for cell function and survival.

The observed variation in focal adhesion formation in the hAD-MSCs across the studied electrospun mats highlights the critical influence of scaffold composition and morphology on cell behavior. While BSHC, BSHC/HA5, and BSHC/HA10 demonstrated superior focal adhesion formation, other samples may require further optimization to improve their adhesive performance. These findings emphasize the importance of tailoring scaffold design to promote robust cell–material interaction.

- Proliferation of hAD-MSCs on Electrospun Fibrous Mats

The evaluation of cell proliferation is a critical aspect of assessing the biocompatibility and cytotoxicity of biomaterials, as it reveals their ability to support long-term cell viability and function. Proliferation studies not only indicate whether a material provides a favorable environment for cell growth but also detect the leaching of harmful substances that could impair cellular health over time [51]. In this study, we examined the proliferation of hAD-MSCs on electrospun fibrous mats to determine whether these scaffolds support the normal division and expansion or can induce inhibitory or cytotoxic effects that could compromise their suitability for biomedical applications [70].

The proliferation kinetics of hAD-MSCs cultured on the electrospun fibrous mats for up to 9 days are presented in Figure 12.

Figure 12 demonstrates that all electrospun mats, including the base BSHC mat and GAG-functionalized variants (BSHC/HA5, BSHC/HA10, BSHC/CS5, BSHC/CS10, and BSHC/HA5/CS5), significantly stimulate the growth of hAD-MSCs. The number of hAD-MSCs increases approximately 6-fold at the 9th day compared to the 1st day, indicating robust cell proliferation across all scaffolds. Notably, proliferation rates on BSHC and BSHC/GAGs mats are higher than those observed on the control glass, CG (Figure 12, colored vs. gray columns), suggesting these electrospun mats better replicate the native ECM by providing optimal structural morphology and biochemical cues to support cell attachment, spreading, and expansion.

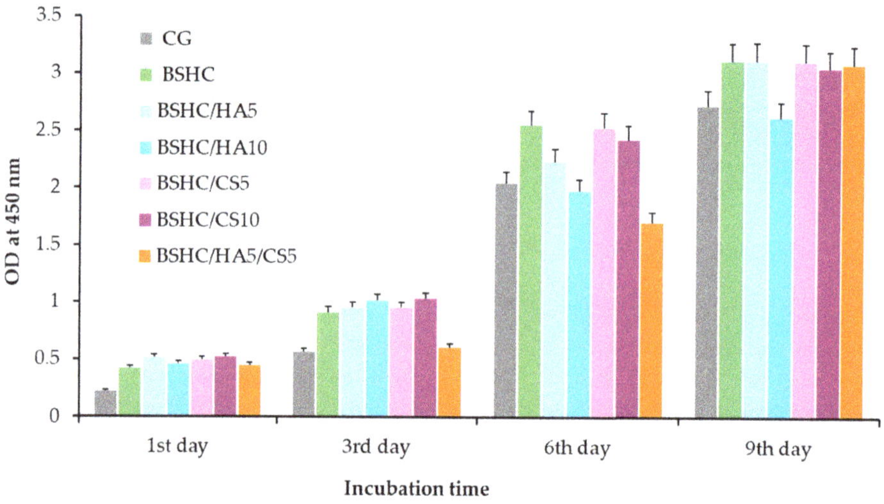

Figure 12. Proliferation of hAD-MSCs cultured for 9 days on control glass (CG); BSHC mat without GAGs; and BSHC mats containing HA (BSHC/HA5 and BSHC/HA10), CS (BSHC/CS5 and BSHC/CS10), or both HA and CS (BSHC/HA5/CS5).

While statistically insignificant differences were observed among mats at individual time points (days 1, 3, 6, and 9), minor variations may arise from the different composition and fiber architecture. For example, some mats may promote faster initial cell attachment, while others may support sustained proliferation longer. Despite these variations, the overall trend indicates that all tested mats support long-term cell growth without cytotoxicity and position them as promising candidates for applications requiring stable cell expansion and tissue maturation.

The results from the hAD-MSCs behavior study demonstrate that electrospun BSHC and BSHC/GAG mats support hAD-MSCs adhesion, spreading, and proliferation without cytotoxicity. Our key findings reveal the structural influence, compositional effects, adhesion-proliferation dynamics, and biocompatibility of the studied electrospun BSHC and BSHC/GAGs fibrous mats and confirm what is known from various scientific reports [25,51,67,71–78].

The findings can be summarized as follows:

1. The structural influence: Optimal cell spreading occurs on linear micro/submicron fiber mats (BSHC, BSHC/HA5, and BSHC/HA10); the fiber architecture (linear vs. branched) significantly impacts focal adhesion formation; mats with linear fibers provide superior mechanical support for cell attachment.
2. Compositional effects: HA-containing mats (BSHC/HA5) show exceptional cell spreading, rivaling fibronectin controls; GAG incorporation influences initial adhesion but not long-term proliferation; all compositions maintain cytoskeletal integrity and actin expression.
3. Adhesion–Proliferation Dynamics: While adhesion rates vary among mats, all support long-term proliferation (6-fold increase by day 9), an inverse relationship exists between the adhesion density and spreading area, and the composition/structure differentially regulates attachment versus spreading behavior.
4. Biocompatibility: Rapid confluence was achieved within 24 h across all mats; no indirect cytotoxicity was observed, and all mats supported sustained growth comparable to controls.

These findings position BSHC-based electrospun mats as promising scaffolds for tissue engineering and regenerative medicine. The linear fiber mats (particularly HA-containing variants) offer an optimal balance of structural and biochemical cues for stem cell maintenance. While adhesion rates may vary, all compositions demonstrate excellent biocompatibility and proliferation support, making them suitable substrates for further differentiation studies.

4. Conclusions

The electrospinning of BSHC from benign solvents, such as phosphate buffers and their mixtures with ethanol, as well as acetic acid/ethanol solutions, was found to be unfeasible for concentrations ranging from 5 wt.% to 25 wt.%. This held true across a wide range of operational conditions, including voltages between 15 and 30 kV, flow rates from 1 to 100 μL/min, nozzle diameters between 0.5 and 0.9 mm, and tip-to-collector distances from 15 to 18 cm.

New electrospun BSHC and BSHC/GAG mats with distinct structural morphologies (micron- and submicron-diameter linear and branched fibers) were successfully prepared from HFP solutions. These mats were created both in the absence of GAGs and in the presence of hyaluronic acid (HA), chondroitin sulfate (CS), or a combination of both HA and CS.

The influence of different compositions and structural morphologies of the electrospun BSHC and BSHC/GAG mats was evaluated using both direct and indirect methods on human adipose-derived mesenchymal stem cells (hAD-MSCs). Parameters such as cell adhesion, spreading, focal adhesion contact formation, and proliferation were assessed. The electrospun BSHC and BSHC/GAG mats were found to support the long-term viability and proliferation of hAD-MSCs without inducing cytotoxicity. Additionally, the mats exhibited excellent hAD-MSC adhesiveness.

The behavior of hAD-MSCs on the BSHC and BSHC/GAG electrospun mats indicated that these mats do not induce indirect cytotoxicity and support hAD-MSCs' adhesion and proliferation effectively. Therefore, BSHC and BSHC/GAG mats are promising candidates as substrates for cell adhesion and proliferation, and potentially for stem cell differentiation applications.

Supplementary Materials: The following supporting information can be downloaded at: https://www.mdpi.com/article/10.3390/jfb16060219/s1.

Author Contributions: Conceptualization, T.G.V.; methodology, T.G.V., D.N.G. and P.D.D.; software, D.N.G. and N.K.; validation, T.G.V.; formal analysis, D.N.G. and N.K.; investigation, T.G.V., D.N.G., P.D.D., N.K., M.K.-M. and K.H.-P.; resources, T.G.V. and D.N.G.; data curation, T.G.V.; writing—original draft preparation, T.G.V. and D.N.G.; writing—review and editing, N.K.; visualization, D.N.G. and N.K.; supervision, T.G.V.; funding acquisition, D.N.G. All authors have read and agreed to the published version of the manuscript.

Funding: This research was funded by Bulgarian National Scientific Fund grant number KP-06-COST 11/2023 and grant number KP-06-COST 15/2023.

Institutional Review Board Statement: Not applicable.

Informed Consent Statement: Not applicable.

Data Availability Statement: The original contributions presented in this study are included in the article. Further inquiries can be directed to the corresponding author.

Acknowledgments: This manuscript is based upon work from COST action CA 20126 NETPORE and COST Action 21145 EURESTOP, supported by COST (European Cooperation in Science and Technology).

Conflicts of Interest: The authors declare no conflicts of interest.

Abbreviations

The following abbreviations are used in this manuscript:

hAD-MSCs	Human adipose-derived mesenchymal stem cells.
BSHC	Bovine split hide collagen.
DMEM	Dulbecco's Modified Eagle Medium.
MSCs	Mesenchymal stem cells.
ECM	Extracellular matrix.
GAG	Glycosaminoglycan.
SEM	Scanning electron microscopy.
HFP	Hexafluoropropanol.
BPS	Standard phosphate buffer.
CG-FN	Cover glass fibronectin coated.
GA	Glutar aldehyde.
CS	Chondroitin sulfate.
CG	Cover glass.

References

1. Lu, W.P.; Guo, Y.; Lu, W.P.; Guo, Y. Electrospinning of Collagen and Its Derivatives for Biomedical Applications. In *Novel Aspects of Nanofibers*; IntechOpen: London, UK, 2018; ISBN 978-1-78923-075-8.
2. Bianchi, S.; Bernardi, S.; Simeone, D.; Torge, D.; Macchiarelli, G.; Marchetti, E. Proliferation and Morphological Assessment of Human Periodontal Ligament Fibroblast towards Bovine Pericardium Membranes: An In Vitro Study. *Materials* **2022**, *15*, 8284. [CrossRef]
3. Kai, D.; Jin, G.; Prabhakaran, M.P.; Ramakrishna, S. Electrospun Synthetic and Natural Nanofibers for Regenerative Medicine and Stem Cells. *Biotechnol. J.* **2013**, *8*, 59–72. [CrossRef]
4. Blackstone, B.N.; Gallentine, S.C.; Powell, H.M. Collagen-Based Electrospun Materials for Tissue Engineering: A Systematic Review. *Bioengineering* **2021**, *8*, 39. [CrossRef]

5. Hu, X.; Wu, Z.; Zhang, Z.; Yao, H.; Wang, D.-A. Type II Collagen Scaffolds for Tissue Engineering. *Commun. Mater.* **2024**, *5*, 149. [CrossRef]
6. Kulkarni, M.A.; Gangal, A.R.; Khare, S.S.; Mundada, H.G.; Pujari, R.R. Electrospun Collagen Based Nanofibrous Mats for Wound Healing: An Integrative Review. *Biosci. Biotechnol. Res. Asia* **2022**, *19*, 515–528. [CrossRef]
7. Drobota, M.; Gradinaru, L.M.; Vlad, S.; Bargan, A.; Butnaru, M.; Angheloiu, M.; Aflori, M. Preparation and Characterization of Electrospun Collagen Based Composites for Biomedical Applications. *Materials* **2020**, *13*, 3961. [CrossRef]
8. Roche, S.; Ronzière, M.-C.; Herbage, D.; Freyria, A.-M. Native and DPPA Cross-Linked Collagen Sponges Seeded with Fetal Bovine Epiphyseal Chondrocytes Used for Cartilage Tissue Engineering. *Biomaterials* **2000**, *22*, 9–18. [CrossRef]
9. Luo, X.; Guo, Z.; He, P.; Chen, T.; Li, L.; Ding, S.; Li, H. Study on Structure, Mechanical Property and Cell Cytocompatibility of Electrospun Collagen Nanofibers Crosslinked by Common Agents. *Int. J. Biol. Macromol.* **2018**, *113*, 476–486. [CrossRef]
10. Zong, X.; Bien, H.; Chung, C.-Y.; Yin, L.; Fang, D.; Hsiao, B.S.; Chu, B.; Entcheva, E. Electrospun Fine-Textured Scaffolds for Heart Tissue Constructs. *Biomaterials* **2005**, *26*, 5330–5338. [CrossRef] [PubMed]
11. Hapach, L.A.; VanderBurgh, J.A.; Miller, J.P.; Reinhart-King, C.A. Manipulation of in Vitro Collagen Matrix Architecture for Scaffolds of Improved Physiological Relevance. *Phys. Biol.* **2015**, *12*, 061002. [CrossRef]
12. Zeugolis, D.I.; Khew, S.T.; Yew, E.S.Y.; Ekaputra, A.K.; Tong, Y.W.; Yung, L.-Y.L.; Hutmacher, D.W.; Sheppard, C.; Raghunath, M. Electro-Spinning of Pure Collagen Nano-Fibres—Just an Expensive Way to Make Gelatin? *Biomaterials* **2008**, *29*, 2293–2305. [CrossRef]
13. Dong, B.; Arnoult, O.; Smith, M.E.; Wnek, G.E. Electrospinning of Collagen Nanofiber Scaffolds from Benign Solvents. *Macromol. Rapid Commun.* **2009**, *30*, 539–542. [CrossRef]
14. Arnoult, O. A Novel Benign Solution for Collagen Processing. Ph.D. Thesis, Case Western Reserve University, Cleveland, OH, USA, 2010.
15. Visser, D.; Rogg, K.; Fuhrmann, E.; Marzi, J.; Schenke-Layland, K.; Hartmann, H. Electrospinning of Collagen: Enzymatic and Spectroscopic Analyses Reveal Solvent-Independent Disruption of the Triple-Helical Structure. *J. Mater. Chem. B* **2023**, *11*, 2207–2218. [CrossRef]
16. Melo, P.; Montalbano, G.; Boggio, E.; Gigliotti, C.L.; Dianzani, C.; Dianzani, U.; Vitale-Brovarone, C.; Fiorilli, S. Electrospun Collagen Scaffold Bio-Functionalized with Recombinant ICOS-Fc: An Advanced Approach to Promote Bone Remodelling. *Polymers* **2022**, *14*, 3780. [CrossRef]
17. Caliari, S.R.; Ramirez, M.A.; Harley, B.A.C. The Development of Collagen-GAG Scaffold-Membrane Composites for Tendon Tissue Engineering. *Biomaterials* **2011**, *32*, 8990–8998. [CrossRef]
18. Harley, B.A.C.; Gibson, L.J. In Vivo and in Vitro Applications of Collagen-GAG Scaffolds. *Chem. Eng. J.* **2008**, *137*, 102–121. [CrossRef]
19. Yang, P.; Lu, Y.; Gou, W.; Qin, Y.; Tan, J.; Luo, G.; Zhang, Q. Glycosaminoglycans' Ability to Promote Wound Healing: From Native Living Macromolecules to Artificial Biomaterials. *Adv. Sci.* **2024**, *11*, 2305918. [CrossRef]
20. Tierney, C.M.; Haugh, M.G.; Liedl, J.; Mulcahy, F.; Hayes, B.; O'Brien, F.J. The Effects of Collagen Concentration and Crosslink Density on the Biological, Structural and Mechanical Properties of Collagen-GAG Scaffolds for Bone Tissue Engineering. *J. Mech. Behav. Biomed. Mater.* **2009**, *2*, 202–209. [CrossRef]
21. Murphy, C.M.; Haugh, M.G.; O'Brien, F.J. The Effect of Mean Pore Size on Cell Attachment, Proliferation and Migration in Collagen–Glycosaminoglycan Scaffolds for Bone Tissue Engineering. *Biomaterials* **2010**, *31*, 461–466. [CrossRef] [PubMed]
22. Lungu, A.; Albu, M.G.; Florea, N.M.; Stancu, I.C.; Vasile, E.; Iovu, H. The Influence of Glycosaminoglycan Type on the Collagen-Glycosaminoglycan Porous Scaffolds. *Dig. J. Nanomater. Biostructures* **2011**, *6*, 1867–1875.
23. Law, J.X.; Liau, L.L.; Saim, A.; Yang, Y.; Idrus, R. Electrospun Collagen Nanofibers and Their Applications in Skin Tissue Engineering. *Tissue Eng. Regen. Med.* **2017**, *14*, 699–718. [CrossRef]
24. Anderegg, U.; Halfter, N.; Schnabelrauch, M.; Hintze, V. Collagen/Glycosaminoglycan-Based Matrices for Controlling Skin Cell Responses. *Biol. Chem.* **2021**, *402*, 1325–1335. [CrossRef]
25. Tören, E.; Mazari, A.A. Needless Electrospun Collagen/Hyaluronic Acid Nanofibers for Skin Moisturization: Research. *Polym. Adv. Technol.* **2024**, *35*, e6434. [CrossRef]
26. Yannas, I.V.; Tzeranis, D.S.; So, P.T.C. Regeneration Mechanism for Skin and Peripheral Nerves Clarified at the Organ and Molecular Scales. *Curr. Opin. Biomed. Eng.* **2018**, *6*, 1–7. [CrossRef]
27. Menard, C.; Mitchell, S.; Spector, M. Contractile Behavior of Smooth Muscle Actin-Containing Osteoblasts in Collagen-GAG Matrices in Vitro: Implant-Related Cell Contraction. *Biomaterials* **2000**, *21*, 1867–1877. [CrossRef]
28. Pfeiffer, E.; Vickers, S.M.; Frank, E.; Grodzinsky, A.J.; Spector, M. The Effects of Glycosaminoglycan Content on the Compressive Modulus of Cartilage Engineered in Type II Collagen Scaffolds. *Osteoarthr. Cartil.* **2008**, *16*, 1237–1244. [CrossRef]

29. Caliari, S.R.; Harley, B.A.C. The Effect of Anisotropic Collagen-GAG Scaffolds and Growth Factor Supplementation on Tendon Cell Recruitment, Alignment, and Metabolic Activity. *Biomaterials* **2011**, *32*, 5330–5340. [CrossRef]
30. Todd, E.A.; Mirsky, N.A.; Silva, B.L.G.; Shinde, A.R.; Arakelians, A.R.L.; Nayak, V.V.; Marcantonio, R.A.C.; Gupta, N.; Witek, L.; Coelho, P.G. Functional Scaffolds for Bone Tissue Regeneration: A Comprehensive Review of Materials, Methods, and Future Directions. *J. Funct. Biomater.* **2024**, *15*, 280. [CrossRef]
31. Dewey, M.J.; Kolliopoulos, V.; Ngo, M.T.; Harley, B.A.C. Glycosaminoglycan Content of a Mineralized Collagen Scaffold Promotes Mesenchymal Stem Cell Secretion of Factors to Modulate Angiogenesis and Monocyte Differentiation. *Materialia* **2021**, *18*, 101149. [CrossRef]
32. Keogh, M.B.; O'Brien, F.J.; Daly, J.S. Substrate Stiffness and Contractile Behaviour Modulate the Functional Maturation of Osteoblasts on a Collagen–GAG Scaffold. *Acta Biomater.* **2010**, *6*, 4305–4313. [CrossRef]
33. Tierney, C.M.; Jaasma, M.J.; O'Brien, F.J. Osteoblast Activity on Collagen-GAG Scaffolds Is Affected by Collagen and GAG Concentrations. *J. Biomed. Mater. Res. Part A* **2009**, *91A*, 92–101. [CrossRef]
34. Murphy, C.M.; Duffy, G.P.; Schindeler, A.; O'brien, F.J. Effect of Collagen-Glycosaminoglycan Scaffold Pore Size on Matrix Mineralization and Cellular Behavior in Different Cell Types. *J. Biomed. Mater. Res. Part A* **2016**, *104*, 291–304. [CrossRef]
35. Osmond, M.J.; Krebs, M.D.; Pantcheva, M.B. Human Trabecular Meshwork Cell Behavior Is Influenced by Collagen Scaffold Pore Architecture and Glycosaminoglycan Composition. *Biotechnol. Bioeng.* **2020**, *117*, 3150–3159. [CrossRef]
36. O'Brien, F.J.; Harley, B.A.; Yannas, I.V.; Gibson, L.J. The Effect of Pore Size on Cell Adhesion in Collagen-GAG Scaffolds. *Biomaterials* **2005**, *26*, 433–441. [CrossRef]
37. Li, Z.; Hu, X.; Zhong, J.F. Mesenchymal Stem Cells: Characteristics, Function, and Application. *Stem Cells Int.* **2019**, *2019*, 8106818. [CrossRef]
38. Tanabe, S. Role of Mesenchymal Stem Cells in Cell Life and Their Signaling. *World J. Stem Cells* **2014**, *6*, 24–32. [CrossRef]
39. Koczkowska, M.; Kostecka, A.; Zawrzykraj, M.; Myszczyński, K.; Skoniecka, A.; Deptuła, M.; Tymińska, A.; Czerwiec, K.; Jąkalski, M.; Zieliński, J.; et al. Identifying Differentiation Markers Between Dermal Fibroblasts and Adipose-derived Mesenchymal Stromal Cells (AD-MSCs) in Human Visceral and Subcutaneous Tissues Using Single-cell Transcriptomics. *Stem Cell Res. Ther.* **2025**, *16*, 64–73. [CrossRef]
40. Alcayaga-Miranda, F.; Cuenca, J.; Khoury, M. Antimicrobial Activity of Mesenchymal Stem Cells: Current Status and New Perspectives of Antimicrobial Peptide-Based Therapies. *Front. Immunol.* **2017**, *8*, 339. [CrossRef]
41. Zuk, P.A.; Zhu, M.; Mizuno, H.; Huang, J.; Futrell, J.W.; Katz, A.J.; Benhaim, P.; Lorenz, H.P.; Hedrick, M.H. Multilineage Cells from Human Adipose Tissue: Implications for Cell-Based Therapies. Available online: https://www.liebertpub.com/doi/10.1089/107632701300062859 (accessed on 7 April 2025).
42. Kern, S.; Eichler, H.; Stoeve, J.; Klüter, H.; Bieback, K. Comparative Analysis of Mesenchymal Stem Cells from Bone Marrow, Umbilical Cord Blood, or Adipose Tissue. *Stem Cells* **2006**, *24*, 1294–1301. [CrossRef] [PubMed]
43. Lee, S.Y.; Kwon, B.; Lee, K.; Son, Y.H.; Chung, S.G. Therapeutic Mechanisms of Human Adipose-Derived Mesenchymal Stem Cells in a Rat Tendon Injury Model. *Am. J. Sports Med.* **2017**, *45*, 1429–1439. [CrossRef] [PubMed]
44. Melief, S.M.; Zwaginga, J.J.; Fibbe, W.E.; Roelofs, H. Adipose Tissue-Derived Multipotent Stromal Cells Have a Higher Immunomodulatory Capacity Than Their Bone Marrow-Derived Counterparts. *Stem Cells Transl. Med.* **2013**, *2*, 455–463. [CrossRef]
45. Doshi, J.; Reneker, D.H. Electrospinning Process and Applications of Electrospun Fibers. *J. Electrost.* **1995**, *35*, 151–160. [CrossRef]
46. Castilla-Casadiego, D.A.; Rivera-Martínez, C.A.; Quiñones-Colón, B.A.; Almodóvar, J. Electrospun Collagen Scaffolds. In *Electrospun Biomaterials and Related Technologies*; Almodovar, J., Ed.; Springer International Publishing: Cham, Switzerland, 2017; pp. 21–55, ISBN 978-3-319-70049-6.
47. Huang, G.P.; Shanmugasundaram, S.; Masih, P.; Pandya, D.; Amara, S.; Collins, G.; Arinzeh, T.L. An Investigation of Common Crosslinking Agents on the Stability of Electrospun Collagen Scaffolds. *J. Biomed. Mater. Res. Part A* **2015**, *103*, 762–771. [CrossRef]
48. Campiglio, C.E.; Marcolin, C.; Draghi, L.; Campiglio, C.E.; Marcolin, C.; Draghi, L. Electrospun ECM Macromolecules as Biomimetic Scaffold for Regenerative Medicine: Challenges for Preserving Conformation and Bioactivity. *AIMS Mater. Sci.* **2017**, *4*, 638–669. [CrossRef]
49. Yannas, I.V.; Lee, E.; Orgill, D.P.; Skrabut, E.M.; Murphy, G.F. Synthesis and Characterization of a Model Extracellular Matrix That Induces Partial Regeneration of Adult Mammalian Skin. *Proc. Natl. Acad. Sci. USA* **1989**, *86*, 933–937. [CrossRef]
50. Zhong, S.; Teo, W.E.; Zhu, X.; Beuerman, R.; Ramakrishna, S.; Yung, L.Y.L. Formation of Collagen−Glycosaminoglycan Blended Nanofibrous Scaffolds and Their Biological Properties. *Biomacromolecules* **2005**, *6*, 2998–3004. [CrossRef]
51. Tian, H.; Chen, Y.; Ding, C.; Li, G. Interaction Study in Homogeneous Collagen-Chondroitin sulfate Blends by Twodimensional Infrared Spectroscopy. *Carbohydr. Polym.* **2012**, *89*, 542–550. [CrossRef]

52. *ISO 10993-5:2009*; International Standard: Biological Evaluation of Medical Devices—Part 5: Tests for In Vitro Cytotoxicity. International Organization for Standardization: Geneva, Switzerland, 2009.
53. Zhang, L.; Webster, T.J. Nanotechnology and Nanomaterials: Promises for Improved Tissue Regeneration. *Nano Today* **2009**, *4*, 66–80. [CrossRef]
54. Fröhlich, E.; Salar-Behzadi, S. Toxicological Assessment of Inhaled Nanoparticles: Role of in Vivo, Ex Vivo, in Vitro, and in Silico Studies. *Int. J. Mol. Sci.* **2014**, *15*, 4795–4822. [CrossRef]
55. Clift, M.J.D.; Rothen-Rutishauser, B.; Brown, D.M.; Duffin, R.; Donaldson, K.; Proudfoot, L.; Guy, K.; Stone, V. The Impact of Different Nanoparticle Surface Chemistry and Size on Uptake and Toxicity in a Murine Macrophage Cell Line. *Toxicol. Appl. Pharmacol.* **2008**, *232*, 418–427. [CrossRef]
56. Armour, A.D.; Powell, H.M.; Boyce, S.T. Fluorescein Diacetate for Determination of Cell Viability in Tissue-Engineered Skin. *Tissue Eng. Part C Methods* **2008**, *14*, 89–96. [CrossRef]
57. Dalton, C.J.; Lemmon, C.A. Fibronectin: Molecular Structure, Fibrillar Structure and Mechanochemical Signaling. *Cells* **2021**, *10*, 2443. [CrossRef]
58. Pankov, R.; Yamada, K.M. Fibronectin at a Glance. *J. Cell Sci.* **2002**, *115*, 3861–3863. [CrossRef]
59. Aota, S.; Yamada, K.M. Fibronectin and Cell Adhesion: Specificity of Integrin–Ligand Interaction. In *Advances in Enzymology and Related Areas of Molecular Biology*; John Wiley & Sons, Ltd.: Hoboken, NJ, USA, 1995; pp. 1–21. ISBN 978-0-470-12316-4.
60. Baer, P.C.; Geiger, H. Adipose-Derived Mesenchymal Stromal/Stem Cells: Tissue Localization, Characterization, and Heterogeneity. *Stem Cells Int.* **2012**, *2012*, 812693. [CrossRef]
61. Geiger, B.; Bershadsky, A.; Pankov, R.; Yamada, K.M. Transmembrane Crosstalk between the Extracellular Matrix and the Cytoskeleton. *Nat. Rev. Mol. Cell Biol.* **2001**, *2*, 793–805. [CrossRef]
62. Barnes, C.P.; Sell, S.A.; Boland, E.D.; Simpson, D.G.; Bowlin, G.L. Nanofiber Technology: Designing the next Generation of Tissue Engineering Scaffolds. *Adv. Drug Deliv. Rev.* **2007**, *59*, 1413–1433. [CrossRef]
63. Matthews, J.A.; Wnek, G.E.; Simpson, D.G.; Bowlin, G.L. Electrospinning of Collagen Nanofibers. *Biomacromolecules* **2002**, *3*, 232–238. [CrossRef]
64. Denchai, A.; Tartarini, D.; Mele, E. Cellular Response to Surface Morphology: Electrospinning and Computational Modeling. *Front. Bioeng. Biotechnol.* **2018**, *24*, 155. [CrossRef]
65. Schnell, E.; Klinkhammer, K.; Balzer, S.; Brook, G.; Klee, D.; Dalton, P.; Mey, J. Guidance of Glial Cell Migration and Axonal Growth on Electrospun Nanofibers of Poly-ε-Caprolactone and a Collagen/Poly-ε-Caprolactone Blend. *Biomaterials* **2007**, *28*, 3012–3025. [CrossRef]
66. Gourlay, C.W.; Ayscough, K.R. The Actin Cytoskeleton: A Key Regulator of Apoptosis and Ageing? *Nat. Rev. Mol. Cell Biol.* **2005**, *6*, 583–589. [CrossRef]
67. Paim, Á.; Tessaro, I.C.; Cardozo, N.S.M.; Pranke, P. Mesenchymal Stem Cell Cultivation in Electrospun Scaffolds: Mechanistic Modeling for Tissue Engineering. *J. Biol. Phys.* **2018**, *44*, 245–271. [CrossRef]
68. Bershadsky, A.D.; Balaban, N.Q.; Geiger, B. Adhesion-Dependent Cell Mechanosensitivity. *Annu. Rev. Cell Dev. Biol.* **2003**, *19*, 677–695. [CrossRef]
69. Parsons, J.T.; Horwitz, A.R.; Schwartz, M.A. Cell Adhesion: Integrating Cytoskeletal Dynamics and Cellular Tension. *Nat. Rev. Mol. Cell Biol.* **2010**, *11*, 633–643. [CrossRef]
70. Huang, Z.-M.; Zhang, Y.-Z.; Kotaki, M.; Ramakrishna, S. A Review on Polymer Nanofibers by Electrospinning and Their Applications in Nanocomposites. *Compos. Sci. Technol.* **2003**, *63*, 2223–2253. [CrossRef]
71. Zhao, Y.; Sun, Q.; Huo, B. Focal Adhesion Regulates Osteogenic Differentiation of Mesenchymal Stem Cells and Osteoblasts. *Biomater. Transl.* **2021**, *2*, 312–322. [CrossRef]
72. Zaidel-Bar, R.; Itzkovitz, S.; Ma'ayan, A.; Iyengar, R.; Geiger, B. Functional Atlas of the Integrin Adhesome. *Nat. Cell Biol.* **2007**, *9*, 858–867. [CrossRef]
73. Chung, E.; Yoon, G.; (Matthew) Suh, H. A Collagen-Hyaluronic Acid Matrix for Stem Cell Culture. In *Biological, Physical and Technical Basics of Cell Engineering*; Artmann, G., Artmann, A., Zhubanova, A., Digel, I., Eds.; Springer: Singapore, 2018. [CrossRef]
74. Murphy, C.M.; Matsiko, A.; Haugh, M.G.; Gleeson, J.P.; O'Brien, F.J. Mesenchymal Stem Cell Fate is Regulated by the Composition and Mechanical Properties of Collagen-Glycosaminoglycan Scaffolds. *J. Mech. Behav. Biomed. Mater.* **2012**, *11*, 53–62. [CrossRef]
75. Amann, E.; Wolff, P.; Breel, E.; Van Griensven, M.; Balmayor, E.R. Hyaluronic Acid Facilitates chondrogenesis and Matrix Deposition of Human Adipose Derived Mesenchymal Stem Cells and Human Chondrocytes Co-cultures. *Acta Biomater.* **2017**, *52*, 130–144. [CrossRef]
76. He, X.; Wang, S.; Yu, X.; Zhou, X. Bone Marrow Mesenchymal Stem Cells Response on Collagen/hyaluronan/chondroitin Scaffold Enriched with Gentamicin-loaded Gelatin Microparticles for Skin Ttissue Engineering. *J. Biomater. Appl.* **2023**, *38*, 134–145. [CrossRef]

77. Somaiah, C.; Kumar, A.; Mawrie, D.; Sharma, A.; Patil, S.D.; Bhattacharyya, J.; Swaminathan, R.; Jaganathan, B.G. Collagen Promotes Higher Adhesion, Survival and Proliferation of Mesenchymal Stem Cells. *PLoS ONE* **2015**, *10*, e0145068. [CrossRef]
78. Kennedy, K.M.; Bhaw Luximon, A.; Jhurry, D. Cell-matrix mechanical interaction in electrospun polymeric scaffolds for tissue engineering: Implications for scaffold design and performance. *Acta Biomater.* **2017**, *50*, 41–55. [CrossRef] [PubMed]

Disclaimer/Publisher's Note: The statements, opinions and data contained in all publications are solely those of the individual author(s) and contributor(s) and not of MDPI and/or the editor(s). MDPI and/or the editor(s) disclaim responsibility for any injury to people or property resulting from any ideas, methods, instructions or products referred to in the content.

Journal of
Functional Biomaterials

Review

Head and Neck 3D Bioprinting—A Review on Recent Advancements in Soft Tissue 3D Bioprinting and Medical Applications

Iosif-Aliodor Timofticiuc [1], Ana Caruntu [2,3], Christiana Diana Maria Dragosloveanu [4,5,*], Andreea-Elena Scheau [6,*], Ioana Anca Badarau [1], Argyrios Periferakis [1,7,8], Serban Dragosloveanu [9,10], Andreea Cristiana Didilescu [11] Constantin Caruntu [1,12] and Cristian Scheau [1,6]

[1] Department of Physiology, The "Carol Davila" University of Medicine and Pharmacy, 8 Eroii Sanitari Boulevard, 050474 Bucharest, Romania
[2] Department of Oral and Maxillofacial Surgery, "Carol Davila" Central Military Emergency Hospital, 010825 Bucharest, Romania
[3] Department of Oral and Maxillofacial Surgery, Faculty of Dental Medicine, Titu Maiorescu University, 031593 Bucharest, Romania
[4] Department of Ophthalmology, Faculty of Dentistry, The "Carol Davila" University of Medicine and Pharmacy, 020021 Bucharest, Romania
[5] Department of Ophthalmology, Clinical Hospital for Ophthalmological Emergencies, 010464 Bucharest, Romania
[6] Department of Radiology and Medical Imaging, "Foisor" Clinical Hospital of Orthopaedics, Traumatology and Osteoarticular TB, 021382 Bucharest, Romania
[7] Akadimia of Ancient Greek and Traditional Chinese Medicine, 16675 Athens, Greece
[8] Elkyda, Research & Education Centre of Charismatheia, 17675 Athens, Greece
[9] Department of Orthopaedics and Traumatology, The "Carol Davila" University of Medicine and Pharmacy, 050474 Bucharest, Romania
[10] Department of Orthopaedics, "Foisor" Clinical Hospital of Orthopaedics, Traumatology and Osteoarticular TB, 021382 Bucharest, Romania
[11] Department of Embryology and Microbiology, Faculty of Dentistry, The "Carol Davila" University of Medicine and Pharmacy, 050474 Bucharest, Romania
[12] Department of Dermatology, "Prof. N.C. Paulescu" National Institute of Diabetes, Nutrition and Metabolic Diseases, 011233 Bucharest, Romania
* Correspondence: christiana.dragosloveanu@umfcd.ro (C.D.M.D.); andreea.ghergus@gmail.com (A.-E.S.)

Abstract: Recent developments in 3D bioprinting offer innovative alternative solutions to classical treatments for head and neck defects. Soft tissues in an anatomical area as diverse in composition as the head and neck are complex in terms of structure and function. Understanding how cellular interaction underlies functionality has led to the development of bioinks capable of mimicking the natural morphology and roles of different human parts. Moreover, from the multitude of recently developed materials, there are now many options for building scaffolds that potentiate the activity of these cells. The fidelity and accuracy of the utilized techniques ensure maximum precision in terms of model construction. Emerging technologies will allow for improved control of the scaffold, facilitating optimal results in the treatment of various pathologies, without concerns about the availability of donors, immunological response, or any other side effects that traditional treatments withhold. This paper explores the current landscape of bioprinted scaffolds and their applications in the head and neck region, with a focus on the properties and use of natural and synthetic bioinks in the attempt to replicate the biomechanical features of native tissues. Customization capabilities that support anatomical precision and biofunctionality are also addressed. Moreover, regulatory requirements, as well as current challenges related to biocompatibility, immune response, and vascularization, are critically discussed in order to provide a comprehensive overview of the pathway to clinical application.

Keywords: 3D bioprinting; head and neck; soft tissue; bionics; tissue engineering; physiology; vascularized scaffolds

1. Introduction

The head and neck region presents high complexity, mainly due to the vast vascular and neural networks integrated into a biomechanical system that ensures a multitude of functions, including sensorial and motor control [1,2]. The soft tissue components of this region play an important role in the integrative sensorial system, which highly contributes to the adaptation of humans in their natural environment [3]. The system consists of complex organs and supporting elements [4,5].

Defects in this anatomical region, with its numerous structures with complex functionality, can greatly impact patients' lives, either psychologically or physically. Causes of these defects can include oncologic resections, other surgical interventions, congenital pathologies, genetic disorders, or post-traumatic tissue loss [6–8]. Most times, the complexity of treatments requires a multidisciplinary approach to restore the functionality and aesthetics of the affected elements. Increasingly complicated transplantations are attempted, but functional results can still be further improved, and supplementary rehabilitation methods are required [9–11]. Recent research focuses on the development of new technologies and treatments aimed at individualized solutions with optimized outcomes for patients.

3D bioprinting is an interdisciplinary branch of the 3D printing industry developed specifically for medical applications [12]. Numerous applications already exist, and the potential of the field is ever-increasing [13–17]. The materials used in this field are carefully designed in terms of mechanical and bioactive properties and include mixes with added cultured cells, also known as bioinks [18,19]. Therefore, while maintaining cell viability and compatibility in a specific material, 3D models or scaffolds are built to serve precise functions mimicking physiological functions [18,19]. Numerous types of technologies have been developed to serve the purposes of 3D bioprinting, such as micrometer precision, controlled deposition of cells, high functionality, increased viability, and bioactive properties [20,21].

The most relevant methods developed in 3D bioprinting are (1) droplet-based bioprinting, composed of inkjet-bioprinting and laser-induced forward transfer, which ensures high resolution and fast printing due to the controlled cell deposition using droplets of cell-laden biomaterial [22]; (2) extrusion-based bioprinting, highly used in current studies due to a large range of available biomaterials with layer-by-layer printed bioinks [23]; (3) laser-assisted bioprinting, which emerged from droplet-based bioprinting [24]; and (4) stereolithographic bioprinting, which uses photopolymerization, similar to digital light processing (DLP) or stereolithography (SLA) 3D printing techniques [25].

One of the advancements desired by the medical and engineering industries is limiting the use of human donors for patient treatments and creating parts that can mimic the functions of various human elements with high specificity. The simpler term for this is creating bionic parts. Bionics is a field based on building elements that can mimic the form or function of something that already exists in nature to serve human evolution and development [26]. Adapting this definition in our paper, we highlight the recent therapeutic advancements made possible by 3D bioprinting, with inspiration from human parts and their functionality (Figure 1). Alongside presenting and integrating the achievements of 3D bioprinting in head and neck soft tissues, we focus on the main challenges and drawbacks of this method, while exploring potential workarounds and future perspectives.

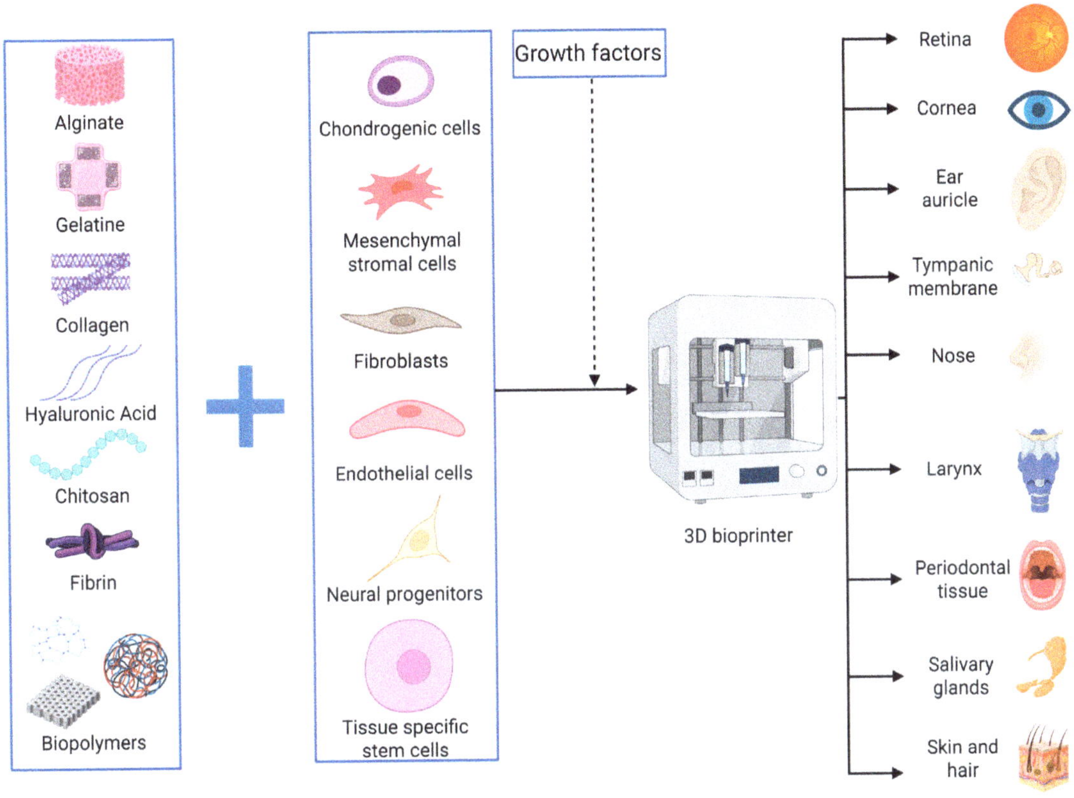

Figure 1. Summary of the results obtained by bioprinting soft tissues of the head and neck. Relevant materials or cells used in this field are presented. Created in BioRender. Timofticiuc, I. (2025) https://BioRender.com/e6j0vbk, accessed on 3 May 2025.

2. Head and Neck Soft Tissue 3D (Bio)Printing

The main defects affecting different parts of the head and neck region are reviewed below, highlighting the limitations of traditional treatments. 3D bioprinting represents an opportunity for the development of specific treatments, as well as the supply of, theoretically, an unlimited amount of biocompatible tissue. In current practice, when designing and building soft tissue scaffolds, the mechanical characteristics are extremely important. Presently available biomaterials, as developed and described in recent studies, may be incorporated into bioprinted elements meant to replace soft tissues of the head and neck. The similitude in mechanical properties to the anatomical structures targeted for restoration is one of the strong arguments in this regard (Tables 1 and 2).

Table 1. Mechanical properties of the natural human tissues and organs of the head and neck.

Physiological Tissue/Organ	Young's Modulus (Stiffness)	Tensile Strength	Reference
Cornea	Central Cornea: 70–100 kPa Peripheral Cornea: 30–60 kPa	380–650 kPa	[27,28]
Retina	431 kPa	100 kPa (porcine models)	[29,30]
Choroid	4620 kPa (4.62 MPa)	300 kPa	[29,31]
Sclera	Anterior Sclera: 17,000–43,000 kPa (17–43 MPa) Equatorial Sclera: 9000–25,000 kPa (9–25 MPa) Posterior Sclera: 7000–18,000 kPa (7–18 MPa)	Not mentioned	[32]
Ear Auricle	4500–5900 kPa (4.5–5.9 MPa)	3460 kPa (3.46 MPa)	[33,34]
Tympanic Membrane	20,000–40,000 kPa (20–40 MPa)	25,000–32,200 kPa (25–32.2 MPa)	[35,36]
Nose	2100–3300 kPa (2.1–3.3 MPa)	1640–2140 kPa (1.64–2.14 MPa)	[37,38]
Periodontal Tissue	10–31 kPa/900–1200 kPa	Not mentioned	[39,40]
Salivary Glands	Parotid glands: 18.4 kPa Submandibular glands: 15.9 kPa	Not mentioned	[41]
Skin	Back of the ear: 40 kPa Chin: 20 kPa	Not mentioned	[42]
Larynx	8.6 kPa	1000 kPa (1 MPa)	[43,44]
Trachea	1000–15,000 kPa (1–15 MPa)	1200–2500 kPa (1.2–2.5 MPa)	[45,46]
Esophagus	Not mentioned	1200–2800 kPa (1.2–2.8 MPa)	[47]

Table 2. Mechanical properties of the hydrogels and other biomaterials that could be used as scaffold materials for different medical applications.

Material	Young's Modulus (Stiffness)	Tensile Strength	Advantages	Disadvantages	Refs
			Hydrogels		
Alginate	<1.5 kPa	Up to 1830 kPa (1.83 MPa)	-Ease of gelation; -Low shear stress; -Compatible in mixes with other materials;	-Low mechanical strength; -Cytotoxicity; -Low cell adhesion;	[48–54]
Gelatin	81 kPa	24 kPa	-Bioactive; -Thermosensitive; -Low shear stress;	-Low mechanical strength; -Thermal instability;	[55–61]
GelMA (Gelatin Methacryloyl)	29.2–43.2 kPa/Up to 200–1000 kPa (Influenced by the % of methacrylation)	2800–3800 kPa (2.8–3.8 MPa)	-Photopolymerization (suitable for DLP- or SLA); -Adjustable mechanical properties (different concentrations of methacrylate can be added); -Great support for cells;	-Low viscosity; -Sensible to the microenvironmental conditions; -Cytotoxicity; -Limited mechanical stability;	[62–70]
Collagen	120–250 kPa	40 kPa	-Good cell adhesion; -Good mechanical properties, which can be adjusted with various methods of crosslinking;	-Low mechanical strength; -Fast degradation rate (this could also represent an advantage if used in fast-degradable scaffolds); -Thermal instability;	[60,64,71–76]
Fibrin	15–150 kPa	1.6–10 kPa	-Cell migration and adhesion; -Adjustable mechanical properties; -Mimics the natural extracellular matrix;	-Fast degradation rate (this could also represent an advantage if used in fast-degradable scaffolds); -Costly; -Sensible to the microenvironmental conditions;	[77–83]
Hyaluronic Acid	24 kPa	63 kPa	-Wettability; -Dynamic cross-linking; -Promotes cell adhesion, cell viability, and cell mobility;	-Costly; -Limited mechanical properties; -Sensible to the microenvironmental conditions;	[84–90]

Table 2. *Cont.*

Material	Young's Modulus (Stiffness)	Tensile Strength	Advantages	Disadvantages	Refs
Chitosan	7900–92,000 kPa (7.9–92 MPa)	3600–12,100 kPa (3.6–12.1 MPa)	-Easy gelation; -Bioactive; Adjustable mechanical properties;	-Limited printability; -Hard to obtain from natural sources; -Limited printability; -Limited mechanical stability;	[91–96]
Agarose	100–300 kPa	Not mentioned	-Exhibits rheological properties; -Thermosensitive; -Promotes cell viability;	-Limited mechanical stability; -Sensible to the microenvironmental conditions; -Costly;	[97–100]
Silk Fibroin	Up to 6300–6700 kPa (6.3–6.7 MPa)	80–700 kPa	-Good mechanical properties; -Promotes cell adhesion, mobility, and viability;	-Different properties if extracted from different sources; -Sensible to the microenvironmental conditions; -Low wettability;	[101–106]
Other Biomaterials					
Polyethylene Glycol (PEG)	16.5–89.5 kPa	Not mentioned	-Adjustable mechanical properties; -High wettability;	-Limited cell adhesion; -Potential toxicity; -Costly; -Limited mechanical strength;	[107–111]
Polylactic Acid (PLA)	3.5 GPa (3500 MPa/ 3,500,000 kPa)	60,000 kPa (60 MPa)	-Thermal and mechanical stability; -Wettability (this could be so intense that it could represent a disadvantage)	-Costly; -Limited cell adhesion; -Fast degradation rate (this could also represent an advantage if used in fast-degradable scaffolds);	[112–119]
Polycaprolactone (PCL)	25,000–120,000 kPa (25–120 MPa)	Not mentioned	-Good mechanical properties; -Adjustable degradation rate; -Promotes cell adhesion and viability;	-Limited wettability; -Thermal instability; -Costly;	[120–125]
Poly (lactic-co-glycolic acid) (PLGA)	32,100–36,100 kPa (32.1–36.1 MPa)	1600–2000 kPa (1.6–2 MPa)	-Adjustable mechanical properties; -Adjustable release of various drugs;	-Hydrophobic; -Fast degradation rate (this could also represent an advantage if used in fast-degradable scaffolds); -It can generate inflammatory response;	[83,120,126–129]

The head consists of numerous complex components, with structural, sensitive, motor, or sensorial roles. Bioprinting is a great option due to the possibility of creating high-precision models and multilayered constructs, where various types of cells can be introduced differently in each layer of the models, thus creating more precise builds that can mimic the natural tissues or organs more closely [130–132]. In the following sections, we highlight the main advancements in the 3D bioprinting of eyes, ear components, nose, salivary glands, oral parts, skin, and hair, while addressing the main limitations and potential solutions to current problems. Vascularization and neural control are critical aspects for fully functional models and will be discussed in the final section of this paper.

2.1. Corneal 3D (Bio)Printing

Considering all the possible causes of blindness worldwide, corneal opacity stands out as a distinct subgroup. Even though such causes represent less than 5% of the defects that lead to blindness, many of them, such as a number of diabetes-induced corneal changes [133], are treatable, and corneal transplantation is the most used method to achieve full vision recovery [134,135]. The transplantation techniques have evolved, from large and multilayered grafts in 1956 to single-layered grafts or partial grafts in the 21st century, offering safer and faster recovery. However, the number of patients in need far exceeds the number of donors, with a current ratio of approximately 70 to 1 [134,136].

Despite donor shortage, there exist alternatives, and the most promising for the future of corneal repair is the FRESH method, a 3D bioprinting protocol used by Isaacson et al. in 2018 to build the first fully 3D bioprinted corneal stroma [137]. The challenge in 3D printing a corneal stroma is obtaining good transparency and optimal architecture of the build, thus ensuring all the physical phenomena of image formation [138].

A common mixture used in corneal 3D bioprinting is based on three materials: collagen, alginate, and gelatin [139,140]. Collagen is used as a material in 3D printing and does not present ideal properties; collagen-based builds perform weakly in terms of mechanical behavior; however, in combination with other transparent materials such as alginates and gelatin, the performance can be improved, especially in cornea bioprinted constructs, where transparency is essential [139,141–143].

Unexpectedly, utilizing three transparent materials does not necessarily result in a transparent final 3D-printed product. Several attempts in the last 10 years showed that transparency is perhaps the hardest characteristic to achieve. The reasons for this significant drawback appear during the printing process. Due to the complicated dome shape of the cornea, the 3D printed layers collapse and create micro gaps filled with air or with insufficiently processed biomaterial, thus affecting light transmittance due to opaque spots or gaps with differing refractive indices [140].

A team of researchers addressed the main challenges of 3D printing corneal parts, i.e., the high costs, the difficulty in printing with viable cells, and the low transparency. A partial thickness corneal construct using a low-cost modified 3D printer was built. Optimal transparency was achieved using a concave mold as support to prevent layers from collapsing, therefore moving away from the FRESH technique, which created porous builds often filled with air bubbles [140].

From our perspective, this method, which was only used for partial grafts, can be scaled up to full corneal constructs. However, scaling up could bring other technical issues due to the printing process, increasing costs. Furthermore, the team did not use viable cells in this method, relying on the post-operative cellularization of the construct—an aspect that might not be as feasible in full corneal stroma 3D-printed builds [140].

An example that highlights the need for viable cells and very detailed techniques in 3D bioprinting large models can be seen in a study where a cell-laden corneal stroma was printed. Using a combination of technologies and Human Corneal Keratocytes (HCKs), full corneal stroma equivalents were able to ensure optimal transparency (75–90%) while maintaining their mechanical properties and integrity [139].

Notably, the FRESH method was used with the same three transparent materials mentioned before, but the performance was improved by combining 3D bioprinting, micro-transfer molding, and optimized bioinks [139]. In our opinion, this study highlights the importance of sophisticated techniques and the use of viable cells, ensuring a physiologically similar microenvironment that could maintain the viability of the construct over time.

However, issues still remain regarding the in vivo interaction between biomaterials and viable cells, which might have uncertain consequences [139]. Under that consideration, it was observed that using different materials, such as a decellularized extracellular matrix/gelatin methacryloyl (CECM-GelMA) bioinks, a corneal matrix with no viable cells and with no support structure can be built without affecting the transparency, nor the structural integrity of the construct [144]. It was demonstrated that the mix of substances in this bioink provides attachment sites for patient cells while also reducing inflammation [144]. However, it is uncertain how the fourth dimension affects these types of builds, and a major challenge is determining how long the 3D-printed corneal stroma equivalents can perform optimally. Furthermore, it is essential to know whether replacement is possible once defects appear, and more importantly, how many times these revisions can be performed.

2.2. *Retina 3D (Bio)Printing*

The retina presents a complex architecture [145]. To understand the possibility of 3D bioprinting a retinal equivalent, we must first address the main challenges from the perspective of printing technology and the availability of printable materials. It is known that the retina has a multilayered structure, with various types of cells [146]. From this observation, we can see two conditions that need to be fulfilled in the final 3D construct: (1) bioprinting the cells in an extracellular matrix (ECM) equivalent similar to the retinal microenvironment in terms of chemical, physical, and physiological properties; and the necessity of (2) designing a heterogeneously multilayered build.

2.2.1. Bioprinting Retinal Cells and ECM-Equivalent Conditions

The cellular composition of the retina is mainly made up of photoreceptors (PRs), horizontal cells, bipolar cells, amacrine cells, and ganglion cells [147]. The first step for an optimal bioprinting process is ensuring the cell's availability and ability to survive, both in vitro and during the mechanical stress of the printing phase. PRs can be obtained by differentiation from retinal progenitor cells, but the cells cannot survive isolated in vitro [148]. Therefore, PRs require the assistance of the retinal ECM composition to develop and survive in vitro [149].

It was demonstrated that maintaining the viability of PRs is difficult in vitro with ECM equivalents that do not mimic the physical properties of the retinal ECM [148]. The retinal microenvironment contains mainly hyaluronic acid [150]. A team of researchers tried to achieve an environment suitable for photoreceptor development using methacrylate hyaluronic acid [148]. They managed to achieve the necessary stiffness for PR development (10–20 kPa) and enhanced cell viability and development by adding retinal pigment epithelium to complement the HA-based gel [148].

In our view, bioprinting should be developed towards the possibility of building structures that require as few viable components as possible, to provide real and sustainable advantages in clinical applications. An interesting avenue is the possibility of bioprinting the retinal pigment epithelium (RPE), a component known to offer optimal circumstances for apoptosis, and for physiological activities of retinal cells [151]. Due to a paucity of studies where other components are used for PR bioprinting, RPE remains, at least for now, the only solution that transforms the artificial ECM equivalent suitable for retinal bioprinting.

RPE consists mainly of lipids (>50%) [152]; thus, bioprinting these substances represents the next main step to be taken for the future of bioprinting RPE. Lipids, mainly because of their thermoplastic behavior, are not yet suitable for the most commonly used bioprinting methods, especially those that use layer-by-layer technology [153]. However, there is a possibility to adapt lipids to 3D bioprinting by using droplet-based deposition as the main form of "layering" [153]. That being said, from our perspective, the faster way to achieve optimal PR bioprinting is by advancing the bioprinting of RPE, rather than isolating or developing other biomaterials that could allow the survival and development of PRs.

Regarding the other cellular components of the retina, data is scarce regarding the 3D bioprinting of functional parts of the eye. The steps needed to achieve 3D printability in these types of cells, thus completing the structure of a retina in regard to cellular composition, are, in our view: (1) obtaining and developing amacrine cells, horizontal cells, and ganglion cells in vitro; (2) testing the mechanical stress they can sustain; (3) identifying or developing biomaterials that support their viability; (4) isolating a 3D bioprinter compatible with the obtained mixture; (5) developing scaffolds viable for an undetermined period; (6) creating solutions to anatomically and functionally interconnect these cells; and (7) connecting the cellular network with the blood supply and ensuring neovascularization.

2.2.2. Designing a Heterogeneously Multilayered Build

Building 3D scaffolds using bioinks for drug testing was a major achievement, which ensured a better understanding of how different substances interact with the retinal microenvironment [154,155]. In an apparently unconnected line of research using other cellular lines (Y79, ARPE 19), for a scaffold used for drug testing, researchers also revealed the possibility of building multi-layered constructs with different cell types in the composition of each layer—one basal layer of ARPE 19 cells topped up with two distinct layers of Y79 cells [154]. This breakthrough opened numerous possibilities for developing multilayered builds in other applications.

However, the ultimate goal is to build a structure able to mimic its living counterpart. One of the options is to design superimposed cell-laden hydrogel layers. However, this comes with many challenges. To start with, due to the high density of cells in a viscous surrounding, the shear stress increases dramatically, which affects cell viability [156]. In addition, the layers can often lack uniformity, hence mimicking tissue behavior becomes extremely difficult [156]. The other solution in constructing multilayered scaffolds is by printing the cells without the surrounding hydrogels, thus allowing the cells to create their ECM [156]. Nevertheless, drawbacks occur when the layers with different types of cells need to be stabilized to form a mechanically resistant structure [156].

It is reasonable to consider that a solution could reside in building a structure with layers consisting of highly porous degradable biomaterials, thus ensuring the stability of the structure and the position of the cells in the scaffold, with the ECM produced by the cells replacing the degradable artificial surroundings over time. Adopting this mixed solution could support the mechanical behavior in the initial phase of the print. A porous construct could help preserve the viability of the cells by reducing the mechanical stress, while also offering them space to produce their own ECM. By degrading over time, the "natural" ECM could ultimately fully replace all artificial environments in a controlled manner.

2.3. Exploring the Feasibility of Developing a Bionic Eye

In essence, the components of the eyeball have the highest complexity in terms of bioprinting. In order to predict the possibility of producing a bionic eye in the future (Figure 2), we must analyze the possibility of printing the other components indispensable for ocular functions and consider their integration into a functional and interdependent system. The structures requiring replication in the outer layer of the eye are the sclera, a fibrous tissue, and the cornea, which is covered by the conjunctiva as a transparent membrane [157]. Moving to the middle layer, the iris, ciliary body, and choroid are structures that are in continuity and functionally connected with each other [157], which may be affected together in some pathologies, like cancer [158]. The retina, the innermost layer, has been discussed in extenso above [157]. Additionally, solutions equivalent to the vitreous humor, aqueous humor, and crystalline lens need to be addressed [157].

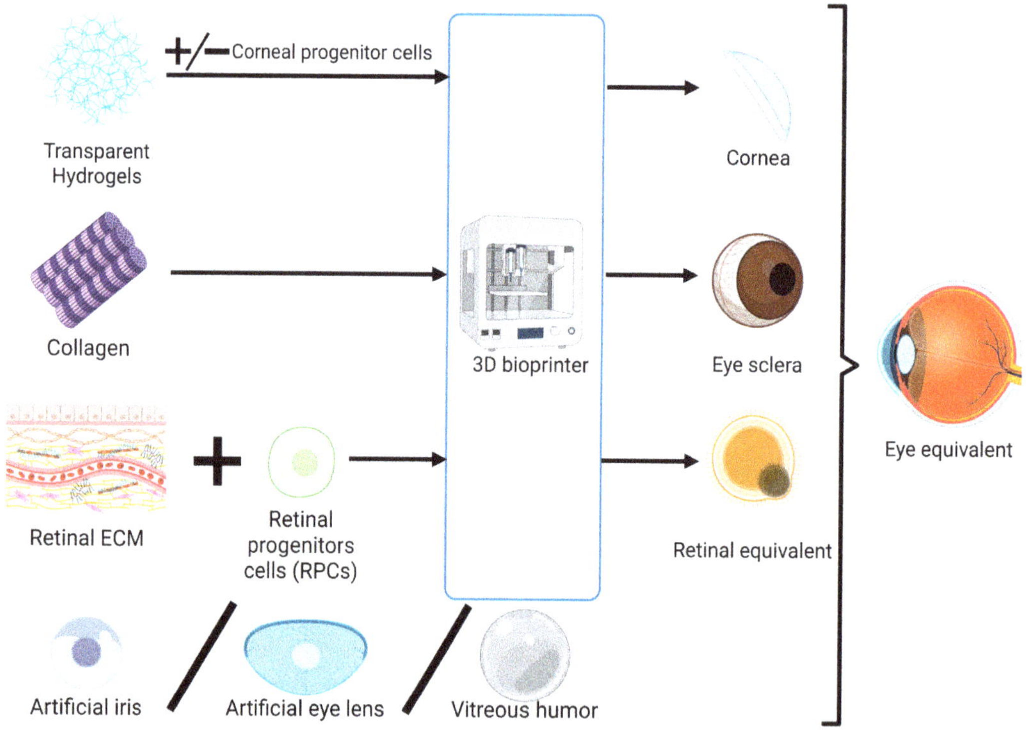

Figure 2. Schematic representation of the process of 3D bioprinting eye components and their integration in an eye equivalent (bionic eye). Created in BioRender. Timofticiuc, I. (2025) https://BioRender.com/d6q5va5, accessed on 21 June 2025.

Regarding the iris and scleral constructs, data is very scarce on bioprinting or 3D printing studies in which models are built that result in a final product with at least minimal functionality. However, there are a few reports of ocular prostheses created to fulfill an aesthetic role, where realistic sclera and iris were produced [159,160].

The sclera is essential for ensuring the biomechanical properties of the eyeball, such as maintaining intraocular pressure for optimal image formation and light transmission [161]. However, from a metabolic standpoint, the sclera does not exhibit activities essential to the main function of the eye. Therefore, in 3D printing a sclera, only mechanical performance is essential [161]. The main component of the sclera is collagen (28% of wet weight, with water comprising 68%), with collagen type I representing around 95% of all types of collagens found in the sclera [161,162]. A 3D-printed sclera needs to be able to withstand pressures of around 12.8 ± 2.7 mm Hg, the mean value for intraocular pressure, requiring an optimal modulus of elasticity around 750 kPa [163]. Collagen type I bioprinting is already possible, but the achieved elasticity is around 200–250 kPa [164]. A collagen ink was also used to create a high-resolution print containing type III collagen, with a resulting elasticity modulus reaching only 679 kPa [165].

These findings are good explanations for why limited advancements have been made in 3D printing a mechanically functional sclera. However, newly developed materials and mixtures could balance the stiffness and the elasticity needed for a potential sclera construct. In our view, a solution for reaching high elasticity while also preserving structural strength could involve perfecting a printing method for collagen type I with improved mechanical features (which will be closer to mimicking the real structure of the sclera) or printing collagen type III in mixtures with an elastic biomaterial. One example in this direction could be MeTro, a recombinant human tropoelastin modified by adding methacryloyl groups, thus resulting in a substance that enhances elasticity but also provides mechanical stability [166].

With regard to the iris or the crystalline lens, there are already available methods that do not necessarily involve 3D bioprinting but require a process of manufacturing that has already proved highly efficient for patients. Some examples in this direction include the silicone iris prosthesis ArtificialIris (HumanOptics, Erlangen, Germany), used as an innovative option for the treatment of various types of iris defects [167], and artificial lenses used in cataract surgeries [168,169]. Given that these are already highly functional and cost-efficient approaches for replacing the iris and the lens, bioprinting them would not bring additional value and could be considered of lesser importance.

The vitreous humor plays a crucial role, in complementarity with the sclera, in the mechanical strength of the eye. It also plays an important role in light transmittance, having a refractive index of 1336, similar to that of the cornea [170]. The vitreous humor consists mainly of water (approximately 90%), but also contains important biomechanical components such as collagen type II and hyaluronic acid [171–173]. Both hyaluronic acid and collagen type II are printable [164,174], so a printed vitreous humor could be developed if the kinematic viscosity of 300–2000 cSt could be reached [175]. The challenge is ensuring a higher viscosity in the central part of the humor [170].

In our view, the flow of aqueous humor in the eye is most probably not achievable, due to the challenges of ensuring drainage and exchange of nutritional substances [176]. Using a biomaterial rich in water and glucose that would also maintain transparency and optimal transmittance of light could be the solution, but, to our knowledge, no reported materials fit this description [177].

Finally, after addressing the main components of the eye and the possibility of building them using 3D printing, we still have to face the main challenge, which is integrating all the structures described above in an interdependent system and ensuring their supply of nutrients and oxygen via vascularization. These represent drawbacks not only for the bionic eye but for any artificially-made functional organ. Further research into understanding and controlling neovascularization, as well as developing bioinks with

VEGF-stimulating materials, may offer solutions for the future of bionic organs. As such, the possibility of printing choroid or conjunctiva remains more distant, due to their heavy reliance on vascularization [178,179].

2.4. Ear Auricle 3D Bioprinting

In order to print an auricle or pinna that mechanically and metabolically mimics a human ear, a stable and bioactive cartilaginous structure first needs to be obtained [180]. The outer ear plays an important role in the formation of sounds due to its complex shape and consistency [181]. Therefore, besides the need to create an optimal turnover of the cells contained in the final print, the shape and the mechanical properties are also very important. There are numerous approaches for creating a stable bioink that serves the purpose of creating a chondrogenic scaffold. A group of researchers developed a new bioink by generating ECM derived from microtia tissue [182]. The affected human ear was implanted subcutaneously in mice, and from that, a microtissue was developed, which further became the source for the cells contained in a bioink used in the 3D bioprinting of a normal ear [182]. Hence, the team managed to develop a bioink capable of creating a model that performed well in vivo while also presenting great mechanical properties, such as optimal stiffness and elasticity [182]. However, due to its invasive character, other methods need to be identified. An innovative technique was developed in this regard, called Digital Near-Infrared Photopolymerization (DNP)-based 3D printing. Relying on the properties used in DLP or SLA, such as photopolymerization under UV light [174], but using Near-Infrared Red light, which is more deeply penetrating, it became possible to induce polymerization processes in the subcutaneous layer of a mouse, thus making it a non-invasive method [183]. With this approach, the research team created a human ear model under the skin of a mouse by injecting the animal with hydrogel mixed with chondrocytes. The light was directed using a digital micromirror device to activate the polymerization process. The next step would be to extract the metabolically active ear from the animal model and surgically attach it to the patient. At this stage, however, a drawback appears: the cells added to the build require resources to ensure proper turnover, and methods for ensuring cell viability need to be developed before transitioning to patient trials.

Considering that the cartilaginous tissues are avascular [184], a good approach to ensuring nutrients and oxygen supply might be creating a highly porous structure, which can then be connected to the patient's vascularization. Furthermore, to ensure good aesthetic final results, the final build should be covered with bioprinted skin.

Another important aspect of the possibility of creating accessible bioprinted prostheses is the cost, and creating more accessible printers could be the solution to this issue. For example, by only making slight modifications to a classical thermoprinter, an outstanding bioprinter was obtained and used to print an equivalent of a human bioactive and cell-laden ear [185].

One of the areas where further optimization would be beneficial is the materials used for creating the cell-laden scaffold with chondrogenic properties. It was shown that stem cells and chondrocytes require soft materials for optimal development, a feature which appears to contradict the implicit rigidity and hardness of the final product [186]. Good results can be obtained by mixing hydrogels with other bioactive materials such as hydroxyapatite, but further research in the area of developing other materials is needed [186].

Another challenge in the generation of cartilaginous structures is the limited time during which chondrogenic-promoting cells maintain their activity and properties [186]. Human mesenchymal stem cells or human chondrocytes are commonly utilized in creating cartilage, but they mostly lose their chondrogenic properties after the fifth passage [187–189]. A solution that could be attempted is the usage of human amniotic fluid-derived stem cells, which can maintain their chondrogenic properties for a longer time and are also bioprintable [190,191].

To summarize, while great results have been reported in bioprinting human-like outer ears, the possibility of applying these constructs in humans remains minimal and costly, even if achievable. Furthermore, the currently existing biomaterials and utilized cells do not yield consistently effective results, indicating that additional research is required.

2.5. Tympanic Membrane 3D (Bio)Printing

The most common tympanic membrane defects are perforations [192]. Around 55,000 surgeries are performed annually and are mostly represented by tympanoplasty [193]. The success rate reported in tympanoplasty surgery varies from under 50% to up to 100% in some cases [194]. These particular statistics suggest that this kind of surgery is strongly dependent on the dimension and shape of the defect, but also on the practical abilities of the surgeon. The main challenge for surgical teams is related to the hand-crafting of grafts, and in cases of unusual morphologies of tympanic membrane defects, the results can be less predictable [195].

Bioprinting patient-specific grafts for defects of the tympanic membrane is now a suitable option for modern tympanoplasty, with a higher success rate. A team of researchers explored this direction by printing GelMA implants, which were later tested in vivo on animal models. An accurate model was built after scanning the tympanic lesion, with precise dimensions and shape. Supportive "butterfly" structures for anchoring the grafts without sutures were also designed [194]. The team reported the achievement of a construct with good mechanical properties, and the supportive structures were able to anchor the grafts even at pressures above 5 kPa [194].

However, it is known that the human tympanic membrane, similar to those of other mammals, can withstand an ultimate stress of over 1MPa [196,197]. Although the possibility of applying patient-specific grafts with no sutures is innovative and opens new possibilities, the mechanical properties could be further improved. One solution to increase the tolerated stress of materials compatible with 3D printing involves decreasing the spacing between filament lines contained in the final product [198]. Therefore, by adding different biomaterials to the final mixture or adapting the temperature or light exposure time, reinforced hydrogels can be obtained to withstand mechanical stress as high as 0.4 MPa [198].

There are a few reported studies where reinforced hydrogels were obtained, but their builds exhibit characteristics that deviate from the optimal mechanical and vibratory properties of the eardrum. For example, a tympanic membrane made of PCL, a material with great elastic and mechanical properties, managed to withstand forces of around 142 kPa [199]. Although this value is much more suitable for behaving like a real eardrum, a graft three times thicker (300 μm) than the physiological thickness of the tympanic membrane was required, which can seriously affect the elastic properties of the eardrum [199].

To address the challenge of implementing elasticity and optimal vibrational properties in the graft, a different approach was needed. As suggested above, reinforced hydrogels can be the perfect candidate to obtain desirable mechanical properties. Thus, a scaffold consisting of three polymers (PLA, PCL, and polydimethylsiloxane-PDMS) was built and filled post-printing with hydrogel. The construct exhibited superior features, both acoustically and mechanically [200]. The final thickness of the graft was around 700 μm—seven times thicker than the human eardrum; however, by designing a radial scaffold with large areas that were later filled with a thin layer of <50 μm of hydrogel, the authors managed to overcome the low elastic properties of other grafts [200]. Still, due to the high density of the graft, acceptable acoustic results were reported only between 1000 and 6000 Hz, while under 1000 Hz, sound transmission dropped significantly [200].

Another solution to reinforce hydrogels for better mechanics was the addition of KerMA (methacrylated keratine) to the printing inks [201]. Considering that keratine presents good mechanical properties, the designed grafts were able to withstand stress up to 0.37 MPa [201]. Furthermore, this is the only study, to our knowledge, in which the graft could be filled with human stem cells, making it a promising approach and the first with reported results regarding its bioprintability [201].

Printing an eardrum is a very demanding task due to its viscoelastic and mechanical properties. While numerous advancements have been made, no constructs performing as well as a human eardrum have been developed. Printing grafts specific to the dimensions of the defect could optimize the surgical time and complexity, but significant scientific improvements are still required in this regard. As previously discussed, further development of printing technologies may allow for more accurate and thinner structures while preserving functional mechanical strength.

Furthermore, none of the articles reviewed here report the elastic modulus of their constructs, an aspect that, in our view, is important for future research, given that the tympanic membrane has a Young's modulus between 34 and 59 MPa [202]. Additionally, hybrid approaches could represent potential solutions, for instance, continuing the clinical use of cadaveric tympanic membranes while reducing the risk of infection by coating them with nanoparticles filled with antibiotics [201].

2.6. Nose 3D (Bio)Printing

This section addresses the possibility of manufacturing a bionic nose. There are three main components of interest, indispensable for a minimal function of a nose equivalent: the cartilage, the nasal epithelium, and the olfactory system sensitive to odors [203]. The nose cartilage plays a double role: cosmetic and functional, i.e., ensuring the optimal velocity and flow of air for good temperature regulation, allowing enough time for the interaction of receptors and air particles, as well as air filtration [204].

Several bioinks have been researched to mimic the nose cartilage. One of them stands out due to the approach used for crosslinking the hydrogels—horseradish peroxidase added to a mix of living cells, glucose, alginates, and cellulose nanofibers [205]. The innovation lies in the function of the peroxidase to catalyze conjugation reactions, thus ensuring optimal crosslinking of gels without altering cellular viability [205]. Cellulose nanofibers have the advantage of superior control of the bioink viscosity, a characteristic relevant both to ensuring an optimal printing process and to the development and survival of the cells [206,207].

Furthermore, this type of nanofiber can mimic the structure of cartilage collagen, enhancing its mechanical stability. However, when relying only on alginates and nanofibers, higher concentrations are required in order to obtain such a complex and robust structure as the nose [205]. These particular aspects directly affect cell viability, limiting survival to one week—drawbacks compounded by the already increased cytotoxicity induced by the H_2O_2 generated by the peroxidase [205].

Other ways to enhance the robustness of the final build are needed. An alternative could be the use of GelMA as the main scaffold material, a gel that ensures high mechanical stability without increasing the early death rate of the cells added to the bioinks [208]. PEGDMA (polyethylene glycol dimethacrylate) can also be added to the mix, with a twofold advantage: increasing the softness of the 3D constructs and ensuring a suitable cell attachment with prolonged viability [209,210].

Other biocompatible materials can be used as equivalents to the nasal cartilage, and collagen type I is a great candidate [211]. However, when choosing optimal mixes for bioprinting cartilages, four main aspects must be considered: (1) using low temperatures that the cells can tolerate while ensuring good printability; (2) designing or creating pores with different biomaterial concentrations to support cellular viability; (3) obtaining a balance for the needed mechanical stability without affecting the cells during printing or afterward due to increased stiffness; and (4) decreasing viscosity as much as possible without affecting the deposition of the materials in the final constructs, thereby ensuring proper cell development and prolonged viability [212–214].

Another crucial contributor to the functionality of the nose is the nasal epithelium. Its roles include filtering harmful particles, creating a barrier against pathogens and other antigens or allergens, and secreting mucus to support the aforementioned properties. An artificial nasal epithelium was created to meet these characteristics [215]. Using human nasal progenitors, five major epithelial cell populations were obtained: basal, suprabasal, goblet, club, and ciliated cells [215]. This multilayered development of various cell types was obtained with the use of droplet-based bioprinting [216]. It was postulated that this technique minimizes the mechanical stress typically experienced by cells during layered printing, thus supporting proper epithelial architecture development [215].

A main focus of recent literature has been obtaining individual parts of different organs. However, the necessity of creating a scaffold for their attachment is often left out. This could represent a good direction for future research, specifically for attaching the nasal epithelium to the nasal cartilage. A viable solution could be a scaffold able to produce ECM, while also maintaining the viability of the cells on both sides, i.e., the chondrocytes in the nasal cartilage and the epithelial cells in the mucosa equivalent. Therefore, researching a good candidate for this type of ECM where the TGF-β1 concentration is controlled is needed [217].

Developing a system that can perceive the sense of smell through bioprinting can be a difficult task due to the complexity of the neuronal connections and the receptive system [218]. Digital sensors already exist, which are usually designed on top of biocompatible polymers that are used in experimental settings [208]. However, numerous challenges remain. Cellular viability in the aforementioned studies was not tested beyond 6 weeks, so it is unclear what the cellular behavior will be, especially when interacting with the biocompatible scaffold. Given the presence of metabolic activity, there is no alternative to "clean" the viable cells in situ without properly developing neovascularization in the scaffold. Additionally, it is uncertain whether designing bionic organs in a hybrid system with digital parts will create any immunological responses or increased cytotoxicity in the scaffold itself. These are a few aspects that need to be further researched to get closer to the goal of replacing human body parts.

2.7. Periodontal Tissue 3D Bioprinting

The periodontium protects and supports the teeth, and is formed of two soft tissues (the gingiva and the periodontal ligament) and two hard tissues (the alveolar bone and the cementum) [219]. Diseases affecting the periodontium are considered important oral pathologies; periodontitis in particular affects 1 in 5 adults [220,221]. Left untreated, periodontitis leads to tooth loss [222], and the chronic inflammation does not allow a proper integration or sustainability of dental implants, if used as a treatment for tooth loss [223]. In contrast with traditional, non-surgical treatments of periodontitis, recent studies have employed a treatment based on stem cell-based regeneration of the periodontal tissues [224–228].

Proper proliferation and bioactivity of stem cells can be achieved through incorporation in biomaterials-based scaffolds via 3D bioprinting. As discussed in earlier sections of this paper, GelMA and sodium alginate mixes are frequently used in bioprinting applications due to their great compatibility with printers. In a study on periodontal tissue regeneration, bioactive glass microspheres, growth factors, and stem cells were added to the biomaterials mix in an attempt to simulate the ECM bioactive properties of the periodontium [229].

The glass microspheres promote osteogenesis and angiogenesis, while the GelMA-sodium alginate hydrogel simulates the mechanical properties of the periodontium, allowing stem cells to reconstruct both soft and hard tissues [230–234]. Repairing both types of periodontal tissues at the same time has proven challenging via traditional methods and even through novel approaches using only stem cells and hydrogels. By adding growth factors such as BMP-2 and PDGF, adequate simultaneous healing was achieved [229]. Although the mix of substances was crucial for fast healing, optimal water retention, and metabolite flow, the researchers reported low mechanical strength of the scaffold when tested in vivo on beagle dogs. Thus, an alternative to the utilized hydrogels needs to be tested in order to improve the final mechanical characteristics of the prints. Biomaterials such as hydroxyapatite [235,236], polycaprolactone [237–239], poly-L-lactic acid [240,241], or collagen [242–244] are great candidates to replace or to be added to GelMA multicomponent hydrogels.

However, materials that exhibit enhanced mechanical properties could affect the viability of stem cells or the efficiency of the printing process [245]. A solution to balance the challenges of using stiff biomaterials could be the design of scaffolds with numerous pores, favoring the proper development of stem cells, while also allowing for the development of optimal vascularization.

As mentioned earlier, treating patients with dental implants without addressing the associated periodontitis is highly inefficient. Defense mechanisms against oral bacteria are impaired if the periodontal ligament is missing [246]. A solution to this issue is the coating of dental implants with human periodontal ligament stem cells. It was shown that these stem cells perform well on titanium dental implants, a material with good properties when used in dental applications [174,246]. The periodontal ligament stem cells promote the generation of fibrous tissues, cementogenesis, and enhanced aesthetics [246]. Human periodontal ligament stem cells also demonstrated good results, and 3D-bioprinted tissue models of the alveolar bone–periodontal ligament biointerface showed enhanced physical, chemical, and rheological properties [247].

One of the most recent approaches in tissue engineering and regenerative medicine is the development of extracellular vesicles, also known as exosomes. Usually referred to as a "cell-free" tissue engineering method, when bioprinted, it can potentiate stem cells in the printed scaffold [248,249]. Bioprinting exosomes has also been used in the treatment of periodontal defects, with the main advantage being the enhancement of stem cell activity without modifying the mechanical properties of the final product [250].

While the repair of periodontal tissue defects using 3D bioprinting has shown promising results, translation to human patients has not yet been achieved. Mechanical stability of the scaffolds remain a critical issue to be considered in parallel with the enhancement of stem cells' period of activity. Additionally, scaffold vascularization optimization is essential, especially for periodontal ligament scaffolds, where metabolite flow is crucial. Further development and testing of bioinks in human patients are necessary for implementing the tissue engineering results presented above.

2.8. Salivary Glands 3D Bioprinting

The parotid, submandibular, and sublingual glands, along with the numerous minor salivary glands found in the oral mucosa, ensure many functions such as antimicrobial activity, and facilitating swallowing, mastication, and speech [251,252]. Many pathologies (autoimmune disorders—Sjögren's syndrome, sialadenitis) or therapies (either medication-based or radiation therapy) can lead to xerostomia, which seriously affects the patient's quality of life [253–255]. Traditional treatments for xerostomia in patients include the use of pilocarpine to stimulate saliva production and surgical interventions to unblock the main ducts. In more severe cases, when gland resection is performed, artificial saliva substitutes are prescribed to improve xerostomia symptoms [256,257].

It can be noticed that none of these common therapeutic approaches for various salivary gland defects attempt to restore the internal architecture or functions of the glands. Self-regeneration in these structures is limited in most adults [258]. In recent years, stem-cell therapies have been attempted with partial success. While salivary gland stem cells can help restore the functionality of the affected areas, there is still an inability to control cell placement, thus resulting in a chaotic architecture [258]. There is a need for regenerative approaches in which good architecture and functionality can be achieved, and an innovative approach is via 3D bioprinting.

Studies on the value of the elastic modulus of the major salivary glands in humans indicate that the optimal interval of functionality is at around 10–20 kPa. These values are required for maintaining the integrity of the acinar structures of the glands, while also not collapsing under the muscular tension involved in the mastication process [259–261]. Therefore, besides ensuring biocompatibility, functionality, and bioactivity of the salivary scaffold, an important characteristic is reaching these mechanical values without affecting the process of 3D bioprinting, nor the viability or proliferation of the stem cells used in bioprinting mixtures.

GelMA is a strong candidate for creating the microarchitecture of salivary glands that ensures the functionality of a gland equivalent. While mimicking the microenvironment, it also offers a favorable mechanical balance, an optimal stiffness that can be adjusted by changing the concentration of methacryloyl added in the hydrogel, and an optimal elastic modulus ranging from 1 to 20 kPa, suitable for the acinar and ductal system of the salivary glands [262]. A recent study used GelMA and alginates to further increase the mechanical stability of the bioink in the process of printing, aiming to create an innovative coaxial design with intersecting hollow tubes. This design allowed the primary salivary gland cells derived from mouse salivary glands to create acinar-like spheroids with enhanced functionality [262].

In theory, as long as the mechanical properties of different biomaterials are achieved without affecting printer functionality or cell viability, any substance can be used for these applications. Thus, other bioactive materials could serve as promising candidates: hyaluronic acid-based hydrogels, which present good mechanical stability for acinar-like structures [263]; polyethylene glycol hydrogels, which offer tunable mechanical properties while promoting cell viability [264]; or fibrin hydrogels, with tunable elastic properties, and the ability to promote cell migration and adhesion [265].

Another study used gelled egg yolk plasma, a natural compound, as a gel for salivary gland scaffold generation [266]. Similarly to the other hydrogels mentioned, its role is to mimic the natural ECM of the salivary glands. Rich in vitamins, lipoproteins, and ECM-like proteins, egg yolk plasma is a good candidate from a bioactive standpoint. Although its mechanical characteristics were not reported, based on the proteins of egg yolk, the elastic modulus is estimated to range between 1 and 5 kPa, which could be a suitable value if further enhanced with other materials or by adjusting the thermal gelation process [266,267]. Compared to other materials, an advantage of using egg yolk plasma is that it contains natural growth factors such as TGF-β, which can promote cell proliferation and viability within the bioink mix; in this case, salivary epithelial cells were used [266,268].

The stem cells used in the aforementioned studies were human or animal salivary gland stem cells. However, other candidates exist for simulating acinar-like structures, such as dental pulp stem cells (DPSCs). It has been observed that when DPSCs are exposed to fibroblast growth factors, epidermal growth factors, and bone morphogenetic proteins, they undergo acinar-like differentiation [269]. While these cells have only been tested in animal models, they could represent a good alternative for gland tissue engineering and may be further integrated into various hydrogels such as GelMA, HA, or fibrin hydrogels, as reviewed earlier.

Although the outlook of regenerative medicine for salivary glands is promising, an entire gland has yet to be created. Ensuring good vascularization of these structures is essential for proper salivary secretion and the flow of metabolites and nutrients, which also enhance cell viability. Furthermore, hybrid salivary glands would require neural control and communication between the major glands through neural and hormonal signaling [258]. The incorporation of different neovascularization factors into the prints or the design of prevascularized scaffolds could be an achievable approach in the near future, but designing appropriate neural control mechanisms requires further study.

2.9. Skin and Hair 3D Bioprinting

The significant progress of 3D bioprinting technology may enable the creation of a bionic head and neck complex. Numerous pathologies—such as thermal or chemical burns, traumatic events, or even genetic disorders—severely impact patients, and restoring aesthetics and functionality, including facial expressions or other receptive functions, remains highly challenging [270,271].

Skin grafts remain the gold standard for restoring skin integrity. However, the lack of sweat gland function, neural control, and hair follicles, as well as the presence of the immune response against the graft or associated infections, represent major disadvantages of this method [272]. Through 3D bioprinting, multi-layered printing has become more controlled, allowing for a better balance in the cell–biomaterial arrangement and advancing the regenerative aspects of skin treatment [273,274].

An important achievement in this direction is represented by a bi-layered skin model tested in nude mice. Epidermal stem cells and skin-derived precursors were added to properly stimulate hair follicle regeneration and sebaceous gland formation, as well as to to promote angiogenesis [275]. These cells were added into a multi-hydrogel mix made of GelMA and hyaluronic acid methacrylate (HAMA), chosen for its mechanical strength and elastic properties [275]. While the results are excellent, especially due to the capacity of the skin model to allow functionality and formation of glands and hair (Figure 3), some aspects need to be discussed regarding its application in humans, especially in the head and neck area.

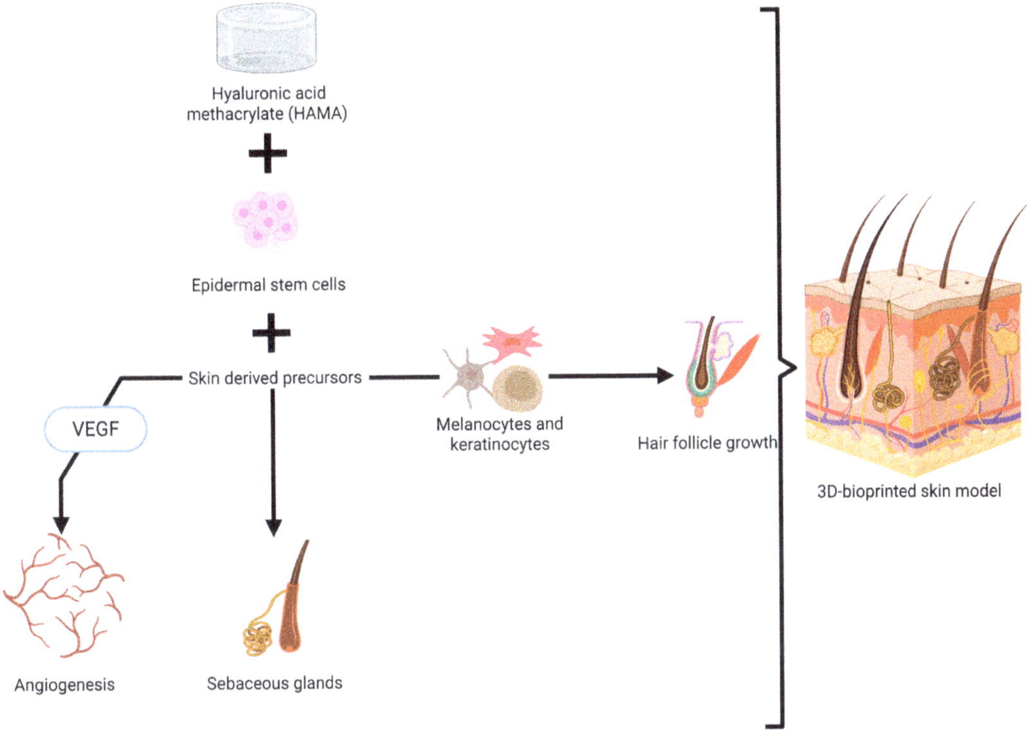

Figure 3. Visual representation of 3D printing of skin with sebaceous glands and hair follicle growth [275–277]. Created in BioRender. Timofticiuc, I. (2025) https://BioRender.com/kt50osr, accessed on 21 June 2025.

Ma et al. ensured mechanical strength by adding 0.5% HAMA to the bioink; however, the elastic modulus and other mechanical characteristics were not reported [275]. We need to consider that the average range of skin stiffness is 13.2–33.4 kPa, which varies greatly in different areas of the body [278]. For example, the chin has a low stiffness of around 20 kPa with a high percentage of elasticity (50.6%), while the back of the ear presents a stiffness of 40 kPa [42]. Furthermore, there are different mechanical properties between each layer of the skin; the dermis has a Young's modulus of around 40 kPa, while the epidermis can reach up to 4 MPa [279].

Furthermore, unlike other regions, the skin of the head and neck not only requires functionality, mechanical strength, and elasticity, but also a cosmetic appearance, which can be difficult to control in models where stem cells activate independently. In addition, the growth of facial hair is different from other parts of the body [280]; thus, creating models for this area is more challenging. Neovascularization in these artificial models may be improved by adding VEGF, or incorporating other stem cells and growth factors—potentially representing a promising direction for future research.

Other scaffolds have also been designed, and improved methods for controlling hair follicle formation and hair growth have been identified, such as the use of mesenchymal stem cells [281]. Through the combination of epithelial and mesenchymal stem cells with collagen droplets, a model with a high density of cells was obtained [281]. This promoted small contractions in the collagen hydrogel, mimicking the natural tensions found in tissues.

This characteristic supports better development of integrative elements of the models, such as vascularization or ECM [281].

Another study highlights the importance of incorporating other types of stem cells for adequate hair follicle generation, such as keratinocytes and melanocytes. These cells contribute to the consolidation of the hair follicle while also playing a role in hair pigmentation, making the skin models more realistic and aesthetic [282]. Integrating sweat gland stem cells in skin models is also possible, which completes the elements required for an aesthetic and functional skin model [283]. More efficient methods of printing, such as robot-assisted techniques, have also been tested when building models based on epidermal and mesenchymal stem cells [284,285].

Controlling such a high number of stem cells in a model is very difficult, especially due to their high metabolite production, which requires good vascularization. Moreover, studies investigating their interactive behavior for long periods of time are lacking, so aspects regarding the immunological response and resistance over time should be further researched. The subcutaneous tissue also warrants further investigation, and some promising results have already been reported using adipose stem cells for generating this layer [286]. However, additional studies may reveal the complexity of creating multilayered models, while also preserving the aesthetic aspects and different mechanical characteristics of each area of the head and neck.

2.10. Exploring the Feasibility of a Bionic Neck

A bionic neck would require crafting a model mimicking all the functions of the neck, while also ensuring integration with other anatomical regions in terms of vascularization, neural control, endocrine control, and muscular and osteoligamentous integration for stability and structural integrity. We have addressed the soft and functional components of this region, i.e., the functions performed by the larynx, trachea, pharynx, the initial part of the esophagus, and also the endocrine elements: thyroid and parathyroid glands [5]. However, the number of bioprinting studies in this anatomical area is very limited. Except for the larynx and trachea, where some research has been conducted, other elements of the neck have not yet been properly studied regarding bioprinting methods, alternative regenerative treatments, or integration into human patients.

2.11. Larynx 3D Bioprinting

The larynx presents a complex neuromuscular architecture that balances its vital functions: the passage of air, protection of the airway during swallowing, and phonation, the latter standing out due to its psychological and social implications [287–289]. Many pathologies, either congenital malformations or acquired defects from traumatic events, inflammation, cancer, or allergic conditions, can affect the larynx's capacity to perform its critical functions [290–293].

Depending on the pathology affecting the larynx, many treatments can be performed, but each one presents challenges and room for improvement. Muscle-related defects can be treated pharmacologically, but the uncertainty regarding the administered dosages and the frequent dysphagia as a side-effect calls for an alternative to be sought [294,295]. Surgical reconstruction of the larynx can also be performed, but many of the interventions are associated with increased complication rates, prolonged healing, and high mortality [296,297]. Furthermore, nerve-related defects, which often result in voice loss, can be treated using recent reinnervation techniques, but with high risks and often yield only partial restoration of phonation [298,299].

The main areas of progress in 3D bioprinting of larynx models have been in relation to recreating its complex structures. Maintaining cell viability in the scaffold while ensuring the scaffolds have good mechanical strength represents a major challenge. The structure of the larynx is firm, but increasing the rigidity of the bioinks requires higher concentrations and viscosities of the hydrogels [300]. The study performed by Galliger et al. demonstrated how cooling systems can increase the mechanical characteristics of the final print required for a larynx equivalent, without affecting the printing process [301]. Additionally, the team developed a mix using decellularized extracellular matrix microparticles of porcine origin to further optimize the printability at lower temperatures, of around 15 °C.

Although GelMA or any other methacryloyl-based bioinks allow a high rate of cell viability post-printing (around 90%), in the study of Galliger et al., no cells or growth factors were used [23]. Obtaining superior mechanical properties and optimizing printing techniques for a simpler print process are promising developments; however, it is unclear if increasing internal forces of the scaffold to better mimic the larynx structure is supported by the cells. Furthermore, increasing the concentrations of the materials used in the scaffolds may decrease pore size, negatively impacting cell viability and function [300].

Other methods have also been proposed to meet the mechanical needs of an artificial larynx model. By integrating a mix of hydrogels (GelMA-glycidyl-methacrylated HA) in a polycaprolactone scaffold, increased structural support can be achieved [302]. By adding a biopolymer such as PCL, a controlled and prolonged degradability of the scaffold is ensured [174,302]. This aspect allows sufficient time for the added stem cells to develop and proliferate in the scaffold, while also replacing the PCL. Over time, a natural chondrocyte mechanical structure that supports the hybrid hydrogel scaffold is created, offering improved integration of the model into a potential patient [302].

However, as observed in the aforementioned study, a larynx model was built only focusing on its cartilaginous constitution. Musculoskeletal differentiation is also important if an accurate model is desired, and this can be achieved by introducing bone marrow-derived mesenchymal stem cells, as shown by McMillan et al. [303]. This reiterates the idea of more in-depth research on the generation of multilayer structures in which superior control can be achieved in terms of cellular composition and mechanical properties.

Features such as clinical translation, scaling up models for integration in human patients, the viability of the cells, and identifying what types of cells better perform under mechanical conditions as found in the larynx, need to be addressed in future research.

2.12. Other Neck Elements and Future Possibilities

Anatomically, the larynx is formed of cartilaginous components (thyroid, cricoid, and arytenoid cartilages) supported by ligaments and muscles [304]. The trachea is also formed of cartilaginous structures, differently shaped to ensure airway passage, mobility, and elasticity during different neck movements or swallowing [305]. Thus, for the cartilaginous part of the trachea, a similar approach to that of the larynx can be undertaken [306,307].

Including chondrogenic stem cells in the model and using compatible hydrogels or polymers with superior mechanical properties could lead to great results. Several studies have explored cartilage and smooth muscle formation for engineering tracheal tissue [308]. A polycaprolactone scaffold with added human mesenchymal stem cells was used to create a trachea model with cartilaginous and muscular components that presented mechanical properties similar to the physiological ones [308].

Hydrogel mixes have also been designed to mimic the extracellular matrix of the trachea, ensuring good chondrogenic differentiation. An innovative approach for bioprinting soft tissues was represented by a bioink made of methacrylated HA, polyethylene glycol succinic acid ester (PEG-NHS), and methacryloyl-modified dermal acellular matrix, which promoted vascularized fibrogenesis, thus addressing vascularization—one of the main drawbacks [309]. Adding fibroblast and chondrocyte stem cells to this mix resulted in obtaining a perfect balance, leading to the formation of mechanically stable C rings and a functional membranous structure, i.e., the vascularized fibrous rings [309].

Furthermore, through chemical reactions obtained by adding PEG-NHS, a stable bond was formed between these rings, achieving an anatomical tracheal equivalent [309]. However, translating these results to humans presents a high risk, mainly due to the need to ensure good airway passage. Moreover, functionalizing the trachea has not yet been achieved, and this could represent the next objective for future research in airway passage bioprinting.

The esophagus and the pharynx are mainly formed of smooth and skeletal muscle [310,311]. Without taking into consideration the neural control that needs to be obtained in these organs, and their secretory functions [310,311], the main objective of research in 3D bioprinting models for these organs should be obtaining a muscle-like scaffold with regulated contraction. Hence, compatible materials and optimal development of stem cells need to be investigated. Some elements have already been studied, but not specifically for the pharynx or esophagus. For example, microfluidic 3D bioprinting was used to ensure good control over cell placement, allowing for human smooth muscle cells to be bioprinted in a scaffold, and controlled contractions and pharmacological response were obtained [312].

Other types of stem cells can be targeted in the formation of smooth muscle using a decellularized smooth muscle extracellular matrix as a microenvironment for stem cell development [313]. Using this approach, adipose-derived mesenchymal stem cells encapsulated in a hydrogel were able to express smooth muscle proteins (alpha-actin) in a 3D bioprinted scaffold [313]. This is a promising result demonstrating various possibilities for generating smooth muscle from different stem cells, while also addressing the regenerative and contractile properties of 3D bioprinted smooth muscle scaffolds.

Through similar approaches, skeletal muscle can also be generated. Scaffolds with contractile and pharmacological responses were crafted using primary human muscle precursor cells, carefully printed after being coated with hydrogels for mechanical stability and protection of the cells during printing [314]. Results on murine models are also promising, providing support for the translation into human models. A team of researchers managed to apply murine muscular stem cells to generate an entire tibial muscle, which was later grafted into the animal model [315]. Worth mentioning is the use of polymers (PEG) instead of hydrogels for the generation of a mechanically stable scaffold for this model, demonstrating the versatility of biomaterials that can be used for future medical applications [315].

These few studies mentioned above are proof enough that smooth and skeletal muscles can be bioprinted. However, integration of these constructs into a muscle-based organ model is challenging. It is still uncertain how the stem cells added to the bioinks integrate with other elements of an active organism. Their activity, viability, or functionality is not fully controllable. Furthermore, a self-sustainable cycle of life for these scaffolds is not yet ensured because the results do not show how these scaffolds or models act in prolonged use. Additionally, neural control is not easily replaceable using only pharmacological approaches or one-function scaffolds. Metabolite production could cause critical defects in the scaffolds or produce immunological or allergic responses in patients if not controlled correctly under optimal vascularization. These are critical aspects that researchers should take into consideration.

Generally speaking, ensuring functionality represents one of the most challenging aspects of artificially created models or scaffolds. It is uncertain whether the use of stem cells derived from donor organs or tissues is sufficient for generating 3D builds capable of bioactivity in the replicated structures. Both mechanical characteristics and cell viability are critical coordinates when first designing such constructs.

While the possibility of generating a bionic part of the human body exists, it is mandatory to properly document all the potentially obtainable bioinks to reach this ambitious medical objective. Developing more materials, researching the maximal viability of the added stem cells, finding solutions to ensure the cells' cycle of life, and controlling their development are some of the main directions that should be researched in this field.

3. Vascularization Possibilities and Neural Control for 3D Models

The rapid development of materials and 3D bioprinting techniques has led to impressive achievements in creating patient-specific and self-regenerative models or scaffolds. However, the dimensions of scaffolds have rapidly scaled up, leading to challenges in their functionality [316]. One of the significant issues when incorporating cells into these constructs is the need for nutrients and oxygen, necessary for cell viability and development [316].

When small scaffolds, under 0.2 mm in thickness, were initially designed for in vitro testing, the nutrients in the in vitro microenvironment were sufficient to supply the cells with all requirements [317,318]. Due to upscaling, the main concern in recent studies has been the identification of methods to prevent the necrosis that occurs frequently in large and thick scaffolds [317,318].

Various solutions have been proposed to enable metabolite and nutrient flow in 3D bioprinted constructs. One approach relies on building vascular networks in vitro, thus creating pre-vascularized scaffolds to be later grafted and connected to the patient's blood supply. For this purpose, a more precise bioprinting technique called microscale continuous optical bioprinting needs to be employed, creating very high-resolution scaffolds that permit adequate vessel development [319].

Another avenue involves printing endothelial cells directly into the scaffold at the same time as the hydrogel, spontaneously generating in vitro tubular-like structures with the ability to connect to the patient's blood supply when transferred in vivo [319]. Designing pores in the scaffolds facilitates optimal development of the endothelial cells, thus resulting in improved release of VEGF. An optimal prevascularized scaffold can be obtained through this method by creating a ceramic outline, which will result in high water retention—necessary for good cell development—with pores ranging from 80 to 100 μm [320].

Creating pre-vascularized scaffolds does not require, in all cases, the use of stem cells. If more specific and controlled vascular networks are desired, direct printing can be performed using a bioink that will be later removed, also known as sacrificial ink [321]. In the end, an exact tubular-like network is created that can play the role of a vascular network equivalent.

Prevascularized scaffolds can be safe for patients in some aspects. The initial vascular formation is created in a controlled in vitro environment, so any potential defects are easy to observe, and also, prolonged viability tests can be performed. A disadvantage of adopting this solution could be the long period required for the production and testing of the scaffold.

Other studies rely on introducing angiogenesis factors and stem cells that promote vessel formation in the scaffold, in vivo. Bioprinting a co-culture of human adipose tissue-derived microvascular fragments and human dermal fibroblasts can lead to an in vivo self-assembled capillary network that connects to the host's blood supply [318]. Additionally, vascular growth factors can be directly added to the scaffolds to support the formation of vascular networks, enhance stem cell activity, and ensure a more controlled development of vessels [322].

Many approaches to vascularizing 3D bioprinted constructs exist (Figure 4). The best method to vascularize bioprinted models needs to be chosen in order to best fulfill its role in patients.

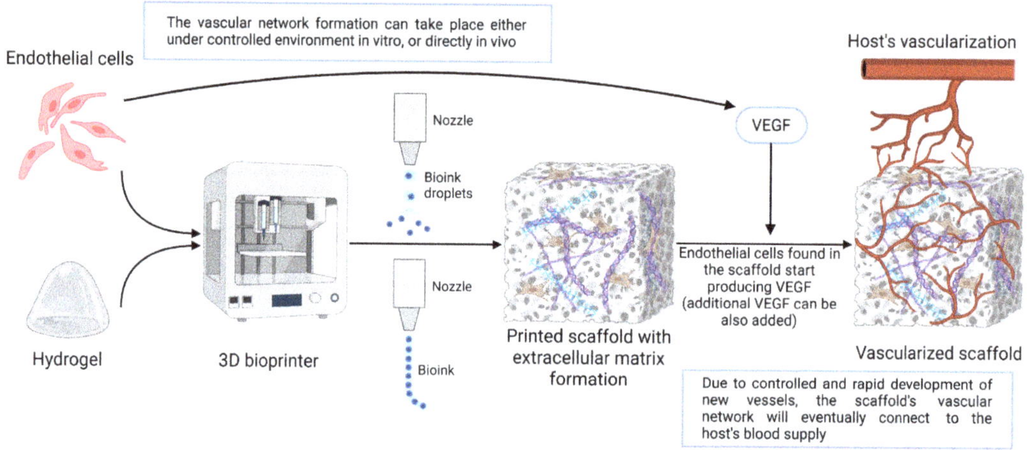

Figure 4. Method for the vascularization of 3D bioprinted constructs. In the initial stages, optimal cells are selected to formulate a bioink based on hydrogel or other materials. Next, through different bioprinting techniques (either droplet-based or classic extrusion), a scaffold or model is obtained, in which extracellular matrix elements form due to the intrinsic activity of the introduced materials and cells. In this case, endothelial cells produce VEGF, a growth factor that underlies the formation of new blood vessels. This process can be controlled in vitro or left to self-regulate in vivo. Finally, the scaffold or model is transferred to patients, where it ends up connecting to the host vasculature. Created in BioRender. Timofticiuc, I. (2025) https://BioRender.com/bintbnp. Date of last access—3 May 2025.

Of course, in most tissues, innervation and associated neural control are of vital importance; the adverse effects of nerve damage have been demonstrated in numerous cases, such as nerve-destroying infectious agents [323–325]. To perfectly complete the intrinsic control of tissue-engineered scaffolds, neural control is the final piece to fully mimic a human tissue or organ, at least from a mechanical functionality point of view. However, studies towards this complex achievement have mainly focused on stimulating the proliferation of neural cells in the scaffolds and designing neural networks capable of responding to chemical or electrical stimuli.

In the study of Joung et. al, neural activity was obtained after neural progenitors were added to the bioinks. After successful printing, these cells were able to differentiate into active neurons [326]. Similar results were obtained by adding pluripotent stem cells to the bioinks, where populations of neural cells with the desired functional characteristics were obtained [327].

More recent studies have managed to create neural tissues with enhanced neural activity and functional neural connectivity. These results were obtained after adding Neurogenin-2 human pluripotent stem cell lines to the hydrogel-based bioinks [328,329]. It was observed that by simply adding specific stem cells, theoretically, any function can be achieved. By respecting the mechanical aspects supported by these cells and printing them in compatible biomaterials, any function can be obtained. Regarding further steps in integrating neural scaffolds into patients—especially brain scaffolds—one of the main challenges is blood supply integration. A simple neovascular proliferation is not sufficient due to the complex vascular network of the brain [330].

However, controlling cellular development, increasing viability, and achieving full functional control remain challenges that need to be further addressed. Neural control is not limited to applying stimuli to the scaffold to obtain any kind of response. Ideally, achieving neural control in bioengineered scaffolds should mean establishing connectivity of neural networks with host nerves, ensuring hierarchical control over the constructs. All these aspects are to be researched in future studies.

4. Summary on 3D Bioprinting Technologies

The main aim of this paper is to review 3D bioprinted soft tissue scaffolds with potential for translation into clinical practice. Most of the constructs presented here are created using extrusion-based bioprinting. There are a few advantages associated with this technique, such as the possibility of printing very high cell-density scaffolds, using various types of different cells, thus allowing for the production of complex cell architectural models [331].

Furthermore, there are various categories of biomaterials compatible with this kind of printing, which can be adapted for the necessities of the cells and the function they will perform [332]. However, these printers can produce high pressures in the moment of the extrusion, generating shear stress that may affect cell viability [331]. Other technologies exist, such as inkjet bioprinting, which is more precise, but are limited to low-viscosity bioinks [333,334].

A distinct group of 3D printing technologies is based on vat-photopolymerization, a group of techniques (DLP and SLA) that uses light to polymerize biocompatible materials [174]. These methods are known for very fast printing, with high precision and resolution, which can be used for printing bioinks; however, additional post-processing steps, such as UV light curing, are often required [174,335]. The main advantages derive from the smooth surfaces that these techniques can produce, and the speed with which this mechanical characteristic is obtained [336,337]. The smooth surfaces and the precisely controlled pore sizes are essential for good cell adhesion and functionality [336,338]; however, the necessity of UV post-processing can affect cell viability [339–341]. Furthermore, with this group of techniques, not many materials are compatible with the added cells [335,342].

These classic printing methods are constantly being developed to expand the range of printable materials, allowing for smaller printed structures, faster printing times, and increased cell viability. Future research will also explore new 3D printing techniques such as in situ bioprinting, which is based on the principle of using bioink directly at the site of the defect [343,344]; multi-material techniques, which allow the creation of highly complex scaffolds [345]; and AI-assisted bioprinting for improved control and precision [346]. Droplet bioprinting using an air-focused platform also shows promise in various medical applications [347,348]. Direct sound printing is another emerging technology that uses ultrasounds to create acoustic cavitation, enabling in situ sonochemical reactions and highly complex prints—even under the skin [349,350].

5. Challenges and Future Perspectives

While it is true that 3D printing has been gaining traction as a realistic therapeutic modality in medicine over the last decade, its regular everyday application has faced a series of bottlenecks [351,352]. Cell viability in the final biomaterial scaffold remains one of the main aspects that need to be addressed. Most of the time, the mechanical characteristics of the build affect the structural integrity of the cells. Shear stress generated by high concentrations of biomaterials such as gelatin drastically decreases the viability of cells—from over 60% to under 20% (at 10% w/v and 20% w/v of gelatin, respectively) [353].

In general, an acceptable cell viability is considered above 70%; thus, enhancing 3D bioprinting techniques and adapting biomaterials are key aspects that every study should consider [354]. Ways of adapting mechanical aspects of the build to maintain high cell viability include increasing pore sizes or decreasing the concentrations of the biomaterials, thereby reducing mechanical stress. However, this also reduces the mechanical integrity of the final build, greatly impacting its functionality [354]. There are other methods to modulate the increase in cell viability, either by incorporating cells into different substances that can have a protective effect, such as graphene oxide [355], or by adapting the extrusion pressure, lowering it to protect the cells from mechanical stress, or even by using special bioinks that generate low shear stress [356].

One of the least studied aspects of cell viability is the maximum time of their survivability. There are optimistic results regarding high rates of cell viability, such as fibroblasts with over 80% viability 48 h post-printing [357], and with prolonged periods of survival, for example, keratinocytes, which presented a few morphological changes 21 days post-printing [358]. However, when translating these results in clinical practice, where a patient-specific scaffold should function for at least a few years, it is uncertain how these cells will perform. As far as we know, there are no reported studies that consider cell viability after 1 year, but considering the emerging ideas on vascularization aspects presented above, optimistic results are expected.

Translating these results on bioprinted scaffolds into clinical practice not only requires an understanding of the period of time during which the added cells can perform, but also how they could possibly interact with the human body, and if they generate any kind of immunological response. Most of the applied cells in the above-mentioned studies are known for their low immunological response [359]. However, some studies report that the mechanical characteristics of the final build, particularly its non-viable aspects, can trigger a nonspecific immunological proinflammatory response (TNL/NF-κB; IL-6/JAK-STAT) [360]. Ensuring adequate vascularization of the build may reduce the immune response [361].

Until recently, only a few clinical studies had examined the usefulness of 3D printed biomaterials in therapy [362]. After the initial successful results, the aim was to implement these techniques into personalized care schemes. In the current setting, point-of-care 3D printing has garnered significant impetus in favor of its implementation [363].

Organizing clinical trials for 3D printing approaches, let alone their regular everyday use, presents considerable challenges. In the last decades, there has been a constant drive for innovation in treatment schemes and solutions [364], which has led to the rapid evolution of the regulatory aspect concerning clinical trials, both from a legal and ethical standpoint [365,366]. However, the introduction of 3D bioprinting is associated with a host of gaps and uncertainties [367,368]. In the pre-clinical bioprinting stage, the results in various applications are encouraging [369–371], and bioprinted constructs have also been used successfully in a research capacity [372–374]. Due to their very nature, 3D printed biomaterials of any type and application are subject to particular constraints and considerations [363]. The two relevant EU regulations are the EU Clinical Trials Regulation [375] and the Medical Devices Regulation [376], which safeguard the rights of the participants and ensure the best possible clinical practices [14]. Nevertheless, given the legal and legislative intricacies associated with 3D printed biomaterials, there are a number of points that must be taken into account [14].

The recent development of materials suitable for 3D printing [377,378] opens even more possibilities for novel approaches both from a research and a treatment perspective [379]. Some interesting future research perspectives may thus be offered. To begin with, as is the case with most artificial implants and materials, there is an increased risk of infection [380,381]; therefore, there is the possibility of natural compound integration with antimicrobial action, aside from antibiotics. Currently, treatment modalities in peri-prosthetic infections center around different antibiotic delivery modalities [382,383]. Natural compounds with multiple different actions, such as kaempferol [384,385], capsaicin [386,387], curcumin [388,389], pinosylvin [390,391], piperine [392,393], and thymol [394,395], could be added during the 3D printing process to reduce the chance of infection or excessive inflammation. Another avenue open is the use of artificial intelligence both for the identification of potential infectious complications [396,397]; however, there still remains a host of other issues to be resolved, including patient rights, privacy, and ethical considerations [398], along with issues pertaining to the issue of Big Data in medicine [399], which is necessary for such applications [400].

Looking ahead, studies should be performed to integrate bioprinted scaffolds first into animal models (in vivo) before translating them to human patients. These studies should follow several important directions: 1. the rate of neovascularization and the density of the vascularized network; 2. how future generations of cells from the initial cell load-up perform (morphological aspect, functional aspects, immunological response, structural integrity, localization in the scaffold), and how many generations of cells can be produced until the effectiveness of the scaffold drops significantly; 3. The long-term degradability of the biomaterial scaffold (e.g., after 1 year, after 5 years); 4. the possibility for revision surgeries (especially if the patients are young).

6. Conclusions

The main directions outlined in this paper relate to the possibility of creating bionic parts in the future. Existing approaches to 3D bioprinting of human components could allow the replacement of traditional treatments and benefit patients. Incorporation of the results in each anatomical component of the head and neck offers researchers a broad vision of what has been achieved and what remains to be researched in order to achieve the ideal of the medical realm, creating a bionic head or neck.

There are numerous challenges to the present scaffolds or models, but possibilities to overcome them have been proposed. The current paper describes the elements that these scaffolds require to be considered a complete equivalent of a certain human part. Furthermore, we described the complexity that needs to be achieved to perfectly mimic a human anatomical component. It is not necessary to create a complete functional replica to obtain great results, since the new constructs already expand numerous limitations of pre-existing treatments.

The promising results described in recent research need to be considered for translation to human patients. These solutions should be accessible, and their advantages should be clearly presented to medical teams and patients alike. Scaling up these models and researching methods to increase the viability of the constructs are critical aspects for the sustainability of these new achievements.

Author Contributions: Conceptualization, I.-A.T., A.C.D., C.C. and C.S.; methodology, A.C., C.D.M.D., I.A.B. and C.S.; validation, A.-E.S., A.P., S.D. and C.C.; formal analysis, C.D.M.D., A.-E.S., A.P. and A.C.D.; investigation, I.-A.T., A.C., C.D.M.D., A.-E.S., I.A.B., S.D. and C.S.; writing—original draft preparation, I.-A.T., A.C., C.D.M.D., A.-E.S., I.A.B., A.P., S.D., A.C.D., C.C. and C.S.;

writing—review and editing, I.-A.T., A.C., A.P., A.C.D. and C.S.; visualization, I.-A.T. and C.C.; supervision, A.C.D., C.C. and C.S. All authors have read and agreed to the published version of the manuscript.

Funding: This research received no external funding.

Institutional Review Board Statement: Not applicable.

Informed Consent Statement: Not applicable.

Data Availability Statement: No new data were created or analyzed in this study. Data sharing is not applicable to this article.

Conflicts of Interest: The authors declare no conflicts of interest.

References

1. Choi, I.S. Functional Vascular Anatomy of the Head and Neck. *Interv. Neuroradiol.* **2003**, *9*, 29–30. [CrossRef] [PubMed]
2. Wali, R.; Halai, T.; Koshal, S. A Review of Head and Neck Clinical Anatomy for the General Dental Practitioner. *Dent. Update* **2020**, *47*, 62–70. [CrossRef]
3. Levin, M.; Baskin, S.M.; Bigal, M.E.; Lipton, R.B.; Markley, H.G.; McGeeney, B.E.; Newman, L.C.; Rapoport, A.M.; Rapoport, M.J.; Shapiro, R.E.; et al. Head Pain Anatomy and Physiology. In *Comprehensive Review of Headache Medicine*; Levin, M., Ed.; Oxford University Press: Oxford, UK, 2008. [CrossRef]
4. Bose, P.; Brockton, N.T.; Dort, J.C. Head and neck cancer: From anatomy to biology. *Int. J. Cancer* **2013**, *133*, 2013–2023. [CrossRef] [PubMed]
5. Kohan, E.J.; Wirth, G.A. Anatomy of the neck. *Clin. Plast. Surg.* **2014**, *41*, 1–6. [CrossRef] [PubMed]
6. Kademani, D.; Mardini, S.; Moran, S.L. Reconstruction of head and neck defects: A systematic approach to treatment. *Semin. Plast. Surg.* **2008**, *22*, 141–155. [CrossRef]
7. Eckardt, A.; Meyer, A.; Laas, U.; Hausamen, J.-E. Reconstruction of defects in the head and neck with free flaps: 20 years experience. *Br. J. Oral. Maxillofac. Surg.* **2007**, *45*, 11–15. [CrossRef]
8. Hurvitz, K.A.; Kobayashi, M.; Evans, G.R. Current options in head and neck reconstruction. *Plast. Reconstr. Surg.* **2006**, *118*, 122e–133e. [CrossRef]
9. Ceradini, D.J.; Tran, D.L.; Dedania, V.S.; Gelb, B.E.; Cohen, O.D.; Flores, R.L.; Levine, J.P.; Saadeh, P.B.; Staffenberg, D.A.; Ben Youss, Z.; et al. Combined Whole Eye and Face Transplant: Microsurgical Strategy and 1-Year Clinical Course. *JAMA* **2024**, *332*, 1551–1558. [CrossRef]
10. Canavero, S.; Lebenstein-Gumovski, M.V.; Kim, C.Y. The rise of transplantation neurosurgery: Spinal cord, eye, brain. *Surg. Neurol. Int.* **2024**, *15*, 478. [CrossRef]
11. Zhou, Y.L.; Chen, G.; Wang, W.J.; Ma, H.; Wen, H.C.; Wang, W. Comparative Efficacy of Masseteric-to-Facial Nerve Transfer with and without Fascia Lata Grafts in Reanimating Facial Paralysis. *Plast. Reconstr. Surg.* **2025**, *155*, 1011–1022. [CrossRef]
12. Tripathi, S.; Mandal, S.S.; Bauri, S.; Maiti, P. 3D bioprinting and its innovative approach for biomedical applications. *MedComm* **2023**, *4*, e194. [CrossRef] [PubMed]
13. Ren, Y.; Yuan, C.; Liang, Q.; Ba, Y.; Xu, H.; Weng, S.; Zhang, Y.; Zuo, A.; Liu, S.; Luo, P.; et al. 3D Bioprinting for Engineering Organoids and Organ-on-a-Chip: Developments and Applications. *Med. Res. Rev.* **2025**. [CrossRef]
14. Periferakis, A.; Periferakis, A.-T.; Troumpata, L.; Dragosloveanu, S.; Timofticiuc, I.-A.; Georgatos-Garcia, S.; Scheau, A.-E.; Periferakis, K.; Caruntu, A.; Badarau, I.A.; et al. Use of Biomaterials in 3D Printing as a Solution to Microbial Infections in Arthroplasty and Osseous Reconstruction. *Biomimetics* **2024**, *9*, 154. [CrossRef]
15. Dong, T.; Li, R.; Lu, Z.; Zhou, J.; Xu, M.T.; Feng, S.; Cui, T.; Gu, N. 3D-bioprinted multifunctional artificial skin patches with synergistic antimicrobial-mechanoadaptive functions for accelerating infected wound healing. *Int. J. Biol. Macromol.* **2025**, *318*, 144844. [CrossRef]
16. Timofticiuc, I.-A.; Dragosloveanu, S.; Caruntu, A.; Scheau, A.-E.; Badarau, I.A.; Garofil, N.D.; Didilescu, A.C.; Caruntu, C.; Scheau, C. 3D Bioprinting in Limb Salvage Surgery. *J. Funct. Biomater.* **2024**, *15*, 383. [CrossRef] [PubMed]
17. Algellay, M.; Sarker, S.D.; Roberts, M.; Bosworth, L.A.; Ehtezazi, T. The potential applications of nanocomposites in 3D-printed drug delivery systems. *J. Pharm. Pharmacol.* **2025**, rgaf028. [CrossRef]
18. Yang, Y.; Jia, Y.; Yang, Q.; Xu, F. Engineering bio-inks for 3D bioprinting cell mechanical microenvironment. *Int. J. Bioprinting* **2022**, *9*, 632. [CrossRef]
19. Gungor-Ozkerim, P.S.; Inci, I.; Zhang, Y.S.; Khademhosseini, A.; Dokmeci, M.R. Bioinks for 3D bioprinting: An overview. *Biomater. Sci.* **2018**, *6*, 915–946. [CrossRef] [PubMed]

20. Zhang, J.; Wehrle, E.; Rubert, M.; Müller, R. 3D bioprinting of human tissues: Biofabrication, bioinks, and bioreactors. *Int. J. Mol. Sci.* **2021**, *22*, 3971. [CrossRef]
21. Hinton, T.J.; Lee, A.; Feinberg, A.W. 3D bioprinting from the micrometer to millimeter length scales: Size does matter. *Curr. Opin. Biomed. Eng.* **2017**, *1*, 31–37. [CrossRef]
22. Taşoğlu, S.; Demirci, U. Bioprinting for Stem Cell Research. *Trends Biotechnol.* **2013**, *31*, 10–19. [CrossRef]
23. Liu, W.; Heinrich, M.A.; Zhou, Y.; Akpek, A.; Hu, N.; Liu, X.; Guan, X.; Zhong, Z.; Jin, X.; Khademhosseini, A.; et al. Extrusion Bioprinting of Shear-Thinning Gelatin Methacryloyl Bioinks. *Adv. Healthc. Mater.* **2017**, *6*, 1601451. [CrossRef] [PubMed]
24. Duan, B.; Hockaday, L.A.; Kang, K.H.; Butcher, J.T. 3D Bioprinting of Heterogeneous Aortic Valve Conduits with Alginate/Gelatin Hydrogels. *J. Biomed. Mater. Res. Part A* **2012**, *101A*, 1255–1264. [CrossRef]
25. Zhang, Y.S.; Arneri, A.; Bersini, S.; Shin, S.R.; Zhu, K.; Goli-Malekabadi, Z.; Aleman, J.; Colosi, C.; Busignani, F.; Dell'erba, V.; et al. Bioprinting 3D Microfibrous Scaffolds for Engineering Endothelialized Myocardium and Heart-on-a-Chip. *Biomaterials* **2016**, *110*, 45–59. [CrossRef] [PubMed]
26. Jiang, M.; Deng, W.; Lin, H. Sustainability through Biomimicry: A Comprehensive Review of Bionic Design Applications. *Biomimetics* **2024**, *9*, 507. [CrossRef]
27. Ramirez-Garcia, M.A.; Sloan, S.R.; Nidenberg, B.; Khalifa, Y.M.; Buckley, M.R. Depth-dependent out-of-plane young's modulus of the human cornea. *Curr. Eye Res.* **2018**, *43*, 595–604. [CrossRef] [PubMed]
28. Randleman, J.B.; Dawson, D.G.; Grossniklaus, H.E.; McCarey, B.E.; Edelhauser, H.F. Depth-dependent cohesive tensile strength in human donor corneas: Implications for refractive surgery. *J. Refract. Surg.* **2008**, *24*, S85–S89. [CrossRef]
29. Qian, X.; Zhang, K.; Liu, Z. A method to determine the mechanical properties of the retina based on an experiment in vivo. *Bio-Med. Mater. Eng.* **2015**, *26*, S287–S297. [CrossRef]
30. Worthington, K.S.; Wiley, L.A.; Bartlett, A.M.; Stone, E.M.; Mullins, R.F.; Salem, A.K.; Guymon, C.A.; Tucker, B.A. Mechanical properties of murine and porcine ocular tissues in compression. *Exp. Eye Res.* **2014**, *121*, 194–199. [CrossRef]
31. Djigo, A.D.; Bérubé, J.; Landreville, S.; Proulx, S. Characterization of a tissue-engineered choroid. *Acta Biomater.* **2019**, *84*, 305–316. [CrossRef]
32. Park, J.; Shin, A.; Jafari, S.; Demer, J.L. Material properties and effect of preconditioning of human sclera, optic nerve, and optic nerve sheath. *Biomech. Model. Mechanobiol.* **2021**, *20*, 1353–1363. [CrossRef]
33. Bos, E.J.; Van Der Laan, K.; Helder, M.N.; Mullender, M.G.; Iannuzzi, D.; Van Zuijlen, P.P. Noninvasive measurement of ear cartilage elasticity on the cellular level: A new method to provide biomechanical information for tissue engineering. *Plast. Reconstr. Surg.–Glob. Open* **2017**, *5*, e1147. [CrossRef]
34. Lem, M.; Pham, J.; Khan, N.; Zaki, D.; Brown, W.; Hu, J.; Ziegler, M.; Athanasiou, K.; Evans, G.R.; Widgerow, A. Biomechanical Characterization of Human Normal Auricular and Microtia Cartilage. *Plast. Reconstr. Surg.–Glob. Open* **2022**, *10*, 150–151. [CrossRef]
35. Luo, H.; Wang, F.; Cheng, C.; Nakmali, D.U.; Gan, R.Z.; Lu, H. Mapping the Young's modulus distribution of the human tympanic membrane by microindentation. *Hear. Res.* **2019**, *378*, 75–91. [CrossRef] [PubMed]
36. Yamamoto, N.; Ishii, T.; Machida, T. Measurement of the mechanical properties of the tympanic membrane with a microtension tester. *Acta Oto-Laryngol.* **1990**, *110*, 85–91. [CrossRef] [PubMed]
37. Griffin, M.; Premakumar, Y.; Seifalian, A.; Szarko, M.; Butler, P. Biomechanical characterisation of the human nasal cartilages; implications for tissue engineering. *J. Mater. Sci. Mater. Med.* **2016**, *27*, 11. [CrossRef] [PubMed]
38. Richmon, J.D.; Sage, A.B.; Wong, V.W.; Chen, A.C.; Pan, C.; Sah, R.L.; Watson, D. Tensile biomechanical properties of human nasal septal cartilage. *Am. J. Rhinol.* **2005**, *19*, 617–622. [CrossRef]
39. Pietrzak, G.; Curnier, A.; Botsis, J.; Scherrer, S.; Wiskott, A.; Belser, U. A nonlinear elastic model of the periodontal ligament and its numerical calibration for the study of tooth mobility. *Comput. Methods Biomech. Biomed. Eng.* **2002**, *5*, 91–100. [CrossRef]
40. Keilig, L.; Drolshagen, M.; Tran, K.l.; Hasan, I.; Reimann, S.; Deschner, J.; Brinkmann, K.t.; Krause, R.; Favino, M.; Bourauel, C. In vivo measurements and numerical analysis of the biomechanical characteristics of the human periodontal ligament. *Ann. Anat.-Anat. Anz.* **2016**, *206*, 80–88. [CrossRef]
41. Ma, H.; Liu, L.; Sun, X.; Liu, M.; Liu, Z. The diagnostic value of shear-wave elastography in the salivary glands of patients with primary Sjögren syndrome. *Int. J. Rheum. Dis.* **2023**, *26*, 1314–1320. [CrossRef]
42. Beatty, M.W.; Wee, A.G.; Marx, D.B.; Ridgway, L.; Simetich, B.; De Sousa, T.C.; Vakilzadian, K.; Schulte, J. Viscoelastic Properties of Human Facial Skin and Comparisons with Facial Prosthetic Elastomers. *Materials* **2023**, *16*, 2023. [CrossRef]
43. Titze, I.R. Mechanical stress in phonation. *J. Voice* **1994**, *8*, 99–105. [CrossRef] [PubMed]
44. Chhetri, D.K.; Zhang, Z.; Neubauer, J. Measurement of Young's modulus of vocal folds by indentation. *J. Voice* **2011**, *25*, 1–7. [CrossRef] [PubMed]

45. Wang, J.-Y.; Mesquida, P.; Lee, T. Young's modulus measurement on pig trachea and bronchial airways. In Proceedings of the 2011 Annual International Conference of the IEEE Engineering in Medicine and Biology Society, Boston, MA, USA, 30 August–3 September 2011; pp. 2089–2092.
46. Rains, J.; Bert, J.; Roberts, C.; Pare, P. Mechanical properties of human tracheal cartilage. *J. Appl. Physiol.* **1992**, *72*, 219–225. [CrossRef]
47. Marcula, K. Determination of the elongation forces of oesophageal tissue in uniaxial tensile test. *Acta Bioeng. Biomech.* **2024**, *26*, 137–143. [CrossRef] [PubMed]
48. Matyash, M.; Despang, F.; Ikonomidou, C.; Gelinsky, M. Swelling and mechanical properties of alginate hydrogels with respect to promotion of neural growth. *Tissue Eng. Part C Methods* **2014**, *20*, 401–411. [CrossRef]
49. Liu, S.; Li, Y.; Li, L. Enhanced stability and mechanical strength of sodium alginate composite films. *Carbohydr. Polym.* **2017**, *160*, 62–70. [CrossRef]
50. Mostolizadeh, S. Alginate, Polymer Purified from Seaweed. In *Alginate-Applications and Future Perspectives*; IntechOpen: London, UK, 2024. [CrossRef]
51. Zimmerling, A.; Zhou, Y.; Chen, D. Synthesis of Alginate/Collagen Bioink for Bioprinting Respiratory Tissue Models. *J. Funct. Biomater.* **2024**, *15*, 90. [CrossRef]
52. Zafaryab, M.; Vig, K. Three-Dimensional Printing of Hydrogel as Skin Substitute and Comparative Evaluation of Melanin Production. *Bioengineering* **2025**, *12*, 270. [CrossRef]
53. Chen, X.; Wu, T.; Bu, Y.; Yan, H.; Lin, Q. Fabrication and Biomedical Application of Alginate Composite Hydrogels in Bone Tissue Engineering: A Review. *Int. J. Mol. Sci.* **2024**, *25*, 7810. [CrossRef]
54. Avnet, S.; Pompo, G.D.; Borciani, G.; Fischetti, T.; Graziani, G.; Baldini, N. Advantages and Limitations of Using Cell Viability Assays for 3D Bioprinted Constructs. *Biomed. Mater.* **2024**, *19*, 025033. [CrossRef] [PubMed]
55. Karimi, A.; Navidbakhsh, M. Material properties in unconfined compression of gelatin hydrogel for skin tissue engineering applications. *Biomed. Eng./Biomed. Tech.* **2014**, *59*, 479–486. [CrossRef]
56. Pérez-Recalde, M.; Pacheco, E.; Aráoz, B.; Hermida, É.B. Effects of Polyhydroxybutyrate-Co-Hydroxyvalerate Microparticle Loading on Rheology, Microstructure, and Processability of Hydrogel-Based Inks for Bioprinted and Moulded Scaffolds. *Gels* **2025**, *11*, 200. [CrossRef] [PubMed]
57. Lameirinhas, N.S.; Teixeira, M.C.; Carvalho, J.P.F.; Valente, B.F.A.; Luís, J.L.; Duarte, I.F.; Pinto, R.J.B.; Oliveira, H.; Oliveira, J.M.; Silvestre, A.J.D.; et al. Biofabrication of HepG2 Cells-Laden 3D Structures Using Nanocellulose-Reinforced Gelatin-Based Hydrogel Bioinks: Materials Characterization, Cell Viability Assessment, and Metabolomic Analysis. *ACS Biomater. Sci. Eng.* **2025**, *11*, 3043–3057. [CrossRef]
58. Wang, Z.; Lin, L.; Li, X.; Zhang, Q.; Mi, X.; Xu, B.; Xu, Y.; Liu, T.; Shen, Y.; Wang, Z.; et al. Improving Thermosensitive Bioink Scaffold Fabrication with a Temperature-Regulated Printhead in Robot-Assisted In Situ Bioprinting System. *ACS Omega* **2024**, *9*, 40618–40631. [CrossRef] [PubMed]
59. Zhou, M.; Lee, B.H.; Tan, L.P. A Dual Crosslinking Strategy to Tailor Rheological Properties of Gelatin Methacryloyl. *Int. J. Bioprinting* **2024**, *3*, 130. [CrossRef]
60. Montanari, M.; Korkeamäki, J.T.; Campodoni, E.; Mohamed-Ahmed, S.; Mustafa, K.; Sandri, M.; Rashad, A. Effects of Magnesium-Doped Hydroxyapatite Nanoparticles on Bioink Formulation for Bone Tissue Engineering. *Acs Appl. Bio Mater.* **2025**, *8*, 535–547. [CrossRef]
61. Denton, O.; Wan, Y.; Beattie, L.; Jack, T.; McGoldrick, P.; McAllister, H.; Mullan, C.; Douglas, C.M.; Shu, W. Understanding the Role of Biofilms in Acute Recurrent Tonsillitis through 3D Bioprinting of a Novel Gelatin-Pegda Hydrogel. *Bioengineering* **2024**, *11*, 202. [CrossRef]
62. Farasatkia, A.; Kharaziha, M.; Ashrafizadeh, F.; Salehi, S. Transparent silk/gelatin methacrylate (GelMA) fibrillar film for corneal regeneration. *Mater. Sci. Eng. C* **2021**, *120*, 111744. [CrossRef]
63. Ferchichi, E.; Stealey, S.; Bogert, P.; Zustiak, S.P. Tunable Gelatin Methacrylate Polyethylene Glycol Diacrylate Hydrogels for Cell Mechanosensing Applications. *Front. Biomater. Sci.* **2024**, *3*, 1408748. [CrossRef]
64. Schlauch, D.; Ebbecke, J.P.; Meyer, J.; Fleischhammer, T.; Pirmahboub, H.; Kloke, L.; Kara, S.; Lavrentieva, A.; Pepelanova, I. Development of a Human Recombinant Collagen for Vat Polymerization-Based Bioprinting. *Biotechnol. J.* **2024**, *19*, e202400393. [CrossRef]
65. González-Martínez, D.; Carpio, M.B.; Tafoya, A.A.; González-Martínez, E.; Hirota, J.A.; Moran-Mirabal, J. GelMA-Carbopol Bioinks with Low Total Solids Content for High-Fidelity Extrusion 3D Bioprinting of Dynamic Tissue Biomimetics. *ChemRxiv* **2024**. [CrossRef]
66. Hwangbo, H.; Chae, S.; Ryu, D.; Kim, G. In Situ Magnetic-Field-Assisted Bioprinting Process Using Magnetorheological Bioink to Obtain Engineered Muscle Constructs. *Bioact. Mater.* **2025**, *45*, 417–433. [CrossRef]

67. Deidda, V.; Ventisette, I.; Langione, M.; Giammarino, L.; Pioner, J.M.; Credi, C.; Carpi, F. 3d-Printable Gelatin Methacrylate-Xanthan Gum Hydrogel Bioink Enabling Human Induced Pluripotent Stem Cell Differentiation Into Cardiomyocytes. *J. Funct. Biomater.* **2024**, *15*, 297. [CrossRef] [PubMed]
68. Zhao, Y. The Effect of Gelatin Methatcryloyl (GelMA) on Adipose-Derived Stem Cells Transplantation and Wound Healing. *Archives* **2025**, *1*. [CrossRef]
69. Potes, M.D.A.; Tilton, M.; Mitra, I.; Liu, X.; Dashtdar, B.; Camilleri, E.T.; Elder, B.D.; Lu, L. 3D Bioprinted Chondrogenic Gelatin Methacrylate-Poly(ethylene Glycol) Diacrylate Composite Scaffolds for Intervertebral Disc Restoration. *Int. J. Extrem. Manuf.* **2024**, *7*, 015507. [CrossRef]
70. Suresh, S.; Shenoy, S.J.; Sekar, J.A.; Harikrishnan, V.; Sabareeswaran, A.; Senthilkumar, M.; Kasoju, N.; Velayudhan, S.; Kumar, P.A. Integration of 3D Printed GelMA in Rat Liver: A Pilot Study. *J. Bioact. Compat. Polym.* **2025**, *440*, 267–281. [CrossRef]
71. Mori, H.; Shimizu, K.; Hara, M. Dynamic viscoelastic properties of collagen gels with high mechanical strength. *Mater. Sci. Eng. C* **2013**, *33*, 3230–3236. [CrossRef]
72. Yang, X.; Lu, Z.; Wu, H.; Li, W.; Zheng, L.; Zhao, J. Collagen-alginate as bioink for three-dimensional (3D) cell printing based cartilage tissue engineering. *Mater. Sci. Eng. C* **2018**, *83*, 195–201. [CrossRef]
73. Jeon, O.; Park, H.; Lee, M.S.; Alsberg, E. In situ cell-only bioprinting of patterned prevascular tissue into bioprinted high-density stem cell-laden microgel bioinks for vascularized bone tissue regeneration. *bioRxiv* **2025**. [CrossRef]
74. Araújo, I.J.d.S.; Perkins, R.S.; Ibrahim, M.M.; Huang, G.T.-J.; Zhang, W. Bioprinting PDLSC-Laden Collagen Scaffolds for Periodontal Ligament Regeneration. *ACS Appl. Mater. Interfaces* **2024**, *16*, 59979–59990. [CrossRef]
75. Chaurasia, P.; Singh, R.; Mahto, S.K. FRESH-based 3D Bioprinting of Complex Biological Geometries Using Chitosan Bioink. *Biofabrication* **2024**, *16*, 045007. [CrossRef] [PubMed]
76. Gómez-Blanco, J.C.; Pagador, J.B.; Galván-Chacón, V.P.; Sánchez-Peralta, L.F.; Matamoros, M.; Marcos, A.; Sánchez-Margallo, F.M. Computational Simulation-Based Comparative Analysis of Standard 3D Printing and Conical Nozzles for Pneumatic and Piston-Driven Bioprinting. *Int. J. Bioprinting* **2024**, *9*, 730. [CrossRef] [PubMed]
77. Rojas-Murillo, J.A.; Simental-Mendía, M.A.; Moncada-Saucedo, N.K.; Delgado-Gonzalez, P.; Islas, J.F.; Roacho-Pérez, J.A.; Garza-Treviño, E.N. Physical, mechanical, and biological properties of fibrin scaffolds for cartilage repair. *Int. J. Mol. Sci.* **2022**, *23*, 9879. [CrossRef] [PubMed]
78. Hense, D.; Strube, O.I. Glutaraldehyde Cross-Linking of Salt-Induced Fibrinogen Hydrogels. *ACS Biomater. Sci. Eng.* **2024**, *10*, 6927–6937. [CrossRef]
79. Haq, A.; Kumari, V.; Kashyap, V.H.; Goel, J. Comparison of Fibrin Glue with Conventional Suturing in Peripheral Nerve Repairs: A Study of Sensory and Motor Outcomes. *Cureus* **2024**, *16*, e63111. [CrossRef]
80. Shpichka, A.; Osipova, D.; Efremov, Y.; Bikmulina, P.; Kosheleva, N.; Lipina, M.; Bezrukov, E.A.; Sukhanov, R.B.; Solovieva, A.B.; Vosough, M.; et al. Fibrin-Based Bioinks: New Tricks from an Old Dog. *Int. J. Bioprinting* **2024**, *6*, 269. [CrossRef]
81. Božek, J.; Kurchakova, O.; Michel, J.; Groß, I.; Gerhards, L.; Zhang, Y.; Brand, I.; Bräuer, A.U. Pneumatic Conveying Inkjet Bioprinting for the Processing of Living Cells. *Biofabrication* **2025**, *17*, 025003. [CrossRef]
82. Stefano, A.B.D.; Urrata, V.; Schilders, K.; Franza, M.; Leo, S.; Moschella, F.; Cordova, A.; Toia, F. Three-Dimensional Bioprinting Techniques in Skin Regeneration: Current Insights and Future Perspectives. *Life* **2025**, *15*, 787. [CrossRef]
83. Lipari, S.; Sacco, P.; Marsich, E.; Donati, I. Silk Fibroin-Enriched Bioink Promotes Cell Proliferation in 3d-Bioprinted Constructs. *Gels* **2024**, *10*, 469. [CrossRef]
84. Levett, P.A.; Hutmacher, D.W.; Malda, J.; Klein, T.J.; Engler, A.J. Hyaluronic acid enhances the mechanical properties of tissue-engineered cartilage constructs. *PLoS ONE* **2014**, *9*, e113216. [CrossRef] [PubMed]
85. Collins, M.N.; Birkinshaw, C. Physical properties of crosslinked hyaluronic acid hydrogels. *J. Mater. Sci. Mater. Med.* **2008**, *19*, 3335–3343. [CrossRef]
86. Salih, A.R.C.; Farooqi, H.M.U.; Amin, H.; Karn, P.R.; Meghani, N.; Nagendran, S. Hyaluronic Acid: Comprehensive Review of a Multifunctional Biopolymer. *Future J. Pharm. Sci.* **2024**, *10*, 63. [CrossRef]
87. Tavakoli, S.; Kocatürkmen, A.; Oommen, O.P.; Varghese, O.P. Ultra-Fine 3D Bioprinting of Dynamic Hyaluronic Acid Hydrogel for in Vitro Modeling. *Adv. Mater.* **2025**, e2500315. [CrossRef] [PubMed]
88. Galocha-León, C.; Antich, C.; Clares-Naveros, B.; Voltes-Martínez, A.; Marchal, J.A.; Gálvez-Martín, P. Design and Characterization of Biomimetic Hybrid Construct Based on Hyaluronic Acid and Alginate Bioink for Regeneration of Articular Cartilage. *Pharmaceutics* **2024**, *16*, 1422. [CrossRef]
89. Nagaraja, K.; Bhattacharyya, A.; Jung, M.; Kim, D.; Khatun, M.R.; Noh, I. 3D Bioprintable Self-Healing Hyaluronic Acid Hydrogel with Cysteamine Grafting for Tissue Engineering. *Gels* **2024**, *10*, 780. [CrossRef]
90. Wang, Y.; Gao, C.; Cheng, S.; Li, Y.; Huang, Y.; Cao, X.; Zhang, Z.; Huang, J. 3D Bioprinting of Double-Layer Conductive Skin for Wound Healing. *Adv. Healthc. Mater.* **2025**, *14*, e2404388. [CrossRef] [PubMed]

91. Ding, B.; Gao, H.; Song, J.; Li, Y.; Zhang, L.; Cao, X.; Xu, M.; Cai, J. Tough and cell-compatible chitosan physical hydrogels for mouse bone mesenchymal stem cells in vitro. *ACS Appl. Mater. Interfaces* **2016**, *8*, 19739–19746. [CrossRef]
92. Senthil, A.; Sakthivel, V.; Jayanthi, P.; Rahmathullah, A.; Ramachandran, S. Valorization of Chitosan for Sustainable Soap Production: Development, Performance Evaluation, and Hygiene Applications. *J. Bioact. Compat. Polym.* **2024**, *40*, 68–83. [CrossRef]
93. García-García, A.; Pérez-Álvarez, L.; Ruiz-Rubio, L.; Larrea-Sebal, A.; Martin, C.; Vilas-Vilela, J.L. Extrusion-Based 3D Printing of Photocrosslinkable Chitosan Inks. *Gels* **2024**, *10*, 126. [CrossRef]
94. Madadian, E.; Ravanbakhsh, H.; Kameni, F.T.; Rahimnejad, M.; Lerouge, S.; Ahmadi, A. In-Foam Bioprinting: An Embedded Bioprinting Technique with Self-Removable Support Bath. *Small Sci.* **2024**, *4*, 2300280. [CrossRef]
95. Khiari, Z. Recent Developments in Bio-Ink Formulations Using Marine-Derived Biomaterials for Three-Dimensional (3D) Bioprinting. *Mar. Drugs* **2024**, *22*, 134. [CrossRef]
96. Smith, D.L.; Rodriguez-Melendez, D.; Montemayor, M.D.; Convento, M.O.; Grunlan, J.C. Fire Resistant Adhesive from Chitosan. *Biomacromolecules* **2025**, *26*, 1141–1148. [CrossRef] [PubMed]
97. Normand, V.; Lootens, D.L.; Amici, E.; Plucknett, K.P.; Aymard, P. New insight into agarose gel mechanical properties. *Biomacromolecules* **2000**, *1*, 730–738. [CrossRef] [PubMed]
98. Hasan, M.M.; Ahmad, A.; Akter, M.Z.; Choi, Y.J.; Yi, H.G. Bioinks for Bioprinting Using Plant-Derived Biomaterials. *Biofabrication* **2024**, *16*, 042004. [CrossRef] [PubMed]
99. Mierke, C.T. Bioprinting of Cells, Organoids and Organs-on-a-Chip Together with Hydrogels Improves Structural and Mechanical Cues. *Cells* **2024**, *13*, 1638. [CrossRef] [PubMed]
100. Neuhäusler, A.; Rogg, K.; Schröder, S.; Spiehl, D.; Zora, H.; Arefaine, E.; Schettler, J.; Hartmann, H.; Blaeser, A. Electrospun Microfibers to Enhance Nutrient Supply in Bioinks and 3d-Bioprinted Tissue Precursors. *Biofabrication* **2024**, *17*, 015038. [CrossRef]
101. Zhu, Z.; Ling, S.; Yeo, J.; Zhao, S.; Tozzi, L.; Buehler, M.J.; Omenetto, F.; Li, C.; Kaplan, D.L. High-strength, durable all-silk fibroin hydrogels with versatile processability toward multifunctional applications. *Adv. Funct. Mater.* **2018**, *28*, 1704757. [CrossRef]
102. Fuest, S.; Salviano-Silva, A.; Maire, C.L.; Xu, Y.; Apel, C.; Grust, A.L.C.; Coste, A.D.; Gosau, M.; Ricklefs, F.L.; Smeets, R. Doping of Casted Silk Fibroin Membranes with Extracellular Vesicles for Regenerative Therapy: A Proof of Concept. *Sci. Rep.* **2024**, *14*, 3553. [CrossRef]
103. Lopes, L.M.; Ottaiano, G.Y.; Guedes, L.d.S.; de Moraes, M.A.; Beppu, M.M. Analyzing the Interactions and Miscibility of Silk Fibroin/Mucin Blends. *J. Appl. Polym. Sci.* **2024**, *141*, e55343. [CrossRef]
104. Rouhani, D.S.B.; Singh, N.K.M.; Chao, J.J.M.; Almutairi, A.; Seradj, M.H.; Badowski-Platz, R.B.; Toranto, J.D.M.; Mofid, M.M.M. Superiority of a Silk Surgical Site Wound Closure Device Over Synthetic Dressings. *Plast. Reconstr. Surg.* **2024**, *154*, 1233–1244. [CrossRef] [PubMed]
105. Di Buduo, C.A.; Lunghi, M.; Kuzmenko, V.; Laurent, P.; Della Rosa, G.; Del Fante, C.; Nogare, D.E.D.; Jug, F.; Perotti, C.; Eto, K.; et al. Bioprinting Soft 3D Models of Hematopoiesis Using Natural Silk Fibroin-Based Bioink Efficiently Supports Platelet Differentiation. *Adv. Sci.* **2024**, *11*, e2308276. [CrossRef] [PubMed]
106. Nettey-Oppong, E.E.; Muhammad, R.; Ali, A.; Jeong, H.-W.; Seok, Y.-S.; Kim, S.W.; Choi, S.H. The Impact of Temperature and Pressure on the Structural Stability of Solvated Solid-State Conformations of Bombyx Mori Silk Fibroins: Insights from Molecular Dynamics Simulations. *Materials* **2024**, *17*, 5686. [CrossRef] [PubMed]
107. Temenoff, J.S.; Athanasiou, K.A.; Lebaron, R.G.; Mikos, A.G. Effect of poly (ethylene glycol) molecular weight on tensile and swelling properties of oligo (poly (ethylene glycol) fumarate) hydrogels for cartilage tissue engineering. *J. Biomed. Mater. Res. An. off. J. Soc. Biomater. Jpn. Soc. Biomater. Aust. Soc. Biomater. Korean Soc. Biomater.* **2002**, *59*, 429–437. [CrossRef]
108. Thaden, O.; Schneider, N.; Walther, T.; Spiller, E.; Taoum, A.; Göpfrich, K.; Campos, D.D. Bioprinting of Synthetic Cell-Like Lipid Vesicles to Augment the Functionality of Tissues after Manufacturing. *Acs Synth. Biol.* **2024**, *13*, 2436–2446. [CrossRef]
109. Töpfer, V.; Melzer, M.; Snowdon, R.J.; Stahl, A.; Matros, A.; Wehner, G. PEG Treatment Is Unsuitable to Study Root Related Traits as It Alters Root Anatomy in Barley (*Hordeum vulgare* L.). *BMC Plant Biol.* **2024**, *24*, 856. [CrossRef]
110. Fazel Anvari-Yazdi, A.; Badea, I.; Chen, X. Biomaterials in Postoperative Adhesion Barriers and Uterine Tissue Engineering. *Gels* **2025**, *11*, 441. [CrossRef]
111. Djunaidi, M.C.; Ayuningrum, D.; Maharani, N.D.; Khabibi, K.; Pardoyo, P.; Raharjo, Y.; Susanto, H.; Filardli, A.M.I. Synthesis of Printed Hollow Fiber Membranes Urea as a Membrane Candidate Hemodialysis. *Indones. J. Chem.* **2024**, *24*, 1583. [CrossRef]
112. Ekinci, A.; Gleadall, A.; Johnson, A.A.; Li, L.; Han, X. Mechanical and hydrolytic properties of thin polylactic acid films by fused filament fabrication. *J. Mech. Behav. Biomed. Mater.* **2021**, *114*, 104217. [CrossRef]
113. Gill, A.; Garnier, S. Floreon Technology, Redefining Polylactic Acid. *Open Access Gov.* **2024**, *44*, 362–363. [CrossRef]
114. Pvr, G.K.; Devi, K.D. Devaki. Influence of Manufacturing Parameters on the Properties of 3D Printed Polylactic Acid Carbon Fiber Components. *Arch. Metall. Mater.* **2025**, *1*, 97–104. [CrossRef]

115. Chetan, C.; James, S. Preliminary Investigation on Biodegradable Ureteral Stents Using 3D Printing. *J. Manuf. Mater. Process.* **2025**, *9*, 52. [CrossRef]
116. Peng, E.; Zou, J.T.; Liu, G.; Xu, J. Research Progress of Polylactic Acid and Its Composites for 3D Printing. In Proceedings of the Fifth International Conference on Mechanical Engineering and Materials (ICMEM 2024), Nanchang, China, 15–16 November 2024. [CrossRef]
117. Palaniyappan, S.; Sivakumar, N.K.; Annamalai, G.; Bodaghi, M.; Saravanamuthukumar, P.; Alageel, O.; Basavarajappa, S.; Hashem, M. Mechanical and Tribological Behaviour of Three-Dimensional Printed Almond Shell Particles Reinforced Polylactic Acid Bio-Composites. *Proc. Inst. Mech. Eng. Part L J. Mater. Des. Appl.* **2024**, *239*, 111–126. [CrossRef]
118. Costa, C.; Marcondes, T.L.; Calvo, W.A.P.; Kodama, Y. Impact of Electron Beam on PBAT/PLA (Ecovio) Bags by Raman Spectroscopy and Thermogravimetry. *Braz. J. Radiat. Sci.* **2025**, *12*, e2774. [CrossRef]
119. Rejiniemon, T.S.; Raishy, H.R.; Bhamadevi, R.; Binobead, M.A.; Aljowaie, R.M.; Ahalliya, R.M. Banana Pseudo-Stem and Cattle Manure for Lactic Acid Production and the Application of Polylactic Acid-Cellulose Silver Nanoparticle-Based Nanocomposite Films in Food Storage. *BioResources* **2024**, *19*, 5654–5671. [CrossRef]
120. Aniagyei, S.E.; Sims, L.B.; Malik, D.A.; Tyo, K.M.; Curry, K.C.; Kim, W.; Hodge, D.A.; Duan, J.; Steinbach-Rankins, J.M. Evaluation of poly (lactic-co-glycolic acid) and poly (dl-lactide-co-ε-caprolactone) electrospun fibers for the treatment of HSV-2 infection. *Mater. Sci. Eng. C* **2017**, *72*, 238–251. [CrossRef] [PubMed]
121. Ramírez-Ruiz, F.; Núñez-Tapia, I.; Piña-Barba, C.; Álvarez-Pérez, M.A.; Guarino, V.; Serrano-Bello, J. Polycaprolactone for Hard Tissue Regeneration: Scaffold Design and in Vivo Implications. *Bioengineering* **2025**, *12*, 46. [CrossRef]
122. Lv, N.; Zhou, Z.; Hou, M.; Hong, L.; Li, H.; Qian, Z.; Gao, X.; Liu, M. Research Progress of Vascularization Strategies of Tissue-Engineered Bone. *Front. Bioeng. Biotechnol.* **2024**, *11*, 1291969. [CrossRef]
123. Luo, W.; Zhang, H.; Wan, R.; Cai, Y.; Liu, Y.; Wu, Y.; Yang, Y.; Chen, J.; Zhang, D.; Luo, Z.; et al. Biomaterials-Based Technologies in Skeletal Muscle Tissue Engineering. *Adv. Healthc. Mater.* **2024**, *13*, e2304196. [CrossRef]
124. Chen, W.; Chen, M.; Chen, S.; Wang, S.; Huang, Z.; Zhang, L.; Wu, J.; Peng, W.; Li, H.; Wen, F. Decellularization of Fish Tissues for Tissue Engineering and Regenerative Medicine Applications. *Regen. Biomater.* **2024**, *12*, rbae138. [CrossRef]
125. Quan, S.; Yang, J.; Huang, S.; Shao, J.; Liu, Y.; Yang, H. Silk Fibroin as a Potential Candidate for Bone Tissue Engineering Applications. *Biomater. Sci.* **2025**, *13*, 364–378. [CrossRef] [PubMed]
126. Pandey, P.; Verma, M.; Lakhanpal, S.; Pandey, S.; Kumar, M.R.; Bhat, M.; Sharma, S.; Alam, M.W.; Khan, F. An Updated Review Summarizing the Anticancer Potential of Poly (Lactic-co-Glycolic Acid)(PLGA) Based Curcumin, Epigallocatechin Gallate, and Resveratrol Nanocarriers. *Biopolymers* **2025**, *116*, e23637. [CrossRef] [PubMed]
127. Bae, M.; Kim, J.J.; Jang, J.; Cho, D.W. 3D Bioprinted Unidirectional Neural Network and Its Application for Alcoholic Neurodegeneration. *Int. J. Extrem. Manuf.* **2025**, *7*, 055003. [CrossRef]
128. Pham, T.N.H.; Dang-Luong, P.-T.; Nguyen, H.-P.; Le-Tuan, L.; Cao, X.T.; Nguyen, T.D.; Tran, V.A.; Quang, H.V. Therapeutic Effect of F127-Folate@PLGA/CHL/Ir780 Nanoparticles on Folate Receptor-Expressing Cancer Cells. *Beilstein J. Nanotechnol.* **2024**, *15*, 954–964. [CrossRef]
129. Guerriero, I.; Pesce, C.; Spanò, R.; Sganga, S.; Tirelli, N.; Mascolo, D.D.; Palange, A.L.; Decuzzi, P. On the Biodegradation of Micropatterned Polymeric Films. *bioRxiv* **2025**. [CrossRef]
130. Lee, G.; Kim, S.J.; Chun, H.; Park, J.K. Multilayered and Heterogeneous Hydrogel Construct Printing System with Crosslinking Aerosol. *Biofabrication* **2021**, *13*, 045027. [CrossRef]
131. Ji, Y.; Yang, Q.; Huang, G.; Shen, M.; Jian, Z.; Thoraval, M.-J.; Lian, Q.; Zhang, X.; Xu, F. Improved Resolution and Fidelity of Droplet-Based Bioprinting by Upward Ejection. *ACS Biomater. Sci. Eng.* **2019**, *5*, 4112–4121. [CrossRef]
132. Fu, Z.; Naghieh, S.; Xu, C.; Wang, C.; Sun, W.; Chen, D. Printability in Extrusion Bioprinting. *Biofabrication* **2021**, *13*, 033001. [CrossRef]
133. LadeaLadea, L.; Zemba, M.; Calancea, M.I.; Călțaru, M.V.; Dragosloveanu, C.D.M.; Coroleucă, R.; Catrina, E.L.; Brezean, I.; Dinu, V. Corneal Epithelial Changes in Diabetic Patients: A Review. *Int. J. Mol. Sci.* **2024**, *25*, 3471. [CrossRef]
134. Gain, P.; Jullienne, R.; He, Z.; Aldossary, M.; Acquart, S.; Cognasse, F.; Thuret, G. Global survey of corneal transplantation and eye banking. *JAMA Ophthalmol.* **2016**, *134*, 167–173. [CrossRef]
135. Wang, E.Y.; Kong, X.; Wolle, M.; Gasquet, N.; Ssekasanvu, J.; Mariotti, S.P.; Bourne, R.; Taylor, H.; Resnikoff, S.; West, S. Global trends in blindness and vision impairment resulting from corneal opacity 1984–2020: A meta-analysis. *Ophthalmology* **2023**, *130*, 863–871. [CrossRef] [PubMed]
136. Moshirfar, M.; Thomson, A.C.; Ronquillo, Y. Corneal endothelial transplantation. In *StatPearls*; StatPearls Publishing: Treasure Island, FL, USA, 2020.
137. Isaacson, A.; Swioklo, S.; Connon, C.J. 3D bioprinting of a corneal stroma equivalent. *Exp. Eye Res.* **2018**, *173*, 188–193. [CrossRef] [PubMed]

138. Lee, M.-H.; Moon, S.-K.; Kim, Y.C. Effect of the corneal nano structure on light transmittance. *Optik* **2017**, *144*, 647–654. [CrossRef]
139. Kutlehria, S.; Dinh, T.C.; Bagde, A.; Patel, N.; Gebeyehu, A.; Singh, M. High-throughput 3D bioprinting of corneal stromal equivalents. *J. Biomed. Mater. Res. Part B Appl. Biomater.* **2020**, *108*, 2981–2994. [CrossRef]
140. Gingras, A.A.; Jansen, P.A.; Smith, C.; Zhang, X.; Niu, Y.; Zhao, Y.; Roberts, C.J.; Herderick, E.D.; Swindle-Reilly, K.E. 3D bioprinting of acellular corneal stromal scaffolds with a low cost modified 3D printer: A feasibility study. *Curr. Eye Res.* **2023**, *48*, 1112–1121. [CrossRef]
141. Meek, K.M. Corneal collagen—Its role in maintaining corneal shape and transparency. *Biophys. Rev.* **2009**, *1*, 83–93. [CrossRef] [PubMed]
142. Sharifi, S.; Islam, M.M.; Sharifi, H.; Islam, R.; Koza, D.; Reyes-Ortega, F.; Alba-Molina, D.; Nilsson, P.H.; Dohlman, C.H.; Mollnes, T.E.; et al. Tuning gelatin-based hydrogel towards bioadhesive ocular tissue engineering applications. *Bioact. Mater.* **2021**, *6*, 3947–3961. [CrossRef]
143. Tarsitano, M.; Cristiano, M.C.; Fresta, M.; Paolino, D.; Rafaniello, C. Alginate-based composites for corneal regeneration: The optimization of a biomaterial to overcome its limits. *Gels* **2022**, *8*, 431. [CrossRef] [PubMed]
144. Zhang, M.; Yang, F.; Han, D.; Zhang, S.-Y.; Dong, Y.; Li, X.; Ling, L.; Deng, Z.; Cao, X.; Tian, J. 3D bioprinting of corneal decellularized extracellular matrix: GelMA composite hydrogel for corneal stroma engineering. *Int. J. Bioprinting* **2023**, *9*, 774. [CrossRef]
145. Chien, Y.; Hsiao, Y.-J.; Chou, S.-J.; Lin, T.-Y.; Yarmishyn, A.A.; Lai, W.-Y.; Lee, M.-S.; Lin, Y.-Y.; Lin, T.-W.; Hwang, D.-K. Nanoparticles-mediated CRISPR-Cas9 gene therapy in inherited retinal diseases: Applications, challenges, and emerging opportunities. *J. Nanobiotechnology* **2022**, *20*, 511. [CrossRef]
146. Eiraku, M.; Takata, N.; Ishibashi, H.; Kawada, M.; Sakakura, E.; Okuda, S.; Sekiguchi, K.; Adachi, T.; Sasai, Y. Self-organizing optic-cup morphogenesis in three-dimensional culture. *Nature* **2011**, *472*, 51–56. [CrossRef]
147. Masland, R.H. Cell populations of the retina: The Proctor lecture. *Investig. Ophthalmol. Vis. Sci.* **2011**, *52*, 4581–4591. [CrossRef]
148. Wang, P.; Li, X.; Zhu, W.; Zhong, Z.; Moran, A.; Wang, W.; Zhang, K.; Chen, S. 3D bioprinting of hydrogels for retina cell culturing. *Bioprinting* **2018**, *12*, e00029. [CrossRef] [PubMed]
149. Baranov, P.; Michaelson, A.; Kundu, J.; Carrier, R.L.; Young, M. Interphotoreceptor matrix-poly (ϵ-caprolactone) composite scaffolds for human photoreceptor differentiation. *J. Tissue Eng.* **2014**, *5*, 2041731414554139. [CrossRef] [PubMed]
150. Liu, Y.; Wang, R.; Zarembinski, T.I.; Doty, N.; Jiang, C.; Regatieri, C.; Zhang, X.; Young, M.J. The application of hyaluronic acid hydrogels to retinal progenitor cell transplantation. *Tissue Eng. Part A* **2013**, *19*, 135–142. [CrossRef]
151. Ramsden, C.M.; Powner, M.B.; Carr, A.-J.F.; Smart, M.J.K.; da Cruz, L.; Coffey, P.J. Stem cells in retinal regeneration: Past, present and future. *Development* **2013**, *140*, 2576–2585. [CrossRef] [PubMed]
152. Berman, E.R.; Schwell, H.; Feeney, L. The retinal pigment epithelium: Chemical composition and structure. *Investig. Ophthalmol. Vis. Sci.* **1974**, *13*, 675–687.
153. Vithani, K.; Goyanes, A.; Jannin, V.; Basit, A.W.; Gaisford, S.; Boyd, B.J. An overview of 3D printing technologies for soft materials and potential opportunities for lipid-based drug delivery systems. *Pharm. Res.* **2019**, *36*, 4. [CrossRef]
154. Shi, P.; Edgar, T.Y.S.; Yeong, W.Y.; Laude, A. Hybrid three-dimensional (3D) bioprinting of retina equivalent for ocular research. *Int. J. Bioprinting* **2017**, *3*, 008. [CrossRef]
155. Vecchiotti, D.; Di Vito Nolfi, M.; Veglianti, F.; Dall'Aglio, F.; Khan, H.N.; Flati, I.; Verzella, D.; Capece, D.; Alesse, E.; Angelucci, A. A 3D Bioprinting Approach to Studying Retinal Müller Cells. *Genes* **2024**, *15*, 1414. [CrossRef]
156. Masaeli, E.; Marquette, C. Direct-write bioprinting approach to construct multilayer cellular tissues. *Front. Bioeng. Biotechnol.* **2020**, *7*, 478. [CrossRef] [PubMed]
157. Cholkar, K.; Dasari, S.R.; Pal, D.; Mitra, A.K. Eye: Anatomy, physiology and barriers to drug delivery. In *Ocular Transporters and Receptors*; Elsevier: Amsterdam, The Netherlands, 2013; pp. 1–36.
158. Păsărică, M.A.; Curcă, P.F.; Dragosloveanu, C.D.M.; Grigorescu, A.C.; Nisipașu, C.I. Pathological and Molecular Diagnosis of Uveal Melanoma. *Diagnostics* **2024**, *14*, 958. [CrossRef] [PubMed]
159. Groot, A.L.; Remmers, J.S.; Hartong, D.T. Three-dimensional computer-aided design of a full-color ocular prosthesis with textured iris and sclera manufactured in one single print job. *3D Print. Addit. Manuf.* **2021**, *8*, 343–348. [CrossRef]
160. Reinhard, J.; Urban, P.; Bell, S.; Carpenter, D.; Sagoo, M.S. Automatic data-driven design and 3D printing of custom ocular prostheses. *Nat. Commun.* **2024**, *15*, 1360. [CrossRef]
161. Boote, C.; Sigal, I.A.; Grytz, R.; Hua, Y.; Nguyen, T.D.; Girard, M.J. Scleral structure and biomechanics. *Progress. Retin. Eye Res.* **2020**, *74*, 100773. [CrossRef] [PubMed]
162. Keeley, F.w.; Morin, J.d.; Vesely, S. Characterization of collagen from normal human sclera. *Exp. Eye Res.* **1984**, *39*, 533–542. [CrossRef]

163. Sit, A.J.; Lin, S.-C.; Kazemi, A.; McLaren, J.W.; Pruet, C.M.; Zhang, X. In vivo noninvasive measurement of young's modulus of elasticity in human eyes: A feasibility study. *J. Glaucoma* **2017**, *26*, 967–973. [CrossRef]
164. Gibney, R.; Ferraris, E. Bioprinting of collagen type I and II via aerosol jet printing for the replication of dense collagenous tissues. *Front. Bioeng. Biotechnol.* **2021**, *9*, 786945. [CrossRef]
165. Gibney, R.; Patterson, J.; Ferraris, E. High-resolution bioprinting of recombinant human collagen type, I.I.I. *Polymers* **2021**, *13*, 2973. [CrossRef]
166. Lee, S.; Sani, E.S.; Spencer, A.R.; Guan, Y.; Weiss, A.S.; Annabi, N. Human-recombinant-Elastin-based bioinks for 3D bioprinting of vascularized soft tissues. *Adv. Mater.* **2020**, *32*, 2003915. [CrossRef]
167. Mayer, C.; Khoramnia, R. Pupil reconstruction with an artificial iris. *Ophthalmol. Z. Dtsch. Ophthalmol. Ges.* **2021**, *119*, 3–12.
168. Kohnen, T.; Baumeister, M.; Kook, D.; Klaproth, O.K.; Ohrloff, C. Cataract surgery with implantation of an artificial lens. *Dtsch. Ärzteblatt Int.* **2009**, *106*, 695. [CrossRef]
169. Lapp, T.; Wacker, K.; Heinz, C.; Maier, P.; Eberwein, P.; Reinhard, T. Cataract surgery—Indications, techniques, and intraocular lens selection. *Dtsch. Ärzteblatt Int.* **2023**, *120*, 377. [CrossRef] [PubMed]
170. Mishra, D.; Gade, S.; Glover, K.; Sheshala, R.; Singh, T.R.R. Vitreous humor: Composition, characteristics and implication on intravitreal drug delivery. *Curr. Eye Res.* **2023**, *48*, 208–218. [CrossRef] [PubMed]
171. Bishop, P.N.; Crossman, M.V.; Mcleod, D.; Ayad, S. Extraction and characterization of the tissue forms of collagen types II and IX from bovine vitreous. *Biochem. J.* **1994**, *299*, 497–505. [CrossRef]
172. Silva, A.F.; Alves, M.A.; Oliveira, M.S. Rheological behaviour of vitreous humour. *Rheol. Acta* **2017**, *56*, 377–386. [CrossRef]
173. Shafaie, S.; Hutter, V.; Brown, M.B.; Cook, M.T.; Chau, D.Y. Diffusion through the ex vivo vitreal body–bovine, porcine, and ovine models are poor surrogates for the human vitreous. *Int. J. Pharm.* **2018**, *550*, 207–215. [CrossRef]
174. Timofticiuc, I.-A.; Călinescu, O.; Iftime, A.; Dragosloveanu, S.; Caruntu, A.; Scheau, A.-E.; Badarau, I.A.; Didilescu, A.C.; Caruntu, C.; Scheau, C. Biomaterials adapted to VAT photopolymerization in 3D printing: Characteristics and medical applications. *J. Funct. Biomater.* **2023**, *15*, 7. [CrossRef]
175. Del Amo, E.M.; Rimpelä, A.-K.; Heikkinen, E.; Kari, O.K.; Ramsay, E.; Lajunen, T.; Schmitt, M.; Pelkonen, L.; Bhattacharya, M.; Richardson, D. Pharmacokinetic aspects of retinal drug delivery. *Progress. Retin. Eye Res.* **2017**, *57*, 134–185. [CrossRef]
176. Brubaker, R.F. The flow of aqueous humor in the human eye. *Trans. Am. Ophthalmol. Soc.* **1982**, *80*, 391.
177. Ansari, R.R.; Bockle, S.; Rovati, L. New optical scheme for a polarimetric-based glucose sensor. *J. Biomed. Opt.* **2004**, *9*, 103–115. [CrossRef] [PubMed]
178. Xie, J.; Gao, Q.; del Prado, Z.N.; Venkateswaran, N.; Mousa, H.M.; Salero, E.; Ye, J.; De Juan-Pardo, E.M.; Sabater, A.L.; Perez, V.L. Establishment of a bi-layered tissue engineered conjunctiva using a 3D-printed melt electrowritten poly-(ε-caprolactone) scaffold. *Int. Ophthalmol.* **2023**, *43*, 215–232. [CrossRef] [PubMed]
179. Maloca, P.M.; Tufail, A.; Hasler, P.W.; Rothenbuehler, S.; Egan, C.; de Carvalho, J.E.R.; Spaide, R.F. 3D printing of the choroidal vessels and tumours based on optical coherence tomography. *Acta Ophthalmol.* **2019**, *97*, e313–e316. [CrossRef] [PubMed]
180. Zopf, D.A.; Flanagan, C.L.; Nasser, H.B.; Mitsak, A.G.; Huq, F.S.; Rajendran, V.; Green, G.E.; Hollister, S.J. Biomechanical evaluation of human and porcine auricular cartilage. *Laryngoscope* **2015**, *125*, E262–E268. [CrossRef]
181. Algazi, V.R.; Duda, R.O.; Satarzadeh, P. Physical and filter pinna models based on anthropometry. In Proceedings of the 122nd Audio Engineering Society Convention, Vienna, Austria, 5–8 May 2007.
182. Xie, X.; Wu, S.; Mou, S.; Guo, N.; Wang, Z.; Sun, J. Microtissue-based bioink as a chondrocyte microshelter for DLP bioprinting. *Adv. Healthc. Mater.* **2022**, *11*, 2201877. [CrossRef]
183. Chen, Y.; Zhang, J.; Liu, X.; Wang, S.; Tao, J.; Huang, Y.; Wu, W.; Li, Y.; Zhou, K.; Wei, X. Noninvasive in vivo 3D bioprinting. *Sci. Adv.* **2020**, *6*, eaba7406. [CrossRef]
184. Jackson, A.R.; Gu, W.Y. Transport properties of cartilaginous tissues. *Curr. Rheumatol. Rev.* **2009**, *5*, 40–50. [CrossRef]
185. Tashman, J.W.; Shiwarski, D.J.; Feinberg, A.W. Development of a high-performance open-source 3D bioprinter. *Sci. Rep.* **2022**, *12*, 22652. [CrossRef]
186. Posniak, S.; Chung, J.H.Y.; Liu, X.; Mukherjee, P.; Gambhir, S.; Khansari, A.; Wallace, G.G. Bioprinting of chondrocyte stem cell co-cultures for auricular cartilage regeneration. *ACS Omega* **2022**, *7*, 5908–5920. [CrossRef]
187. Jiang, Y.; Cai, Y.; Zhang, W.; Yin, Z.; Hu, C.; Tong, T.; Lu, P.; Zhang, S.; Neculai, D.; Tuan, R.S.; et al. Human cartilage-derived progenitor cells from committed chondrocytes for efficient cartilage repair and regeneration. *Stem Cells Transl. Med.* **2016**, *5*, 733–744. [CrossRef]
188. Cruz, I.B.M.d.; Severo, A.L.; Azzolin, V.F.; Garcia, L.F.M.; Kuhn, A.; Lech, O. Regenerative potential of the cartilaginous tissue in mesenchymal stem cells: Update, limitations, and challenges. *Rev. Bras. De Ortop.* **2017**, *52*, 2–10. [CrossRef]
189. Richter, W. Mesenchymal stem cells and cartilage in situ regeneration. *J. Intern. Med.* **2009**, *266*, 390–405. [CrossRef]

190. Skardal, A.; Mack, D.; Kapetanovic, E.; Atala, A.; Jackson, J.D.; Yoo, J.; Soker, S. Bioprinted amniotic fluid-derived stem cells accelerate healing of large skin wounds. *Stem Cells Transl. Med.* **2012**, *1*, 792–802. [CrossRef]
191. Tricomi, B.J.; Dias, A.D.; Corr, D.T. Stem cell bioprinting for applications in regenerative medicine. *Ann. N. Y. Acad. Sci.* **2016**, *1383*, 115–124. [CrossRef]
192. Castelhano, L.; Correia, F.; Colaço, T.; Reis, L.; Escada, P. Tympanic membrane perforations: The importance of etiology, size and location. *Eur. Arch. Oto-Rhino-Laryngol.* **2022**, *279*, 4325–4333. [CrossRef]
193. Sarkar, S. A review on the history of tympanoplasty. *Indian J. Otolaryngol. Head. Neck Surg.* **2013**, *65*, 455–460. [CrossRef] [PubMed]
194. Kuo, C.-Y.; Wilson, E.; Fuson, A.; Gandhi, N.; Monfaredi, R.; Jenkins, A.; Romero, M.; Santoro, M.; Fisher, J.P.; Cleary, K. Repair of tympanic membrane perforations with customized bioprinted ear grafts using chinchilla models. *Tissue Eng. Part A* **2018**, *24*, 527–535. [CrossRef] [PubMed]
195. Bayram, A.; Muluk, N.B.; Cingi, C.; Bafaqeeh, S.A. Success rates for various graft materials in tympanoplasty–A review. *J. Otol.* **2020**, *15*, 107–111. [CrossRef]
196. Liang, J.; Engles, W.G.; Smith, K.D.; Dai, C.; Gan, R.Z. Mechanical properties of baboon tympanic membrane from young to adult. *J. Assoc. Res. Otolaryngol.* **2020**, *21*, 395–407. [CrossRef]
197. Zhou, L.; Feng, M.; Huang, X.; Duan, M. Fatigue analysis of tympanic membrane after ossiculoplasty. *Acta Oto-Laryngol.* **2017**, *137*, 679–685. [CrossRef]
198. Williams, D.; Thayer, P.; Martinez, H.; Gatenholm, E.; Khademhosseini, A. A perspective on the physical, mechanical and biological specifications of bioinks and the development of functional tissues in 3D bioprinting. *Bioprinting* **2018**, *9*, 19–36. [CrossRef]
199. von Witzleben, M.; Stoppe, T.; Zeinalova, A.; Chen, Z.; Ahlfeld, T.; Bornitz, M.; Bernhardt, A.; Neudert, M.; Gelinsky, M. Multimodal additive manufacturing of biomimetic tympanic membrane replacements with near tissue-like acousto-mechanical and biological properties. *Acta Biomater.* **2023**, *170*, 124–141. [CrossRef] [PubMed]
200. Kozin, E.D.; Black, N.L.; Cheng, J.T.; Cotler, M.J.; McKenna, M.J.; Lee, D.J.; Lewis, J.A.; Rosowski, J.J.; Remenschneider, A.K. Design, fabrication, and in vitro testing of novel three-dimensionally printed tympanic membrane grafts. *Hear. Res.* **2016**, *340*, 191–203. [CrossRef] [PubMed]
201. Bedir, T.; Baykara, D.; Yildirim, R.; Calikoglu Koyuncu, A.C.; Sahin, A.; Kaya, E.; Tinaz, G.B.; Insel, M.A.; Topuzogulları, M.; Gunduz, O. Three-dimensional-printed GelMA-KerMA composite patches as an innovative platform for potential tissue engineering of tympanic membrane perforations. *Nanomaterials* **2024**, *14*, 563. [CrossRef]
202. Volandri, G.; Di Puccio, F.; Forte, P.; Carmignani, C. Biomechanics of the tympanic membrane. *J. Biomech.* **2011**, *44*, 1219–1236. [CrossRef]
203. Kern, E.B. The nose structure and function. *Postgrad. Med.* **1975**, *57*, 101–103. [CrossRef]
204. Vertu-Ciolino, D.; Brunard, F.; Courtial, E.-J.; Pasdeloup, M.; Marquette, C.A.; Perrier-Groult, E.; Mallein-Gerin, F.; Malcor, J.-D. Challenges in nasal cartilage tissue engineering to restore the shape and function of the nose. *Tissue Eng. Part B Rev.* **2024**, *30*, 581–595. [CrossRef]
205. Gantumur, E.; Nakahata, M.; Kojima, M.; Sakai, S. Extrusion-based bioprinting through glucose-mediated enzymatic hydrogelation. *Int. J. Bioprinting* **2020**, *6*, 250. [CrossRef]
206. Pitton, M.; Fiorati, A.; Buscemi, S.; Melone, L.; Farè, S.; Contessi Negrini, N. 3D bioprinting of pectin-cellulose nanofibers multicomponent bioinks. *Front. Bioeng. Biotechnol.* **2021**, *9*, 732689. [CrossRef]
207. Shin, S.; Park, S.; Park, M.; Jeong, E.; Na, K.; Youn, H.J.; Hyun, J. Cellulose nanofibers for the enhancement of printability of low viscosity gelatin derivatives. *BioResources* **2017**, *12*, 2941–2954. [CrossRef]
208. Jodat, Y.A.; Kiaee, K.; Jarquin, D.V.; Hernández, R.L.D.l.G.; Wang, T.; Joshi, S.; Rezaei, Z.; de Melo, B.A.G.; Ge, D.; Mannoor, M.S.; et al. A 3D-printed hybrid nasal cartilage with functional electronic olfaction. *Adv. Sci.* **2020**, *7*, 1901878. [CrossRef]
209. Tamura, N.; Hasunuma, K.; Saito, T.; Fujisawa, S. Mechanical and Thermal Properties of Porous Nanocellulose/Polymer Composites: Influence of the Polymer Chemical Structure and Porosity. *ACS Omega* **2024**, *9*, 19560–19565. [CrossRef] [PubMed]
210. Tie, B.S.H.; Manaf, E.; Halligan, E.; Zhuo, S.; Keane, G.; Geever, J.; Geever, L. The Effects of Incorporating Nanoclay in NVCL-NIPAm Hydrogels on Swelling Behaviours and Mechanical Properties. *Nanomaterials* **2024**, *14*, 597. [CrossRef] [PubMed]
211. Lan, X.; Liang, Y.; Vyhlidal, M.; Erkut, E.J.; Kunze, M.; Mulet-Sierra, A.; Osswald, M.; Ansari, K.; Seikaly, H.; Boluk, Y. In vitro maturation and in vivo stability of bioprinted human nasal cartilage. *J. Tissue Eng.* **2022**, *13*, 20417314221086368. [CrossRef] [PubMed]
212. Roseti, L.; Cavallo, C.; Desando, G.; Parisi, V.; Petretta, M.; Bartolotti, I.; Grigolo, B. Three-dimensional bioprinting of cartilage by the use of stem cells: A strategy to improve regeneration. *Materials* **2018**, *11*, 1749. [CrossRef]

213. Petretta, M.; Desando, G.; Grigolo, B.; Roseti, L. Cartilage tissue engineering by extrusion bioprinting: Process analysis, risk evaluation, and mitigation strategies. *Materials* **2021**, *14*, 3528. [CrossRef]
214. Li, M.; Sun, D.; Zhang, J.; Wang, Y.; Wei, Q.; Wang, Y. Application and development of 3D bioprinting in cartilage tissue engineering. *Biomater. Sci.* **2022**, *10*, 5430–5458. [CrossRef]
215. Derman, I.D.; Yeo, M.; Castaneda, D.C.; Callender, M.; Horvath, M.; Mo, Z.; Xiong, R.; Fleming, E.; Chen, P.; EPeeples, M.; et al. High-throughput bioprinting of the nasal epithelium using patient-derived nasal epithelial cells. *Biofabrication* **2023**, *15*, 044103.
216. Gudapati, H.; Dey, M.; Ozbolat, I. A comprehensive review on droplet-based bioprinting: Past, present and future. *Biomaterials* **2016**, *102*, 20–42. [CrossRef]
217. Park, I.H.; Um, J.Y.; Hong, S.M.; Cho, J.S.; Lee, S.H.; Lee, S.H.; Lee, H.M. Metformin reduces TGF-β1–induced extracellular matrix production in nasal polyp–derived fibroblasts. *Otolaryngol.—Head. Neck Surg.* **2014**, *150*, 148–153. [CrossRef]
218. DeMaria, S.; Ngai, J. The cell biology of smell. *J. Cell Biol.* **2010**, *191*, 443–452. [CrossRef]
219. Fiorellini, J.P.; Stathopoulou, P.G. Anatomy of the periodontium. In *Carranza's Clinical Periodontology*, 12th ed.; Elsevier Saunders: St. Louis, MO, USA, 2015; pp. 9–10.
220. Hughes, F.J. Chapter 34-Periodontium and Periodontal Disease. In *Stem Cell Biology and Tissue Engineering in Dental Sciences*; Vishwakarma, A., Sharpe, P., Shi, S., Ramalingam, M., Eds.; Academic Press: Boston, MA, USA, 2015; pp. 433–444.
221. Eke, P.I.; Borgnakke, W.S.; Genco, R.J. Recent epidemiologic trends in periodontitis in the USA. *Periodontology 2000* **2020**, *82*, 257–267. [CrossRef] [PubMed]
222. Ong, G. Periodontal disease and tooth loss. *Int. Dent. J.* **1998**, *48*, 233–238. [CrossRef] [PubMed]
223. Wen, X.; Liu, R.; Li, G.; Deng, M.; Liu, L.; Zeng, X.-T.; Nie, X. History of periodontitis as a risk factor for long-term survival of dental implants: A meta-analysis. *Int. J. Oral. Maxillofac. Implant.* **2014**, *29*, 1271–1280. [CrossRef] [PubMed]
224. Sanz, I.; Alonso, B.; Carasol, M.; Herrera, D.; Sanz, M. Nonsurgical treatment of periodontitis. *J. Evid. Based Dent. Pract.* **2012**, *12*, 76–86. [CrossRef]
225. Cobb, C.M.; Low, S.B.; Coluzzi, D.J. Lasers and the treatment of chronic periodontitis. *Dent. Clin.* **2010**, *54*, 35–53. [CrossRef]
226. Zheng, W.; Wang, S.; Wang, J.; Jin, F. Periodontitis promotes the proliferation and suppresses the differentiation potential of human periodontal ligament stem cells. *Int. J. Mol. Med.* **2015**, *36*, 915–922. [CrossRef]
227. Liu, Y.; Zheng, Y.; Ding, G.; Fang, D.; Zhang, C.; Bartold, P.M.; Gronthos, S.; Shi, S.; Wang, S. Periodontal ligament stem cell-mediated treatment for periodontitis in miniature swine. *Stem Cells* **2008**, *26*, 1065–1073. [CrossRef]
228. Zhang, Z.; Deng, M.; Hao, M.; Tang, J. Periodontal ligament stem cells in the periodontitis niche: Inseparable interactions and mechanisms. *J. Leukoc. Biol.* **2021**, *110*, 565–576. [CrossRef]
229. Miao, G.; Liang, L.; Li, W.; Ma, C.; Pan, Y.; Zhao, H.; Zhang, Q.; Xiao, Y.; Yang, X. 3D bioprinting of a bioactive composite scaffold for cell delivery in periodontal tissue regeneration. *Biomolecules* **2023**, *13*, 1062. [CrossRef]
230. Chen, Z.; Lv, Y. Gelatin/sodium alginate composite hydrogel with dynamic matrix stiffening ability for bone regeneration. *Compos. Part B Eng.* **2022**, *243*, 110162. [CrossRef]
231. Ansari, S.; Pouraghaei Sevari, S.; Chen, C.; Sarrion, P.; Moshaverinia, A. RGD-modified alginate–GelMA hydrogel sheet containing gingival mesenchymal stem cells: A unique platform for wound healing and soft tissue regeneration. *ACS Biomater. Sci. Eng.* **2021**, *7*, 3774–3782. [CrossRef] [PubMed]
232. Ansari, S.; Sarrion, P.; Hasani-Sadrabadi, M.M.; Aghaloo, T.; Wu, B.M.; Moshaverinia, A. Regulation of the fate of dental-derived mesenchymal stem cells using engineered alginate-GelMA hydrogels. *J. Biomed. Mater. Res. Part A* **2017**, *105*, 2957–2967. [CrossRef] [PubMed]
233. Francis, L.; Meng, D.; Knowles, J.C.; Roy, I.; Boccaccini, A.R. Multi-functional P (3HB) microsphere/45S5 Bioglass®-based composite scaffolds for bone tissue engineering. *Acta Biomater.* **2010**, *6*, 2773–2786. [CrossRef]
234. Hossain, K.M.Z.; Patel, U.; Ahmed, I. Development of microspheres for biomedical applications: A review. *Progress. Biomater.* **2015**, *4*, 1–19. [CrossRef]
235. Bendtsen, S.T.; Quinnell, S.P.; Wei, M. Development of a novel alginate-polyvinyl alcohol-hydroxyapatite hydrogel for 3D bioprinting bone tissue engineered scaffolds. *J. Biomed. Mater. Res. Part A* **2017**, *105*, 1457–1468. [CrossRef]
236. Wenz, A.; Borchers, K.; Tovar, G.E.; Kluger, P.J. Bone matrix production in hydroxyapatite-modified hydrogels suitable for bone bioprinting. *Biofabrication* **2017**, *9*, 044103. [CrossRef]
237. Koch, F.; Thaden, O.; Conrad, S.; Tröndle, K.; Finkenzeller, G.; Zengerle, R.; Kartmann, S.; Zimmermann, S.; Koltay, P. Mechanical properties of polycaprolactone (PCL) scaffolds for hybrid 3D-bioprinting with alginate-gelatin hydrogel. *J. Mech. Behav. Biomed. Mater.* **2022**, *130*, 105219. [CrossRef]
238. Mousavi Nejad, Z.; Zamanian, A.; Saeidifar, M.; Vanaei, H.R.; Salar Amoli, M. 3D bioprinting of polycaprolactone-based scaffolds for pulp-dentin regeneration: Investigation of physicochemical and biological behavior. *Polymers* **2021**, *13*, 4442. [CrossRef]

239. Cunniffe, G.M.; Gonzalez-Fernandez, T.; Daly, A.; Sathy, B.N.; Jeon, O.; Alsberg, E.; Kelly, D.J. Three-dimensional bioprinting of polycaprolactone reinforced gene activated bioinks for bone tissue engineering. *Tissue Eng. Part A* **2017**, *23*, 891–900. [CrossRef]
240. Singh, S.; Prakash, C.; Singh, M.; Mann, G.S.; Gupta, M.K.; Singh, R.; Ramakrishna, S. Poly-lactic-Acid: Potential Material for Bio-printing Applications. In *Biomanufacturing*; Prakash, C., Singh, S., Singh, R., Ramakrishna, S., Pabla, B.S., Puri, S., Uddin, M.S., Eds.; Springer International Publishing: Cham, Switzerland, 2019; pp. 69–87.
241. Narayanan, L.K.; Huebner, P.; Fisher, M.B.; Spang, J.T.; Starly, B.; Shirwaiker, R.A. 3D-bioprinting of polylactic acid (PLA) nanofiber–alginate hydrogel bioink containing human adipose-derived stem cells. *ACS Biomater. Sci. Eng.* **2016**, *2*, 1732–1742. [CrossRef] [PubMed]
242. Stepanovska, J.; Supova, M.; Hanzalek, K.; Broz, A.; Matejka, R. Collagen bioinks for bioprinting: A systematic review of hydrogel properties, bioprinting parameters, protocols, and bioprinted structure characteristics. *Biomedicines* **2021**, *9*, 1137. [CrossRef] [PubMed]
243. Lee, J.M.; Suen, S.K.Q.; Ng, W.L.; Ma, W.C.; Yeong, W.Y. Bioprinting of collagen: Considerations, potentials, and applications. *Macromol. Biosci.* **2021**, *21*, 2000280. [CrossRef]
244. Osidak, E.O.; Kozhukhov, V.I.; Osidak, M.S.; Domogatsky, S.P. Collagen as bioink for bioprinting: A comprehensive review. *Int. J. Bioprinting* **2020**, *6*, 270. [CrossRef] [PubMed]
245. Emmermacher, J.; Spura, D.; Cziommer, J.; Kilian, D.; Wollborn, T.; Fritsching, U.; Steingroewer, J.; Walther, T.; Gelinsky, M.; Lode, A. Engineering considerations on extrusion-based bioprinting: Interactions of material behavior, mechanical forces and cells in the printing needle. *Biofabrication* **2020**, *12*, 025022. [CrossRef]
246. Lee, U.-L.; Yun, S.; Cao, H.-L.; Ahn, G.; Shim, J.-H.; Woo, S.-H.; Choung, P.-H. Bioprinting on 3D printed titanium scaffolds for periodontal ligament regeneration. *Cells* **2021**, *10*, 1337. [CrossRef]
247. Vurat, M.T.; Şeker, Ş.; Lalegül-Ülker, Ö.; Parmaksiz, M.; Elçin, A.E.; Elçin, Y.M. Development of a multicellular 3D-bioprinted microtissue model of human periodontal ligament-alveolar bone biointerface: Towards a pre-clinical model of periodontal diseases and personalized periodontal tissue engineering. *Genes Dis.* **2022**, *9*, 1008–1023. [CrossRef]
248. Jiao, K.; Liu, C.; Basu, S.; Raveendran, N.; Nakano, T.; Ivanovski, S.; Han, P. Bioprinting extracellular vesicles as a" cell-free" regenerative medicine approach. *Extracell. Vesicles Circ. Nucleic Acids* **2023**, *4*, 218. [CrossRef]
249. Yerneni, S.S.; Whiteside, T.L.; Weiss, L.E.; Campbell, P.G. Bioprinting exosome-like extracellular vesicle microenvironments. *Bioprinting* **2019**, *13*, e00041. [CrossRef]
250. Han, P.; Raveendran, N.; Liu, C.; Basu, S.; Jiao, K.; Johnson, N.; Moran, C.S.; Ivanovski, S. 3D bioprinted small extracellular vesicles from periodontal cells enhance mesenchymal stromal cell function. *Biomater. Adv.* **2024**, *158*, 213770. [CrossRef]
251. Chibly, A.M.; Aure, M.H.; Patel, V.N.; Hoffman, M.P. Salivary gland function, development, and regeneration. *Physiol. Rev.* **2022**, *102*, 1495–1552. [CrossRef] [PubMed]
252. Miletich, I. Introduction to salivary glands: Structure, function and embryonic development. *Front. Oral. Biol.* **2010**, *14*, 1–20.
253. Villa, A.; Wolff, A.; Aframian, D.; Vissink, A.; Ekström, J.; Proctor, G.; McGowan, R.; Narayana, N.; Aliko, A.; Sia, Y.W. World Workshop on Oral Medicine VI: A systematic review of medication-induced salivary gland dysfunction: Prevalence, diagnosis, and treatment. *Clin. Oral. Investig.* **2015**, *19*, 1563–1580. [CrossRef] [PubMed]
254. Wilson, K.F.; Meier, J.D.; Ward, P.D. Salivary gland disorders. *Am. Fam. Physician* **2014**, *89*, 882–888. [PubMed]
255. Atkinson, J.C.; Wu, A.J. Salivary gland dysfunction: Causes, symptoms, treatment. *J. Am. Dent. Assoc.* **1994**, *125*, 409–416. [CrossRef] [PubMed]
256. Beahm, D.D.; Peleaz, L.; Nuss, D.W.; Schaitkin, B.; Sedlmayr, J.C.; Rivera-Serrano, C.M.; Zanation, A.M.; Walvekar, R.R. Surgical approaches to the submandibular gland: A review of literature. *Int. J. Surg.* **2009**, *7*, 503–509. [CrossRef]
257. Nahlieli, O. Complications of traditional and modern therapeutic salivary approaches. *Acta Otorhinolaryngol. Ital.* **2017**, *37*, 142. [CrossRef]
258. Phan, T.V.; Oo, Y.; Ahmed, K.; Rodboon, T.; Rosa, V.; Yodmuang, S.; Ferreira, J.N. Salivary gland regeneration: From salivary gland stem cells to three-dimensional bioprinting. *SLAS Technol.* **2023**, *28*, 199–209. [CrossRef]
259. Muntean, D.D.; Lenghel, M.L.; Petea-Balea, D.-R.; Ciurea, A.I.; Solomon, C.; Dudea, S.M. Functional evaluation of major salivary glands using viscosity PLUS and 2D shear-wave PLUS elastography techniques in healthy subjects—A pilot study. *Diagnostics* **2022**, *12*, 1963. [CrossRef]
260. Bedewi, M.A.; Elsifey, A.A.; Nassir, E.M.; Elgazzar, H.M.; Alfawaz, A.F.; Hussein, R.S.; Moawad, K.; Bediwy, A.M.; Swify, S.M. Shear wave elastography of the submandibular gland in healthy individuals. *J. Int. Med. Res.* **2020**, *48*, 0300060520979445. [CrossRef]
261. Arda, K.; Ciledag, N.; Aktas, E.; Arıbas, B.K.; Köse, K. Quantitative assessment of normal soft-tissue elasticity using shear-wave ultrasound elastography. *Am. J. Roentgenol.* **2011**, *197*, 532–536. [CrossRef] [PubMed]

262. Yin, Y.; Vázquez-Rosado, E.J.; Wu, D.; Viswananthan, V.; Farach, A.; Farach-Carson, M.C.; Harrington, D.A. Microfluidic coaxial 3D bioprinting of cell-laden microfibers and microtubes for salivary gland tissue engineering. *Biomater. Adv.* **2023**, *154*, 213588. [CrossRef] [PubMed]
263. Pillai, S.; Munguia-Lopez, J.G.; Tran, S.D. Hydrogels for salivary gland tissue engineering. *Gels* **2022**, *8*, 730. [CrossRef]
264. Shubin, A.D.; Felong, T.J.; Graunke, D.; Ovitt, C.E.; Benoit, D.S. Development of poly (ethylene glycol) hydrogels for salivary gland tissue engineering applications. *Tissue Eng. Part A* **2015**, *21*, 1733–1751. [CrossRef]
265. Rose, S.C.; Larsen, M.; Xie, Y.; Sharfstein, S.T. Salivary Gland Bioengineering. *Bioengineering* **2023**, *11*, 28. [CrossRef]
266. Charbonneau, A.M.; Kinsella, J.M.; Tran, S.D. 3D cultures of salivary gland cells in native or gelled egg yolk plasma, combined with egg white and 3D-printing of gelled egg yolk plasma. *Materials* **2019**, *12*, 3480. [CrossRef] [PubMed]
267. Telis-Romero, J.; Thomaz, C.; Bernardi, M.; Telis, V.R.N.; Gabas, A.L. Rheological properties and fluid dynamics of egg yolk. *J. Food Eng.* **2006**, *74*, 191–197. [CrossRef]
268. Anton, M. Egg yolk: Structures, functionalities and processes. *J. Sci. Food Agric.* **2013**, *93*, 2871–2880. [CrossRef]
269. Muallah, D.; Matschke, J.; Kappler, M.; Kroschwald, L.M.; Lauer, G.; Eckert, A.W. Dental Pulp Stem Cells for Salivary Gland Regeneration—Where Are We Today? *Int. J. Mol. Sci.* **2023**, *24*, 8664. [CrossRef]
270. Jeschke, M.G.; van Baar, M.E.; Choudhry, M.A.; Chung, K.K.; Gibran, N.S.; Logsetty, S. Burn injury. *Nat. Rev. Dis. Primers* **2020**, *6*, 11. [CrossRef]
271. Zeiderman, M.R.; Pu, L.L. Contemporary reconstruction after complex facial trauma. *Burn. Trauma.* **2020**, *8*, tkaa003. [CrossRef] [PubMed]
272. Schlottmann, F.; Bucan, V.; Vogt, P.M.; Krezdorn, N. A short history of skin grafting in burns: From the gold standard of autologous skin grafting to the possibilities of allogeneic skin grafting with immunomodulatory approaches. *Medicina* **2021**, *57*, 225. [CrossRef] [PubMed]
273. Böttcher-Haberzeth, S.; Biedermann, T.; Reichmann, E. Tissue engineering of skin. *Burns* **2010**, *36*, 450–460. [CrossRef] [PubMed]
274. Lee, V.; Singh, G.; Trasatti, J.P.; Bjornsson, C.; Xu, X.; Tran, T.N.; Yoo, S.-S.; Dai, G.; Karande, P. Design and fabrication of human skin by three-dimensional bioprinting. *Tissue Eng. Part C Methods* **2014**, *20*, 473–484. [CrossRef]
275. Ma, X.; Zhu, X.; Lv, S.; Yang, C.; Wang, Z.; Liao, M.; Zhou, B.; Zhang, Y.; Sun, S.; Chen, P. 3D bioprinting of prefabricated artificial skin with multicomponent hydrogel for skin and hair follicle regeneration. *Theranostics* **2025**, *15*, 2933. [CrossRef]
276. Tanadchangsaeng, N.; Pasanaphong, K.; Tawonsawatruk, T.; Rattanapinyopituk, K.; Tangketsarawan, B.; Rawiwet, V.; Kongchanagul, A.; Srikaew, N.; Yoyruerop, T.; Panupinthu, N.; et al. 3D bioprinting of fish skin-based gelatin methacryloyl (GelMA) bio-ink for use as a potential skin substitute. *Sci. Rep.* **2024**, *14*, 23240. [CrossRef]
277. Kang, M.S.; Kwon, M.; Lee, S.H.; Kim, W.H.; Lee, G.W.; Jo, H.J.; Kim, B.; Yang, S.Y.; Kim, K.S.; Han, D.W. 3D Printing of Skin Equivalents with Hair Follicle Structures and Epidermal-Papillary-Dermal Layers Using Gelatin/Hyaluronic Acid Hydrogels. *Chem. Asian J.* **2022**, *17*, e202200620. [CrossRef]
278. Boyer, G.; Zahouani, H.; Le Bot, A.; Laquieze, L. In vivo characterization of viscoelastic properties of human skin using dynamic micro-indentation. In Proceedings of the 2007 29th Annual International Conference of the IEEE Engineering in Medicine and Biology Society, Lyon, France, 22–26 August 2007; pp. 4584–4587.
279. Feng, X.; Li, G.-Y.; Ramier, A.; Eltony, A.M.; Yun, S.-H. In vivo stiffness measurement of epidermis, dermis, and hypodermis using broadband Rayleigh-wave optical coherence elastography. *Acta Biomater.* **2022**, *146*, 295–305. [CrossRef]
280. Otberg, N.; Richter, H.; Schaefer, H.; Blume-Peytavi, U.; Sterry, W.; Lademann, J. Variations of hair follicle size and distribution in different body sites. *J. Investig. Dermatol.* **2004**, *122*, 14–19. [CrossRef]
281. Nanmo, A.; Yan, L.; Asaba, T.; Wan, L.; Kageyama, T.; Fukuda, J. Bioprinting of hair follicle germs for hair regenerative medicine. *Acta Biomater.* **2023**, *165*, 50–59. [CrossRef]
282. Motter Catarino, C.; Cigaran Schuck, D.; Dechiario, L.; Karande, P. Incorporation of hair follicles in 3D bioprinted models of human skin. *Sci. Adv.* **2023**, *9*, eadg0297. [CrossRef] [PubMed]
283. Zhang, Y.; Enhejirigala, n.; Yao, B.; Li, Z.; Song, W.; Li, J.; Zhu, D.; Wang, Y.; Duan, X.; Yuan, X. Using bioprinting and spheroid culture to create a skin model with sweat glands and hair follicles. *Burn. Trauma.* **2021**, *9*, tkab013. [CrossRef] [PubMed]
284. Zhao, W.; Chen, H.; Zhang, Y.; Zhou, D.; Liang, L.; Liu, B.; Xu, T. Adaptive multi-degree-of-freedom in situ bioprinting robot for hair-follicle-inclusive skin repair: A preliminary study conducted in mice. *Bioeng. Transl. Med.* **2022**, *7*, e10303. [CrossRef]
285. Chen, H.; Ma, X.; Gao, T.; Zhao, W.; Xu, T.; Liu, Z. Robot-assisted in situ bioprinting of gelatin methacrylate hydrogels with stem cells induces hair follicle-inclusive skin regeneration. *Biomed. Pharmacother.* **2023**, *158*, 114140. [CrossRef]
286. Kang, Y.; Yeo, M.; Derman, I.D.; Ravnic, D.J.; Singh, Y.P.; Alioglu, M.A.; Wu, Y.; Makkar, J.; Driskell, R.R.; Ozbolat, I.T. Intraoperative bioprinting of human adipose-derived stem cells and extra-cellular matrix induces hair follicle-like downgrowths and adipose tissue formation during full-thickness craniomaxillofacial skin reconstruction. *Bioact. Mater.* **2024**, *33*, 114–128. [CrossRef]

287. Gökmen, M.F.; Büyükatalay, Z.Ç.; Beton, S.; Gökcan, M.K.; Dursun, G.; Meço, C.; Küçük, T.B. Functional and Oncological Outcomes of Open Partial Laryngectomy vs. Transoral Laser Surgery in Supraglottic Larynx Cancer. *Turk. Arch. Otorhinolaryngol.* **2021**, *58*, 227–233. [CrossRef] [PubMed]
288. Candelo, E.; Belafsky, P.C.; Corrales, M.; Farwell, D.G.; Gonzales, L.F.; Grajek, M.; Walczak, D.; Strome, M.; Lorenz, R.R.; Tintinago, L.F.; et al. The Global Experience of Laryngeal Transplantation: Series of Eleven Patients in Three Continents. *Laryngoscope* **2024**, *134*, 4313–4320. [CrossRef]
289. Granell, J.; Garrido, L.; Millas, T.; Gutiérrez-Fonseca, R. Management of Oropharyngeal Dysphagia in Laryngeal and Hypopharyngeal Cancer. *Int. J. Otolaryngol.* **2012**, *2012*, 1–9. [CrossRef]
290. Aibara, S.; Okada, M.; Tanaka-Nishikubo, K.; Asayama, R.; Sato, E.; Sei, H.; Aoishi, K.; Takagi, T.; Teraoka, M.; Mukai, N.; et al. Laryngeal Complications after Endotracheal Intubation and Prone Positioning in Patients with Coronavirus Disease 2019. *Laryngoscope Investig. Otolaryngol.* **2022**, *7*, 1909–1914. [CrossRef]
291. Roybal, J.L.; Liechty, K.W.; Hedrick, H.L.; Bebbington, M.; Johnson, M.P.; Coleman, B.G.; Adzick, N.S.; Flake, A.W. Predicting the Severity of Congenital High Airway Obstruction Syndrome. *J. Pediatr. Surg.* **2010**, *45*, 1633–1639. [CrossRef]
292. Leboulanger, N.; Garabédian, É.-N. Laryngo-Tracheo-Oesophageal Clefts. *Orphanet J. Rare Dis.* **2011**, *6*, 81. [CrossRef] [PubMed]
293. Wang, Y.; Chang, G.H.; Yang, Y.H.; Liu, C.Y.; Tsai, Y.T.; Hsu, C.M.; Lee, Y.C.; Lee, L.A.; Yang, P.R.; Tsai, M.S.; et al. Allergic Rhinitis and Laryngeal Pathology: Real-World Evidence. *Healthcare* **2021**, *9*, 36. [CrossRef] [PubMed]
294. Holzer, S.E.S.; Ludlow, C.L. The Swallowing Side Effects of Botulinum Toxin Type a Injection in Spasmodic Dysphonia. *Laryngoscope* **1996**, *106*, 86–92. [CrossRef]
295. Casserly, P.; Timon, C. Botulinum Toxin a Injection Under Electromyographic Guidance for Treatment of Spasmodic Dysphonia. *J. Laryngol. Otol.* **2007**, *122*, 52–56. [CrossRef] [PubMed]
296. Liu, J.; Lv, D.; Deng, D.; Wang, J.; Li, L.; Chen, F. Free Bipaddled Anterolateral Thigh Flap for Simultaneous Reconstruction of Large Larynx and Prelaryngeal Skin Defects after Resection of the Local Recurrent Laryngeal Cancer Invading the Cricoid Cartilage and Prelaryngeal Skin. *Medicine* **2019**, *98*, e14199. [CrossRef] [PubMed]
297. Werner, J.; Scaglioni, M.F.; Ko, S.S.; Rajan, G.P. Preservation of Laryngeal Function through Reconstruction of the Supraglottis and Thyrohyoid Membrane with a Chimeric Anterolateral Thigh Flap after Supraglottic Laryngectomy: A Case Report and Literature Review. *Microsurgery* **2021**, *41*, 462–467. [CrossRef]
298. Smith, M.E.; Roy, N.; Houtz, D. Laryngeal Reinnervation for Paralytic Dysphonia in Children Younger Than 10 Years. *Arch. Otolaryngol.—Head. Neck Surg.* **2012**, *138*, 1161. [CrossRef]
299. Krasnodębska, P.; Szkiełkowska, A. Course of Laryngeal Function Recovery after Reconstruction of the laryngeal Nerves. *J. Hear. Sci.* **2020**, *10*, 51–56. [CrossRef]
300. Zhang, M.; Yang, Y.; Li, M.; Shang, Q.; Xie, R.; Yu, J.; Shen, K.; Zhang, Y.; Cheng, Y. Toughening Double-Network Hydrogels by Polyelectrolytes. *Adv. Mater.* **2023**, *35*, e2301551. [CrossRef]
301. Galliger, Z.; Vogt, C.D.; Helms, H.R.; Panoskaltsis-Mortari, A. Extracellular matrix microparticles improve GelMA bioink resolution for 3D bioprinting at ambient temperature. *Macromol. Mater. Eng.* **2022**, *307*, 2200196. [CrossRef]
302. Park, H.S.; Lee, J.S.; Kim, C.B.; Lee, K.H.; Hong, I.S.; Jung, H.; Lee, H.; Lee, Y.J.; Ajiteru, O.; Sultan, M.T. Fluidic integrated 3D bioprinting system to sustain cell viability towards larynx fabrication. *Bioeng. Transl. Med.* **2023**, *8*, e10423. [CrossRef] [PubMed]
303. McMillan, A.; Hoffman, M.R.; Xu, Y.; Wu, Z.; Thayer, E.; Peel, A.; Guymon, A.; Kanotra, S.; Salem, A.K. 3D bioprinted ferret mesenchymal stem cell-laden cartilage grafts for laryngotracheal reconstruction in a ferret surgical model. *Biomater. Sci.* **2025**, *13*, 1304–1322. [CrossRef]
304. Iqbal, F.; Cornett, E.M. Larynx. In *Basic Anesthesia Review*; Oxford Academic: New York, NY, USA, 2024; pp. 585–587. [CrossRef]
305. Mete, A.; Akbudak, İ.l.H. *Functional Anatomy and Physiology of Airway*; InTech: London, UK, 2018. [CrossRef]
306. Chiesa, I.; Esposito, A.; Vozzi, G.; Gottardi, R.; Maria, C.D. 4D Bioprinted Self-Folding Scaffolds Enhance Cartilage Formation in the Engineering of Trachea. *Adv. Mater. Technol.* **2025**, *10*, 2401210. [CrossRef]
307. Yu, Y.S.; Ahn, C.B.; Son, K.H.; Lee, J.W. Motility Improvement of Biomimetic Trachea Scaffold via Hybrid 3d-Bioprinting Technology. *Polymers* **2021**, *13*, 971. [CrossRef]
308. Ke, D.; Yi, H.; Est-Witte, S.; George, S.; Kengla, C.; Lee, S.J.; Atala, A.; Murphy, S.V. Bioprinted Trachea Constructs with Patient-Matched Design, Mechanical and Biological Properties. *Biofabrication* **2019**, *12*, 015022. [CrossRef] [PubMed]
309. Huo, Y.; Xu, Y.; Wu, X.; Gao, E.; Zhan, A.; Chen, Y.; Zhang, Y.; Hua, Y.; Święszkowski, W.; Zhang, Y.S.; et al. Functional Trachea Reconstruction Using 3D-Bioprinted Native-Like Tissue Architecture Based on Designable Tissue-Specific Bioinks. *Adv. Sci.* **2022**, *9*, e2202181. [CrossRef]
310. Low, V.; Rubesin, S. Contrast evaluation of the pharynx and esophagus. *Radiol. Clin. N. A.* **1993**, *31*, 1265–1291. [CrossRef]
311. Goyal, R.K.; Mashimo, H. Physiology of oral, pharyngeal, and esophageal motility. *GI Motil. Online* **2006**. [CrossRef]

312. Dickman, C.T.; Russo, V.; Thain, K.R.; Pan, S.; Beyer, S.; Walus, K.; Getsios, S.; Mohamed, T.; Wadsworth, S.J. Functional Characterization of 3D Contractile Smooth Muscle Tissues Generated Using a Unique Microfluidic 3D Bioprinting Technology. *Faseb J.* **2019**, *34*, 1652–1664. [CrossRef]
313. Yeleswarapu, S.; Chameettachal, S.; Bera, A.K.; Pati, F. Smooth Muscle Matrix Bioink Promotes Myogenic Differentiation of Encapsulated Adipose-derived Stem Cells. *J. Biomed. Mater. Res. Part A* **2022**, *110*, 1761–1773. [CrossRef]
314. Reyes-Furrer, A.A.; Andrade, S.D.; Bachmann, D.; Jeker, H.; Steinmann, M.W.; Accart, N.; Dunbar, A.; Rausch, M.; Bono, E.; Rimann, M.; et al. 3D Bioprinting of Contractile Human Skeletal Muscle Models Recapitulating Exercise and Pharmacological Responses. *Commun. Biol.* **2021**, *4*, 1183. [CrossRef] [PubMed]
315. Fornetti, E.; Paolis, F.D.; Fuoco, C.; Bernardini, S.; Giannitelli, S.M.; Rainer, A.; Seliktar, D.; Magdinier, F.; Baldi, J.; Biagini, R.; et al. A Novel Extrusion-Based 3D Bioprinting System for Skeletal Muscle Tissue Engineering. *Biofabrication* **2023**, *15*, 025009. [CrossRef] [PubMed]
316. Rouwkema, J.; Khademhosseini, A. Vascularization and angiogenesis in tissue engineering: Beyond creating static networks. *Trends Biotechnol.* **2016**, *34*, 733–745. [CrossRef]
317. Richards, D.; Jia, J.; Yost, M.J.; Markwald, R.R.; Mei, Y. 3D Bioprinting for Vascularized Tissue Fabrication. *Ann. Biomed. Eng.* **2016**, *45*, 132–147. [CrossRef]
318. Zhang, Z.; Xu, C.; Xu, L.; Wan, J.; Cao, G.; Liu, Z.; Ji, P.; Jin, Q.; Fu, Y.; Le, Y.; et al. Bioprinted Dermis with Human Adipose Tissue-derived Microvascular Fragments Promotes Wound Healing. *Biotechnol. Bioeng.* **2023**, *121*, 1406–1420. [CrossRef]
319. Zhu, W.; Qu, X.; Zhu, J.; Ma, X.; Patel, S.; Liu, J.; Wang, P.; Lai, C.S.E.; Gou, M.; Xu, Y.; et al. Direct 3D Bioprinting of Prevascularized Tissue Constructs with Complex Microarchitecture. *Biomaterials* **2017**, *124*, 106–115. [CrossRef]
320. Buranawat, B.; Shaalan, A.; Garna, D.F.; Di Silvio, L. Development of Prevascularized Synthetic Block Graft for Maxillofacial Reconstruction. *J. Funct. Biomater.* **2025**, *16*, 18. [CrossRef] [PubMed]
321. Ouyang, L.; Armstrong, J.P.K.; Chen, Q.; Lin, Y.; Stevens, M.M. Void-Free 3D Bioprinting for in Situ Endothelialization and Microfluidic Perfusion. *Adv. Funct. Mater.* **2019**, *30*, 1908349. [CrossRef]
322. Yeo, M.; Sarkar, A.; Singh, Y.P.; Derman, İ.D.; Datta, P.; Özbolat, İ.T. Synergistic Coupling Between 3D Bioprinting and Vascularization Strategies. *Biofabrication* **2023**, *16*, 012003. [CrossRef]
323. Daia, C.; Scheau, C.; Neagu, G.; Andone, I.; Spanu, A.; Popescu, C.; Stoica, S.I.; Verenca, M.C.; Onose, G. Nerve conduction study and electromyography findings in patients recovering from Covid-19—Case report. *Int. J. Infect. Dis.* **2021**, *103*, 420–422. [CrossRef]
324. Brizzi, K.T.; Lyons, J.L. Peripheral nervous system manifestations of infectious diseases. *Neurohospitalist* **2014**, *4*, 230–240. [CrossRef]
325. De León, A.M.; Garcia-Santibanez, R.; Harrison, T.B. Article Topic: Neuropathies Due to Infections and Antimicrobial Treatments. *Curr. Treat. Options Neurol.* **2023**, *25*, 213–229. [CrossRef] [PubMed]
326. Joung, D.; Truong, V.; Neitzke, C.C.; Guo, S.Z.; Walsh, P.J.; Monat, J.R.; Meng, F.; Park, S.H.; Dutton, J.R.; Parr, A.M.; et al. 3D Printed Stem-Cell Derived Neural Progenitors Generate Spinal Cord Scaffolds. *Adv. Funct. Mater.* **2018**, *28*, 1801850. [CrossRef] [PubMed]
327. Salaris, F.; Colosi, C.; Brighi, C.; Soloperto, A.; Turris, V.d.; Benedetti, M.C.; Ghirga, S.; Rosito, M.; Angelantonio, S.D.; Rosa, A. 3D Bioprinted Human Cortical Neural Constructs Derived from Induced Pluripotent Stem Cells. *J. Clin. Med.* **2019**, *8*, 1595. [CrossRef] [PubMed]
328. Clair-Glover, M.S.; Finol-Urdaneta, R.K.; Maddock, M.; Wallace, E.; Miellet, S.; Wallace, G.G.; Yue, Z.; Dottori, M. Efficient Fabrication of 3D Bioprinted Functional Sensory Neurons Using an Inducible Neurogenin-2 Human Pluripotent Stem Cell Line. *Biofabrication* **2024**, *16*, 045022. [CrossRef]
329. Gu, Q.; Tomaskovic-Crook, E.; Lozano, R.; Chen, Y.; Kapsa, R.M.I.; Zhou, Q.; Wallace, G.G.; Crook, J.M. Functional 3D Neural Mini-Tissues from Printed Gel-Based Bioink and Human Neural Stem Cells. *Adv. Healthc. Mater.* **2016**, *5*, 1429–1438. [CrossRef]
330. Klar, R.M.; Cox, J.C.; Raja, N.; Lohfeld, S. The 3D-McMap Guidelines: Three-Dimensional Multicomposite Microsphere Adaptive Printing. *Biomimetics* **2024**, *9*, 94. [CrossRef]
331. Kotani, T.; Hananouchi, T.; Sakai, S. Enhancing Visible Light-Induced 3D Bioprinting: Alternating Extruded Support Materials for Bioink Gelation. *Biomed. Mater.* **2025**, *20*, 035005. [CrossRef]
332. Lee, J.M.; Sing, S.L.; Zhou, M.; Yeong, W.Y. 3D Bioprinting Processes: A Perspective on Classification and Terminology. *Int. J. Bioprinting* **2024**, *4*, 151. [CrossRef]
333. Lukyanov, D.A.; Levin, O.V. Inkjet Printing with (Semi)conductive Conjugated Polymers: A Review. *Chemengineering* **2024**, *8*, 53. [CrossRef]
334. Mariani, M.; Cervellera, F.; Migliori, E.; Maspero, F.; Bernasconi, R.; Galassi, C.; Magagnin, L.; Lecis, N. Inkjet Printing of Cobalt Ferrite for Hard Ferromagnetic Thick Films Manufacturing. *Adv. Eng. Mater.* **2024**, *26*, 2400371. [CrossRef]

335. Rau, D.A.; Kim, M.; Xu, B.; Cai, L.H. Vat Photopolymerization Printing of Modular Soft Stretchable Low-Cost Elastomers. *Acs Appl. Polym. Mater.* **2025**, *7*, 7566–7574. [CrossRef]
336. Cai, J.; Ma, S.; Yi, W.; Wang, J. Fabrication of High-Density Microarchitected Tungsten via DLP 3D Printing. *Adv. Sci.* **2024**, *11*, e2405487. [CrossRef]
337. Ma, Y.; Shih, C.J.; Bao, Y. Advances in 4D Printing of Biodegradable Photopolymers. *Responsive Mater.* **2024**, *2*, e20240008. [CrossRef]
338. Hisham, M.; Butt, H. Vat Photopolymerization Printing of Functionalized Hydrogels on Commercial Contact Lenses. *Sci. Rep.* **2024**, *14*, 13860. [CrossRef]
339. Gómez-Lizárraga, K.; Ruiz-García, L.; Garduño-Wilches, I.; Aguilar-Frutis, M.; Piña-Barba, C.; Alarcón-Flores, G. Photoluminescence Studies of DoF Effect on the Photopolymerization Reaction of a GelMA-Based Monomers for Tissue Engineering. *Macromol. Chem. Phys.* **2024**, *225*, 2300085. [CrossRef]
340. Piao, C.H.; Du, X.; Xu, Y.; To, S.; Zhu, L.; Zhu, Z. Time-Dependent Volumetric Printing of Precision Lenses through Dynamic Laser Writing. *Int. J. Extrem. Manuf.* **2025**, *7*, 045007. [CrossRef]
341. Daoset, N.; Inglam, S.; Wanchat, S.; Chantarapanich, N. Effects of Post-Processing Curing Parameters and Gamma Irradiation on the Mechanical Properties of Medical Graded Vat Photopolymerization Parts. *Rapid Prototyp. J.* **2024**, *30*, 475–489. [CrossRef]
342. Alam, F.; Alsharif, A.A.; Al-Modaf, F.; El-Atab, N. 3D-Printed Smartwatch Fabricated via Vat Photopolymerization for UV and Temperature Sensing Applications. *ACS Omega* **2024**, *9*, 14830–14839. [CrossRef]
343. Jain, P.; Kathuria, H.; Ramakrishna, S.; Parab, S.; Pandey, M.M.; Dubey, N. In Situ Bioprinting: Process, Bioinks, and Applications. *Acs Appl. Bio Mater.* **2024**, *7*, 7987–8007. [CrossRef]
344. Mirsky, N.A.; Ehlen, Q.T.; Greenfield, J.; Antonietti, M.; Slavin, B.V.; Nayak, V.V.; Pelaez, D.; Tse, D.T.; Witek, L.; Daunert, S.; et al. Three-Dimensional Bioprinting: A Comprehensive Review for Applications in Tissue Engineering and Regenerative Medicine. *Bioengineering* **2024**, *11*, 777. [CrossRef] [PubMed]
345. Ravanbakhsh, H.; Karamzadeh, V.; Bao, G.; Mongeau, L.; Juncker, D.; Zhang, Y.S. Emerging technologies in multi-material bioprinting. *Adv. Mater.* **2021**, *33*, 2104730. [CrossRef] [PubMed]
346. Alhetheel, A.; Alrfaei, B.M.; Mujamammi, A.H.; Hakami, J.; Alshuhri, A. REGECEL (An Oxidized Regenerated Cellulose) Provides Superior Bioactivity Effect on Microorganisms. *Infect. Drug Resist.* **2024**, *17*, 3353–3365. [CrossRef]
347. Yang, C.; Wu, W.; Lao, S.; Zhang, S.; Tang, J.; Pu, X.; Lu, X.; Ye, F.; Zhao, P.; Chen, D.; et al. Controlled preparation of droplets for cell encapsulation by air-focused microfluidic bioprinting. *Int. J. Bioprinting* **2024**, *10*, 1102. [CrossRef]
348. Dai, X.; Zhang, J.; Bao, X.; Guo, Y.; Jin, Y.; Yang, C.; Zhang, H.; Liu, L.; Gao, Y.; Ye, C.; et al. Induction of Tumor Ferroptosis-Dependent Immunity via an Injectable Attractive Pickering Emulsion Gel. *Adv. Mater.* **2023**, *35*, e2303542. [CrossRef]
349. Habibi, M.; Foroughi, S.; Karamzadeh, V.; Packirisamy, M. Direct sound printing. *Nat. Commun.* **2022**, *13*, 1800. [CrossRef]
350. Davoodi, E.; Li, J.; Ma, X.; Najafabadi, A.H.; Yoo, J.; Lu, G.; Sani, E.S.; Lee, S.; Montazerian, H.; Kim, G.; et al. Imaging-guided deep tissue in vivo sound printing. *Science* **2025**, *388*, 616–623. [CrossRef]
351. Velazquez, E.R.; Parmar, C.; Jermoumi, M.; Mak, R.H.; van Baardwijk, A.; Fennessy, F.M.; Lewis, J.H.; De Ruysscher, D.; Kikinis, R.; Lambin, P.; et al. Volumetric CT-based segmentation of NSCLC using 3D-Slicer. *Sci. Rep.* **2013**, *3*, 3529. [CrossRef]
352. Watson, R.A. A low-cost surgical application of additive fabrication. *J. Surg. Educ.* **2014**, *71*, 14–17. [CrossRef]
353. Lipshutz, S.; Kim, Y.; Curtis, M.; Friedrich, L.; Alimperti, S. Development of Multiparametric Bioprinting Method for Generation of 3D Printed Cell-laden Structures. *Biotechnol. Progress.* **2025**, e70016. [CrossRef]
354. Mohammadrezaei, D.; Podina, L.; Silva, J.D.; Kohandel, M. Cell Viability Prediction and Optimization in Extrusion-Based Bioprinting via Neural Network-Based Bayesian Optimization Models. *Biofabrication* **2024**, *16*, 025016. [CrossRef]
355. Ferreras, A.; Matesanz, A.; Mendizabal, J.; Artola, K.; Nishina, Y.; Acedo, P.; Jorcano, J.L.; Ruiz, A.; Reina, G.; Martín, C. Light-Responsive and Antibacterial Graphenic Materials as a Holistic Approach to Tissue Engineering. *Acs Nanosci. Au* **2024**, *4*, 263–272. [CrossRef] [PubMed]
356. Pourchet, L.; Casado-Medina, L.; Richaud-Patín, Y.; Tadevosyan, K.; Morillas-García, A.; Lorenzo, E.; Lazis, I.; Ventura, A.; Litowczenko, J.; Guiu, J.; et al. 3D Bioprinting of Human iPSC-derived Cardiac Constructs with Microvascular Network Support for Improved Graft Survival in Vivo. *Biofabrication* **2025**, *17*, 035010. [CrossRef] [PubMed]
357. Greco, I.; Machrafi, H.; Iorio, C.S. Double-Network Hydrogel 3D BioPrinting Biocompatible with Fibroblast Cells for Tissue Engineering Applications. *Gels* **2024**, *10*, 684. [CrossRef]
358. Andrade, T.A.; da Silva, V.A.; Scheck, K.; Garay, T.; Sharma, R.; Willerth, S.M. 3D Bioprinting a Novel Skin Co-Culture Model Using Human Keratinocytes and Fibroblasts. *J. Biomed. Mater. Res. Part A* **2025**, *113*, e37831. [CrossRef]
359. Soman, S.S.; Vijayavenkataraman, S. Applications of 3D Bioprinted-Induced Pluripotent Stem Cells in Healthcare. *Int. J. Bioprinting* **2024**, *6*, 280. [CrossRef] [PubMed]

360. Liu, Q.; Feng, Y.; Yao, B.; Li, Z.; Kong, Y.; Zhang, C.; Tan, Y.; Song, W.; Jirigala, E.; Li, X.; et al. Effect of Tunable Stiffness on Immune Responses in 3d-Bioprinted Alginate–gelatin Scaffolds. *Int. J. Bioprinting* **2024**, *10*, 2874. [CrossRef]
361. Goldfracht, I.; Machour, M.; Michael, I.; Bulatova, M.; Zavin, J.; Levenberg, S. 3D Bioprinting of Thick Adipose Tissues with Integrated Vascular Hierarchies. *Adv. Funct. Mater.* **2024**, *35*, 2410311. [CrossRef]
362. Witowski, J.; Sitkowski, M.; Zuzak, T.; Coles-Black, J.; Chuen, J.; Major, P.; Pdziwiatr, M. From ideas to long-term studies: 3D printing clinical trials review. *Int. J. Comput. Assist. Radiol. Surg.* **2018**, *13*, 1473–1478. [CrossRef]
363. Parramon-Teixido, C.J.; Rodríguez-Pombo, L.; Basit, A.W.; Worsley, A.; Cañete-Ramírez, C.; Alvarez-Lorenzo, C.; Cabañas-Poy, M.J.; Goyanes, A. A framework for conducting clinical trials involving 3D printing of medicines at the point-of-care. *Drug Deliv. Transl. Res.* **2025**, 1–20. [CrossRef]
364. Periferakis, A.; Bolocan, A.; Ion, D. A review of innovation in medicine. *Technol. Innov. Life Sci.* **2022**, *1*, 42–48. [CrossRef]
365. Bhatt, A. Evolution of clinical research: A history before and beyond james lind. *Perspect. Clin. Res.* **2010**, *1*, 6–10. [CrossRef] [PubMed]
366. Kandi, V.; Vadakedath, S. Clinical Trials and Clinical Research: A Comprehensive Review. *Cureus* **2023**, *15*, e35077. [CrossRef]
367. Li, P.; Faulkner, A.; Medcalf, N. 3D bioprinting in a 2D regulatory landscape: Gaps, uncertainties, and problems. *Law Innov. Technol.* **2020**, *12*, 1–29. [CrossRef]
368. Mladenovska, T.; Choong, P.F.; Wallace, G.G.; O'Connell, C.D. The Regulatory Challenge of 3D Bioprinting. *Regen. Med.* **2023**, *18*, 659–674. [CrossRef] [PubMed]
369. Kim, J.H.; Seol, Y.J.; Ko, I.K.; Kang, H.W.; Lee, Y.K.; Yoo, J.J.; Atala, A.; Lee, S.J. 3D Bioprinted Human Skeletal Muscle Constructs for Muscle Function Restoration. *Sci. Rep.* **2018**, *8*, 12307. [CrossRef]
370. Maiullari, F.; Costantini, M.; Milan, M.; Pace, V.; Chirivì, M.; Maiullari, S.; Rainer, A.; Baci, D.; Marei, H.E.; Seliktar, D.; et al. A multi-cellular 3D bioprinting approach for vascularized heart tissue engineering based on HUVECs and iPSC-derived cardiomyocytes. *Sci. Rep.* **2018**, *8*, 13532. [CrossRef]
371. Apelgren, P.; Amoroso, M.; Säljö, K.; Lindahl, A.; Brantsing, C.; Stridh Orrhult, L.; Gatenholm, P.; Kölby, L. Skin Grafting on 3D Bioprinted Cartilage Constructs in vivo. *Plast. Reconstr. Surg. Glob. Open* **2018**, *6*, e1930. [CrossRef]
372. Bregenzer, M.E.; Horst, E.N.; Mehta, P.; Novak, C.M.; Raghavan, S.; Snyder, C.S.; Mehta, G. Integrated cancer tissue engineering models for precision medicine. *PLoS ONE* **2019**, *14*, e0216564. [CrossRef]
373. Nguyen, D.G.; Funk, J.; Robbins, J.B.; Crogan-Grundy, C.; Presnell, S.C.; Singer, T.; Roth, A.B. Bioprinted 3D Primary Liver Tissues Allow Assessment of Organ-Level Response to Clinical Drug Induced Toxicity in vitro. *PLoS ONE* **2016**, *11*, e0158674. [CrossRef]
374. Skardal, A.; Murphy, S.V.; Devarasetty, M.; Mead, I.; Kang, H.W.; Seol, Y.J.; Shrike Zhang, Y.; Shin, S.R.; Zhao, L.; Aleman, J.; et al. Multi-tissue interactions in an integrated three-tissue organ-on-a-chip platform. *Sci. Rep.* **2017**, *7*, 8837. [CrossRef]
375. European Union Regulation. *Regulation (EU) No 536/2014 of the European Parliament and of the Council of 16 April 2014 on Clinical Trials on Medicinal Products for Human Use, and Repealing Directive 2001/20/EC Text with EEA Relevance*; European Union Regulation: Brussels, Belgium, 2014; pp. 1–76.
376. European Union Regulation. *Regulation (EU) 2017/745 of the European Parliament and of the Council of 5 April 2017 on Medical Devices, Amending Directive 2001/83/EC, Regulation (EC) No 178/2002 and Regulation (EC) No 1223/2009 and Repealing Council Directives 90/385/EEC and 93/42/EEC (Text with EEA Relevance)*; European Union Regulation: Brussels, Belgium, 2017; pp. 1–175.
377. Tappa, K.; Jammalamadaka, U. Novel biomaterials used in medical 3D printing techniques. *J. Funct. Biomater.* **2018**, *9*, 17. [CrossRef] [PubMed]
378. Zhu, Y.; Guo, S.; Ravichandran, D.; Ramanathan, A.; Sobczak, M.T.; Sacco, A.F.; Patil, D.; Thummalapalli, S.V.; Pulido, T.V.; Lancaster, J.N. 3D-Printed Polymeric Biomaterials for Health Applications. *Adv. Healthc. Mater.* **2025**, *14*, 2402571. [CrossRef]
379. Madan, A.; Saini, R.; Satapathy, M.K.; Arora, D. Biomaterials in Clinical Trials and Clinical Applications. In *Biomaterials and Neurodegenerative Disorders*; Kumar, G., Mukherjee, S., Kumar, S., Eds.; Springer Nature Singapore: Singapore, 2025; pp. 205–250.
380. Tsantes, A.G.; Altsitzioglou, P.; Papadopoulos, D.V.; Lorenzo, D.; Romanò, C.L.; Benzakour, T.; Tsukamoto, S.; Errani, C.; Angelini, A.; Mavrogenis, A.F. Infections of Tumor Prostheses: An Updated Review on Risk Factors, Microbiology, Diagnosis, and Treatment Strategies. *Biology* **2023**, *12*, 314. [CrossRef] [PubMed]
381. Hitchman, L.H.; Smith, G.E.; Chetter, I.C. Prosthetic infections and high-risk surgical populations. *Surgery* **2019**, *37*, 38–44. [CrossRef]
382. Le Vavasseur, B.; Zeller, V. Antibiotic Therapy for Prosthetic Joint Infections: An Overview. *Antibiotics* **2022**, *11*, 486. [CrossRef]
383. Harrabi, H.; Meyer, E.; Dournon, N.; Bouchand, F.; Kilu, C.M.; Perronne, V.; Jaffal, K.; d'Anglejan, E.; Duran, C.; Dinh, A. Suppressive Antibiotic Therapy in Prosthetic Joint Infections: A Contemporary Overview. *Antibiotics* **2025**, *14*, 277. [CrossRef]
384. Periferakis, A.; Periferakis, A.T.; Troumpata, L.; Periferakis, K.; Scheau, A.E.; Savulescu-Fiedler, I.; Caruntu, A.; Badarau, I.A.; Caruntu, C.; Scheau, C. Kaempferol: A Review of Current Evidence of Its Antiviral Potential. *Int. J. Mol. Sci.* **2023**, *24*, 16299. [CrossRef]

385. Herrera, T.E.S.; Tello, I.P.S.; Mustafa, M.A.; Jamil, N.Y.; Alaraj, M.; Atiyah Altameem, K.K.; Alasheqi, M.Q.; Hamoody, A.M.; Alkhafaji, A.T.; Shakir, M.N.; et al. Kaempferol: Unveiling its anti-inflammatory properties for therapeutic innovation. *Cytokine* **2025**, *186*, 156846. [CrossRef]
386. Periferakis, A.T.; Periferakis, A.; Periferakis, K.; Caruntu, A.; Badarau, I.A.; Savulescu-Fiedler, I.; Scheau, C.; Caruntu, C. Antimicrobial Properties of Capsaicin: Available Data and Future Research Perspectives. *Nutrients* **2023**, *15*, 4097. [CrossRef]
387. Petran, E.M.; Periferakis, A.; Troumpata, L.; Periferakis, A.-T.; Scheau, A.-E.; Badarau, I.A.; Periferakis, K.; Caruntu, A.; Savulescu-Fiedler, I.; Sima, R.-M.; et al. Capsaicin: Emerging Pharmacological and Therapeutic Insights. *Curr. Issues Mol. Biol.* **2024**, *46*, 7895–7943. [CrossRef] [PubMed]
388. El-Kattan, N.; Ibrahim, M.A.; Emam, A.N.; Metwally, K.; Youssef, F.S.; Nassar, N.A.; Mansour, A.S. Evaluation of the antimicrobial activity of chitosan-and curcumin-capped copper oxide nanostructures against multi-drug-resistant microorganisms. *Nanoscale Adv.* **2025**, *7*, 2988–3007. [CrossRef] [PubMed]
389. Ribeiro, A.; Oliveira, D.; Cabral-Marques, H. Curcumin in Ophthalmology: Mechanisms, Challenges, and Emerging Opportunities. *Molecules* **2025**, *30*, 457. [CrossRef] [PubMed]
390. Periferakis, A.; Periferakis, A.-T.; Troumpata, L.; Periferakis, K.; Georgatos-Garcia, S.; Touriki, G.; Dragosloveanu, C.D.M.; Caruntu, A.; Savulescu-Fiedler, I.; Dragosloveanu, S.; et al. Pinosylvin: A Multifunctional Stilbenoid with Antimicrobial, Antioxidant, and Anti-Inflammatory Potential. *Curr. Issues Mol. Biol.* **2025**, *47*, 204. [CrossRef]
391. Zhang, W.; Wu, X.; Li, W.; Zhang, H.; Wang, Y.; Xu, J.; Li, W.; Qin, Y.; Wu, Z.; Ge, G.; et al. Pinosylvin Inhibits Inflammatory and Osteoclastogenesis via NLRP3 Inflammasome. *Adv. Sci.* **2025**, e01532. [CrossRef]
392. Sharma, M.R.; Sharma, P.K.; Mohan, B.; Bhatti, S.P.; Bendi, A.; Raghav, N. Piperine as Potential Anti-Inflammatory Agent: In-Silico Studies Along with In-Vitro Anti-Cathepsin B Activity, and Serum Protein Protection Validation. *Chem. Biodivers.* **2025**, e202500144. [CrossRef]
393. Periferakis, A.; Troumpata, L.; Periferakis, K.; Adalis, G.; Periferakis, A.; Georgatos-Garcia, S.; Maier, C.; Costache, A.; Garofil, D.; Costache, D. Traditional Ethnomedical and Ethnobotanical Applications and Uses of Piper Nigrum. *RJ Mil. Med.* **2025**, *128*, 286–303. [CrossRef]
394. Gago, C.; Serralheiro, A.; Miguel, M.d.G. Anti-Inflammatory Activity of Thymol and Thymol-Rich Essential Oils: Mechanisms, Applications, and Recent Findings. *Molecules* **2025**, *30*, 2450. [CrossRef]
395. Stojanović, N.M.; Ranđelović, P.J.; Simonović, M.; Radić, M.; Todorović, S.; Corrigan, M.; Harkin, A.; Boylan, F. Essential Oil Constituents as Anti-Inflammatory and Neuroprotective Agents: An Insight through Microglia Modulation. *Int. J. Mol. Sci.* **2024**, *25*, 5168. [CrossRef]
396. Cesaro, A.; Hoffman, S.C.; Das, P.; de la Fuente-Nunez, C. Challenges and applications of artificial intelligence in infectious diseases and antimicrobial resistance. *npj Antimicrob. Resist.* **2025**, *3*, 2. [CrossRef]
397. Gorecki, G.P.; Tomescu, D.R.; Pleș, L.; Panaitescu, A.M.; Dragosloveanu, Ș.; Scheau, C.; Sima, R.M.; Coman, I.S.; Grigorean, V.T.; Cochior, D. Implications of using artificial intelligence in the diagnosis of sepsis/sepsis shock. *Germs* **2024**, *14*, 77–84. [CrossRef] [PubMed]
398. Sarantopoulos, A.; Mastori Kourmpani, C.; Yokarasa, A.L.; Makamanzi, C.; Antoniou, P.; Spernovasilis, N.; Tsioutis, C. Artificial Intelligence in Infectious Disease Clinical Practice: An Overview of Gaps, Opportunities, and Limitations. *Trop. Med. Infect. Dis.* **2024**, *9*, 228. [CrossRef] [PubMed]
399. Periferakis, A.; Busnatu, S.; Nastasa, I.; Stanescu, A.M.A.; Andronic, O. Current bioethical concerns associated with the use of big data in medicine. In Proceedings of the XVII National Conference of Bioethics, Bucharest, Romania, 8–10 December 2022; pp. 65–66.
400. Lokesh, S.; Chakraborty, S.; Pulugu, R.; Mittal, S.; Pulugu, D.; Muruganantham, R. AI-based big data analytics model for medical applications. *Meas. Sens.* **2022**, *24*, 100534. [CrossRef]

Disclaimer/Publisher's Note: The statements, opinions and data contained in all publications are solely those of the individual author(s) and contributor(s) and not of MDPI and/or the editor(s). MDPI and/or the editor(s) disclaim responsibility for any injury to people or property resulting from any ideas, methods, instructions or products referred to in the content.

Journal of
Functional Biomaterials

MDPI

Article

Pre-Vascularized 3-Dimensional Skin Substitutes Promote Angiogenesis and Tissue Repair in a Murine Model of Refractory Skin Ulcers

Shota Tojo [1], Hiromi Miyazaki [2,*], Takami Saiki [2], Yasuyuki Tsunoi [3], Shingo Nakamura [2,*] and Ryuichi Azuma [1]

[1] Department of Plastic and Reconstructive Surgery, National Defense Medical College, Saitama 359-8513, Japan; shota1005@ndmc.ac.jp (S.T.); azuma@ndmc.ac.jp (R.A.)
[2] Division of Biomedical Engineering, National Defense Medical College Research Institute, Saitama 359-8513, Japan; res372@ndmc.ac.jp
[3] Division of Bioinformation and Therapeutic Systems, National Defense Medical College Research Institute, Saitama 359-8513, Japan; ytsunoi@ndmc.ac.jp
* Correspondence: hiromyzk@ndmc.ac.jp (H.M.); snaka@ndmc.ac.jp (S.N.)

Abstract: Restoring blood flow is crucial for treating refractory ulcers. Despite advancements in various biomaterials, none incorporating pre-formed blood vessels have been commercialized. To address this, we developed a pre-vascularized three-dimensional (3D) skin substitute (PV-3D skin) designed to enhance healing when treating refractory ulcers. This study aimed to evaluate the therapeutic role of PV-3D skin transplantation in refractory ulcer models, induced by applying mitomycin C to wounds in severe immunodeficient mice. The wounds were then treated with PV-3D skin, non-vascularized 3D skin, skin grafts, or wound dressings. The PV-3D skin group demonstrated healing dynamics comparable to those of the skin graft group, with similar tissue morphology and wound temperature changes. Furthermore, at day 7 post-transplantation, the PV-3D skin group demonstrated significantly higher hypoxia-inducible factor 1-alpha expression levels compared to the 3D skin group. By day 14, the PV-3D skin group exhibited a significantly larger vascular area compared to the 3D skin group. Notably, PV-3D skin treatment stimulated host-derived angiogenesis, thereby enhancing wound healing and reducing the recurrence of refractory ulcers. These results suggest that PV-3D skin transplantation offers a promising therapeutic approach for refractory ulcers, especially in terms of angiogenesis.

Keywords: angiogenesis; regional blood flow; skin substitutes; skin ulcer; wound healing

1. Introduction

Chronic wounds, including refractory ulcers, pose significant challenges in clinical management, as standard therapies have shown limited efficacy. Conventional treatment typically involves debridement to remove necrotic tissue, followed by non-surgical approaches to promote granulation tissue formation [1,2]. Once sufficient granulation tissue develops, epithelialization usually progresses, leading to wound healing. However, skin grafting or flap procedures may be required in cases of extensive ulcers or when granulation tissue remains immature and non-epithelializing. Conservative treatment approaches include the application of growth factors such as basic fibroblast growth factor (FGF), prostaglandins (PGs), and cyclic adenosine monophosphate (cAMP) derivatives, along with wound dressings like hydrocolloids, polyurethane foams, and negative-pressure wound therapy. Biomaterials such as dermal–epidermal composite substitutes, dried amniotic membranes, and collagen-based extracellular matrices (e.g., small intestinal submucosa) are highly effective for refractory ulcers [3–5]. Despite these therapeutic advancements, chronic wounds often remain challenging to heal due to underlying conditions such as venous stasis, atherosclerosis, radiation-induced tissue damage, prolonged corticosteroid use, or diabetes, all of which impair blood flow and fibroblast function. These factors hinder granulation tissue formation and delay wound healing [6,7]. Thus, optimizing tissue perfusion is crucial for successful treatment [8,9], highlighting the urgent need for advanced biomaterials and regenerative strategies to address these underlying pathophysiological challenges.

We have developed a pre-vascularized three-dimensional (PV-3D) skin substitute with a two-layer structure (epidermal and vascularized dermal layers) [10]. This design aims to overcome the challenge of tissue repair failure caused by insufficient blood flow, a critical factor in the impaired healing of chronic refractory wounds. Notably, no previous studies have reported using vascularized, three-dimensional cultured skin composed entirely of human skin-derived cells without relying on scaffolds such as Matrigel to treat refractory ulcers. If successful, this approach could be a novel and effective therapeutic option for refractory ulcers.

This study aimed to evaluate the therapeutic efficacy of PV-3D skin transplantation in a refractory ulcer mouse model.

2. Materials and Methods

2.1. Cell Culture

Neonatal normal human dermal fibroblasts (NHDFs, CC-2509; Lonza, Basel, Switzerland) were cultured in Dulbecco's modified Eagle's medium (DMEM; Nacalai Tesque, Kyoto, Japan) supplemented with 5% fetal bovine serum (Gibco, Waltham, MA, USA) and 1% antibiotic-antimycotic mixed stock solution (Nacalai Tesque), and were then used until passage 6. Human dermal microvascular endothelial cells (HMVECs, CC-2813; Lonza) were cultured in endothelial growth medium (EGM-2MV, Lonza) and used at passage 5. Neonatal human epidermal keratinocytes (NHEKs, KK-4009; Lifeline Cell Technology, Frederick, TX, USA) were maintained in HuMedia–KG2 medium (KURABO, Osaka, Japan) and used at passage 3. Cells were passaged upon reaching 80–90% confluency using 0.25% trypsin/ethylenediaminetetraacetic acid solution (Thermo Fisher Scientific Inc., Waltham, MA, USA). The culture media were changed every 1–2 days.

2.2. Construction of Vascularized 3D Skin Substitute

In a previous study [10], a 3D skin substitute incorporating a vascular network was constructed using a cell accumulation approach combined with a layer–by–layer cell coating technique [11]. Briefly, NHDFs were alternately suspended with 0.04 mg/mL fibronectin (FN; Sigma-Aldrich, St. Louis, MO, USA) or 0.04 mg/mL gelatin (G; Fujifilm Wako, Osaka, Japan) and washed with phosphate-buffered saline (PBS) between each coating step. After four repetitions of this cycle, cells were again suspended with FN solution and washed with Dulbecco's PBS. To construct a multilayer tissue containing blood vessel network, the coated NHDFs (1×10^7 cells) were cultured alone or co-cultured with HMVECs (2×10^5 cells) in 12-well insert with a 0.4 μm pore size (Corning, Corning, NY, USA) for 24 h. NHEKs (3×10^5 cells) in a 1:1 v/v ratio of DMEM and HuMedia–KG2 were subsequently seeded on the dermal layer coated with 0.04 mg/mL collagen type IV solution (from human placenta, Sigma-Aldrich), submerged in the mixed medium and cultured for 24 h. Epidermal differentiation was induced by replacing the mixed medium with 25 μg/mL L-ascorbic acid (Fujifilm Wako), followed by air-liquid interface cultivation for 5 days. Prior to grafting, a subset of the PV-3D skin constructs was reserved for quality control; pre-vascularization was verified by immunostaining for endothelial markers (e.g., human CD31) to confirm the presence of interconnected capillary-like networks. Representative images are provided in Figure S1B.

2.3. Animal Experiments

Following all ethical regulations, animal experiments in this study received approval from the Animal Ethics Committee of the National Defense Medical College (Approval No. 22062: Study on the Treatment of Refractory Ulcers Using 3D Cultured Skin, 14 March 2023). For application of human derived skin substitutes in transplantation, male mice with severe combined immunodeficiency (SCID), aged 8–9 weeks (25–30 g), were obtained from Charles River Laboratories (Kanagawa, Japan). After induction of inhalation anesthesia with isoflurane, the dorsal hair was shaved, and depilatory cream (Epilat®; Kracie Home Products, Tokyo, Japan) was applied. A full-thickness wound (8 mm in diameter) was created at the center of the back and excised skin was used as graft. To create a refractory wound, mitomycin C (MMC; 20 μg in 20 μL of a solution containing 10% (*v*/*v*) ethanol and 90% (*v*/*v*) ethylene glycol; Fujifilm Wako) was applied to the wound for 10 min [12]. After washing the wound with saline, mice were divided into four groups and treated with either PV-3D skin, 3D skin substitutes (without vasculature), syngeneic skin grafts, or wound dressing. The wound was protected with a silicone-faced dressing (SI–Aid®; ALCARE Co., Tokyo, Japan) and secured with transparent film dressing (Opsite®; Smith & Nephew, Tokyo, Japan) (Figure 1). The wound dressing comprised an inert, non-adherent silicone mesh without any bioactive components such as antimicrobials, drugs, biologics, or growth factors. This dressing was used solely as a neutral contact layer to prevent adhesion and to allow exudate to pass to a secondary absorbent pad. A circular silicone ring was secured to prevent wound contraction during long-term observation of 28 days. At 7, 14, and 28 days post-transplantation, the mice were euthanized; the dorsal skin, including the graft, was then harvested for histological evaluation. During tissue harvesting, each wound was inverted to expose the underside, and was photographed at a fixed distance under standardized conditions.

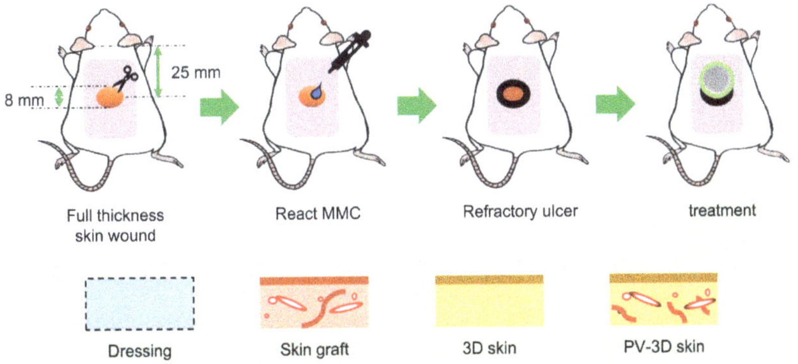

Figure 1. Schematic illustration of the treatment of refractory skin ulcers in mice with severe combined immunodeficiency. MMC: mitomycin C; PV-3D: pre-vascularized three-dimensional.

2.4. Evaluation of Wound State

On postoperative days 3, 7, 10, and 14, the surface temperature surrounding the wounds was assessed using an infrared thermography non-contact thermometer (FLIR ONE® GEN 3; Teledyne FLIR LLC, Wilsonville, OR, USA), and blood flow was measured by laser speckle blood flow imaging system (OMEGAZONE OZ-2; Omegawave Inc., Osaka, Japan).

2.5. Histological Observation and Immunohistochemistry

Tissues were fixed in 10% neutral buffered formalin for 24 h and embedded in paraffin. Hematoxylin and Eosin (H&E) and Trichrome (Abcam, Cambridge, UK) staining were performed according to the manufacturer's recommendations.

For immunohistochemical analysis, sections were stained using primary antibodies after blocking with normal goat serum. Primary antibodies against human leukocyte antigen (HLA)-Class I ABC (1:2500, ab70328, clone EMR8-5; Abcam), human and mouse-specific Fibroblast activation protein (FAP; 1:200, ABIN2929221, clone AA 542-761; antibodies.com, Atlanta, GA, USA), human-specific FAP (1:200, ab314456, clone RM1080; Abcam), human-specific CD31 (1:200, NBP2-15202, clone C31.3; Novus Biologicals, Centennial, CO, USA), mouse-specific CD31 (1:100, DIA-310, clone sz31; dianova, Hamburg, Germany), and hypoxia-inducible factor 1-alpha (HIF-1α; 1:400, ab228649, clone C-terminal; Abcam), Ly6g/6c (1:500, ab25377, clone RB6-8C5; abcam) were used. Before immunostaining with mouse-derived antibodies, we treated the sections with endogenous mouse immunoglobulin G (IgG) blocking reagent (M.O.M.® Immunodetection Kit; Vector Laboratories, Burlingame, CA, USA). The sections were reacted with anti-mouse or anti-rabbit IgG ImmPRESS-HRP reagent (Vector Laboratories) and visualized using DAB (Dako, Glostrup, Denmark). The stained tissue sections were observed under a microscope (BZ-X710; Keyence, Osaka, Japan). Wound size, thickness of the epidermal and dermal layers, and immunoreactivity of the stained areas were quantified using ImageJ software version 1.54p (National Institutes of Health, Bethesda, MD, USA).

2.6. Statistical Analysis

Statistical analyses were performed using GraphPad Prism version 9.5.1 (GraphPad Software, San Diego, CA, USA). Comparisons between two groups were performed using an unpaired t-test, whereas comparisons among multiple groups were performed using ANOVA followed by the Tukey–Kramer test. A linear mixed-effects model was used to analyze wound temperatures, with group, time, and their interaction included as fixed effects, and individual animal ID included as a random effect to account for repeated measurements. Statistical significance was defined as a p-value < 0.05. Data are presented as mean $\pm$ standard error.

3. Results

3.1. PV-3D Skin Accelerated Wound Healing in Refractory Ulcer

In both the PV-3D and 3D skin groups, the wound surface appeared white on day 7 post-grafting. On day 14, disparities in the progression of wound healing became more discernible: the 3D skin group exhibited residual white tissue in the center, whereas the PV-3D skin group displayed signs of crust detachment, revealing underlying red tissue. Conversely, the skin graft group exhibited partial dehiscence and necrosis at the wound edges, whereas the wound dressing group was found to have a translucent wound bed and hypogranulation (Figure 2).

To further assess healing, a thermographic analysis was conducted. On day 3 after grafting, the PV-3D skin group showed significantly higher wound temperatures than the 3D skin group ($p < 0.05$), approaching normal skin temperature. The wound temperature in the wound dressing group was significantly lower than that in the other groups, suggesting a delayed thermal recovery process during the inflammatory phase of wound healing. On day 7, both the PV-3D skin and skin graft groups maintained temperatures similar to that of the surrounding skin, no significant differences observed among the groups after day 10. Importantly, throughout the observation period, no abnormal temperature increases indicative of infection were detected (Figure 3A,B).

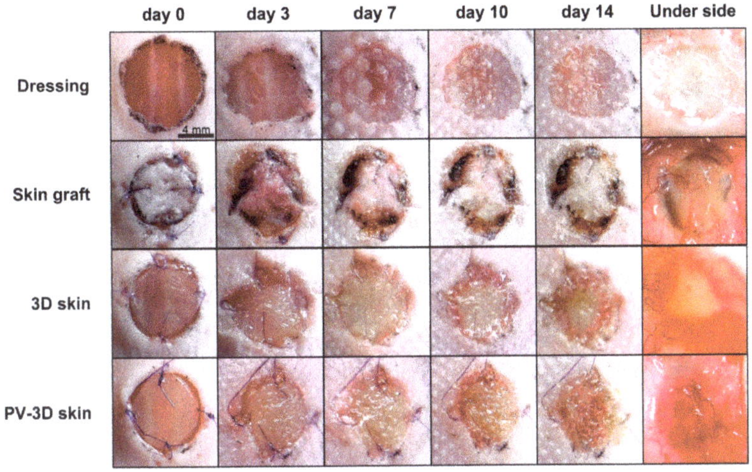

Figure 2. Post-transplantation macroscopic images of the wound area in the recipient mice.

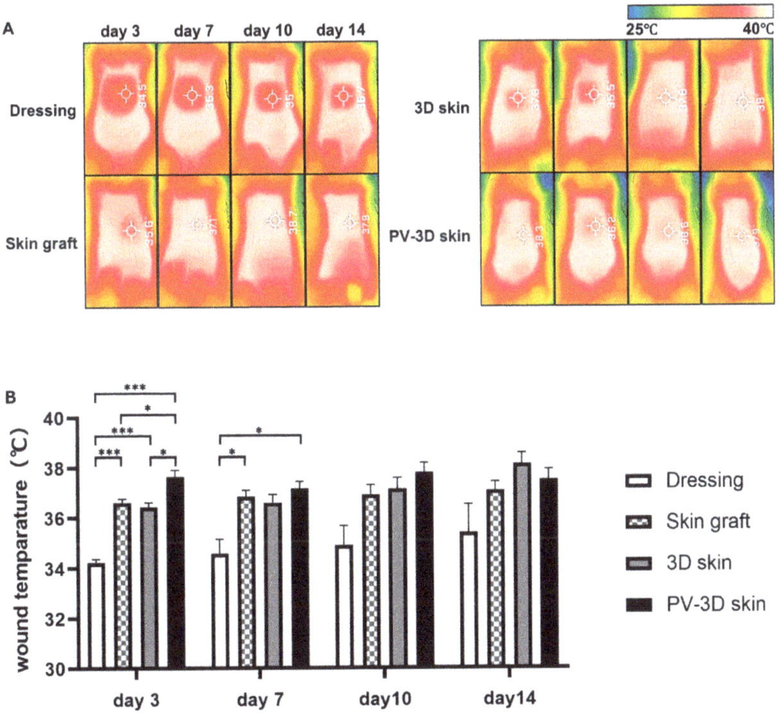

Figure 3. Evaluation of the therapeutic response in experimental models using infrared thermographic imaging. (**A**) Representative observation of the wound temperature. (**B**) Comparative analysis of wound temperatures across groups at each time point (n = 4–8 animals/time point). Data are represented as mean ± standard error of the mean. * $p < 0.05$, *** $p < 0.001$.

Histological observations provided further insights into wound healing. On day 7, the PV-3D and 3D skin groups exhibited signs of engraftment and crust formation, which subsequently underwent detachment. In the skin graft group, hair follicle reduction and marginal necrosis at the wound edges were observed; however, the epidermal and dermal structures remained, suggesting a tendency toward engraftment. Conversely, the wound dressing group exhibited minimal granulation tissue without epithelialization (Figure 4A). On day 14, the dermal and subdermal tissue thickness in the wound dressing group was significantly lower compared to its thickness in other groups ($p < 0.05$); however, the wound diameter did not differ significantly (Figure 4B). A critical marker of wound remodeling, collagen deposition, was also analyzed. On day 14, the PV-3D skin group exhibited widespread collagen deposition, exhibiting a larger collagen-stained area in the Skin graft and PV-3D skin group (Figure 4C). These findings suggest that PV-3D skin enhances extracellular matrix formation, potentially improving wound stability.

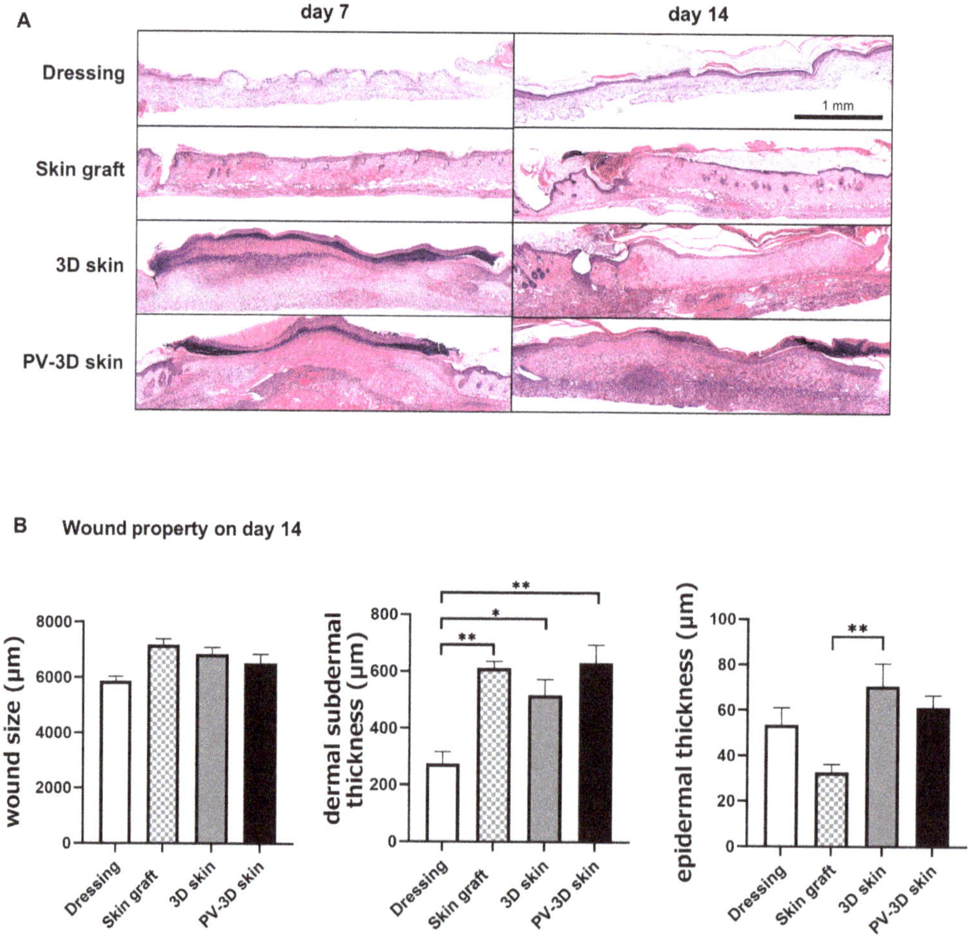

Figure 4. *Cont.*

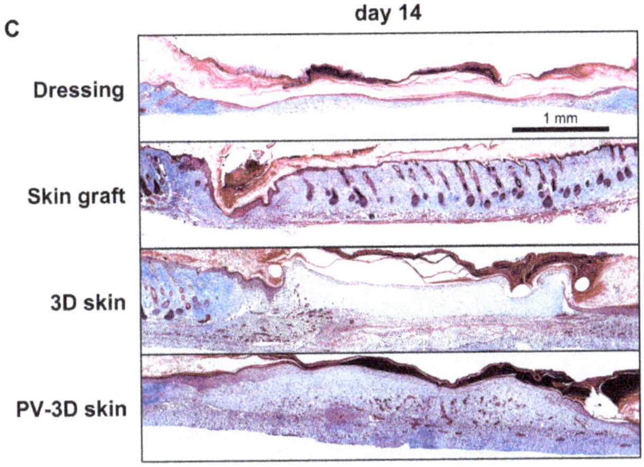

Figure 4. Histological analysis of the skin. (**A**) Hematoxylin and eosin staining performed on days 7 and 14 post-transplantation. (**B**) Quantitative analysis of wound diameter and the thickness of the epidermis, dermis, and subdermal layer on day 14 (n = 6–12 animals). Data are represented as mean ± standard error of the mean. * $p < 0.05$, ** $p < 0.01$. (**C**) Masson's Trichrome staining on day 14.

To investigate the cellular composition of the grafted tissues, immunohistochemical analysis was performed. On day 14 post-grafting, HLA staining was observed in the upper wound area of 3D and PV-3D skin groups, confirming the presence of human-derived cells in the grafted regions (Figure 5A). Subsequent analysis employing FAP staining, which detects human and mouse FAP, revealed pervasive staining throughout the wound area. However, staining with an antibody specific to human FAP revealed positive signals confined to the upper wound region (Figure 5B). Notably, the PV-3D skin group exhibited fewer positive areas than the 3D skin group, suggesting potential disparities in cellular dynamics between these two skin substitutes.

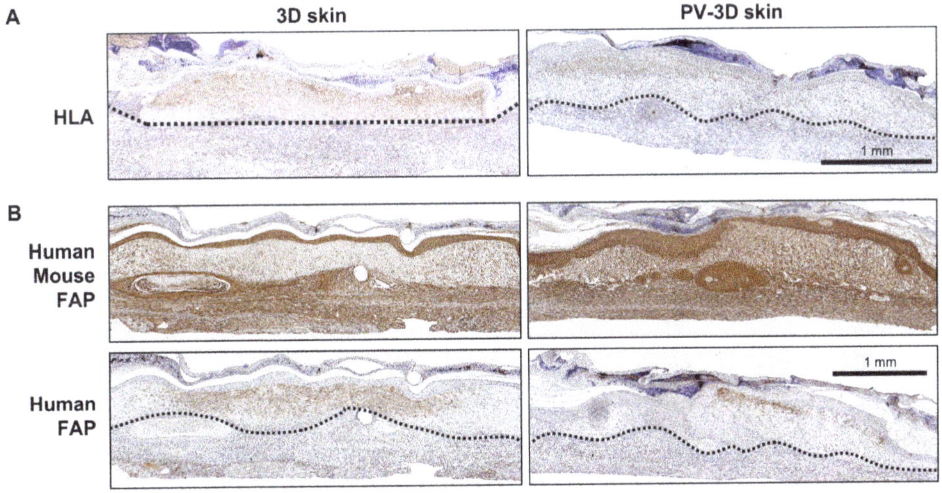

Figure 5. Immunohistochemical analysis of transplanted tissues on day 14. (**A**) Human leukocyte antigen and (**B**) Fibroblast activation protein staining. Dashed black lines indicate the interface between the transplanted tissue and the mouse wound bed.

3.2. Influence of Vascular Structure on Wound Healing

The results of the study indicate that the PV-3D skin substitute demonstrated wound healing dynamics closely aligned with those of normal skin. Given the established role of vascularization in skin graft survival and wound healing, the subsequent investigation focused on the role of vascular structures in the 3D and PV-3D skin groups.

Laser speckle contrast imaging revealed that beginning on day 3 post-grafting, the blood flow in the PV-3D skin group was significantly higher than in the 3D skin group ($p < 0.05$; Figure 6A,B).

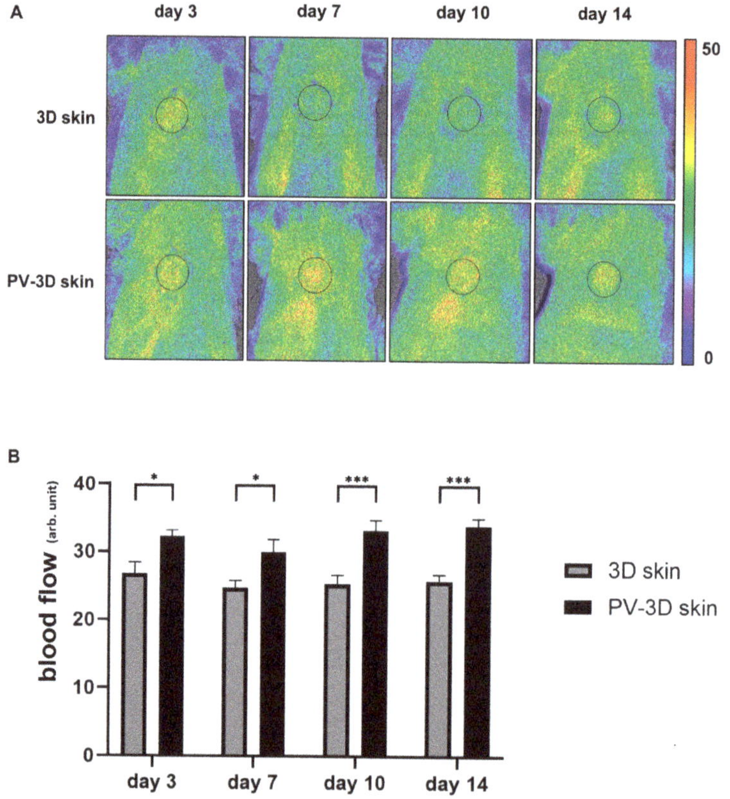

Figure 6. Blood flow monitoring in the pre-vascularized three-dimensional (3D) skin substitute and 3D skin groups following transplantation. (**A**) Representative laser speckle contrast images. (**B**) Quantitative analysis of blood flow changes at each time point ($n = 6$ animals/time point). Data are represented as mean ± standard error of the mean. * $p < 0.05$, *** $p < 0.001$.

To further assess neovascularization, we performed CD31 staining, a marker of endothelial cells. On day 7, the PV-3D skin group exhibited some positive CD31 staining; however, by day 14, no human CD31-positive staining was detectable. On day 7, 3D and PV-3D skin groups exhibited host-derived CD31-positive blood vessels surrounding the grafted tissue. However, by day 14, the PV-3D skin group displayed abundant blood vessels in the wound center, whereas the 3D skin group retained blood vessels primarily around the wound periphery (Figure 7A). Quantitative analysis confirmed that the PV-3D skin group had a significantly larger blood vessel area in the wound center compared to the 3D skin group ($6.4 \pm 0.94 \times 104$ μm^2 vs. $3.4 \pm 0.76 \times 104$ μm^2, $p < 0.05$; Figure 7B).

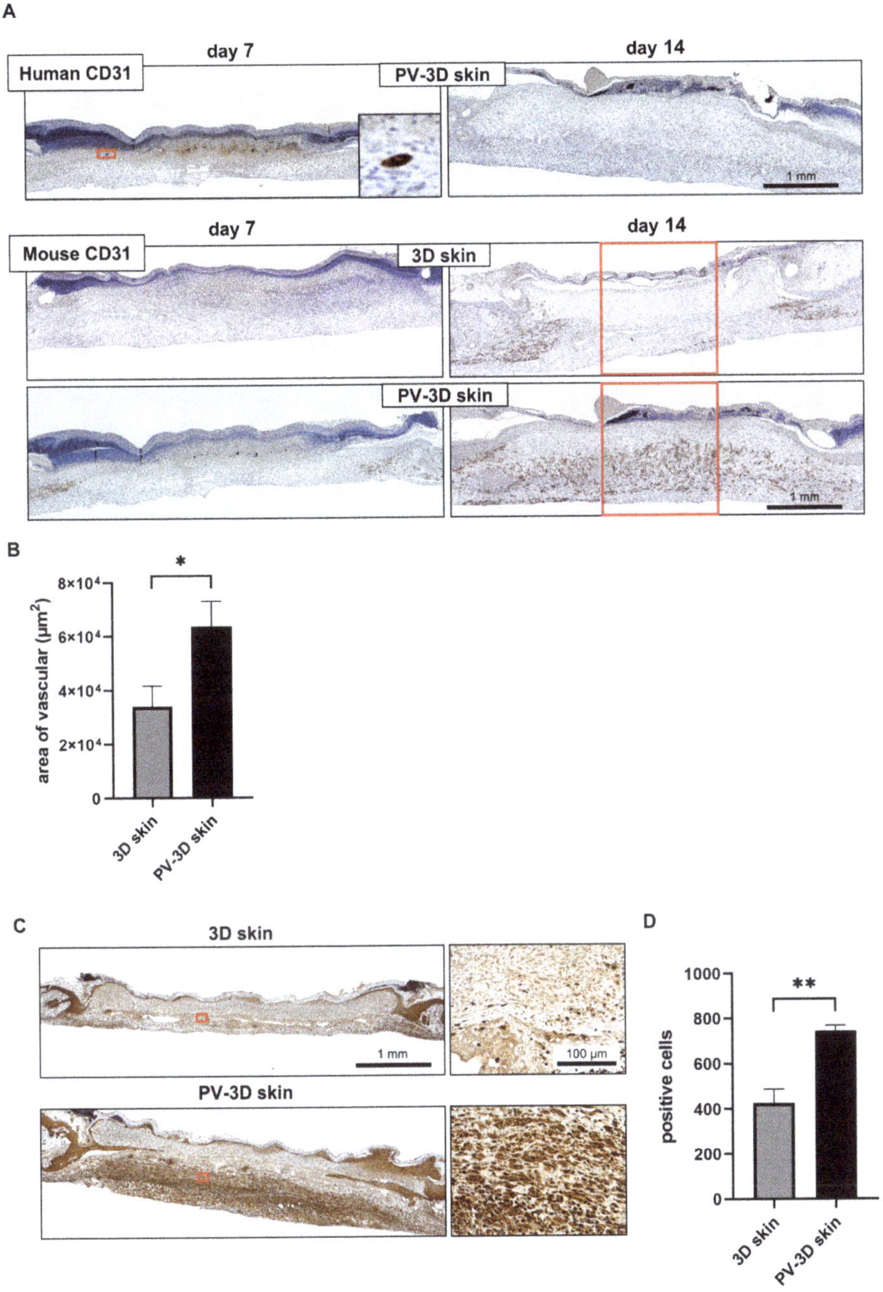

Figure 7. Immunohistochemical analysis at the wound site. (**A**) Human-specific CD31 and mouse-specific CD31 staining (n = 11–12 animals) on days 7 and 14 post-transplantation. (**B**) Quantification of the area occupied by mouse-derived blood vessels on day 14. (**C**) Hypoxia-inducible factor 1-alpha (HIF-1α) staining at the wound site on day 7. (**D**) Quantification of the HIF-1α stained cells (n = 5 animals). Data are represented as mean $\pm$ standard error of the mean. * $p < 0.05$, ** $p < 0.01$.

To further explore the mechanisms underlying these differences, we assessed HIF-1α expression, a key regulator of hypoxia-induced angiogenesis. On day 7 post-grafting, HIF-1α stained cells in the wound center was significantly greater in the PV-3D skin group (743 ± 25.5 cells) than in the 3D skin group (426 ± 61.2 cells, $p < 0.05$; Figure 7C,D). This finding suggests that PV-3D skin fosters a microenvironment conducive to angiogenesis, thereby further substantiating its role in promoting wound healing.

3.3. Long Term Observation of PV-3D Skin

Since wound contraction exerts a significant influence on long-term assessments beyond 2 weeks, experiments were conducted with the application of a silicone ring to prevent contraction. Both 3D skin and PV-3D skin formed eschars macroscopically by 4 weeks (Figure 8A), which subsequently detached from the healed wound bed. Hematoxylin and eosin staining revealed that PV-3D skin was covered with an epidermis of comparable thickness to the surrounding murine epidermis, whereas partial epidermal loss was observed in 3D skin. Collagen staining showed weaker staining in 3D skin, while PV-3Dskin exhibited more intense staining. HLA staining was negative in both PV-3D skin and 3D skin. Immunohistochemistry for Ly6g/6c, a marker of inflammatory cells, demonstrated positive signals within the healing tissue of 3D skin, but no positivity in PV-3D skin. Murine blood vessels were observed throughout the wound area in both 3D skin and PV-3D skin (Figure 8B).

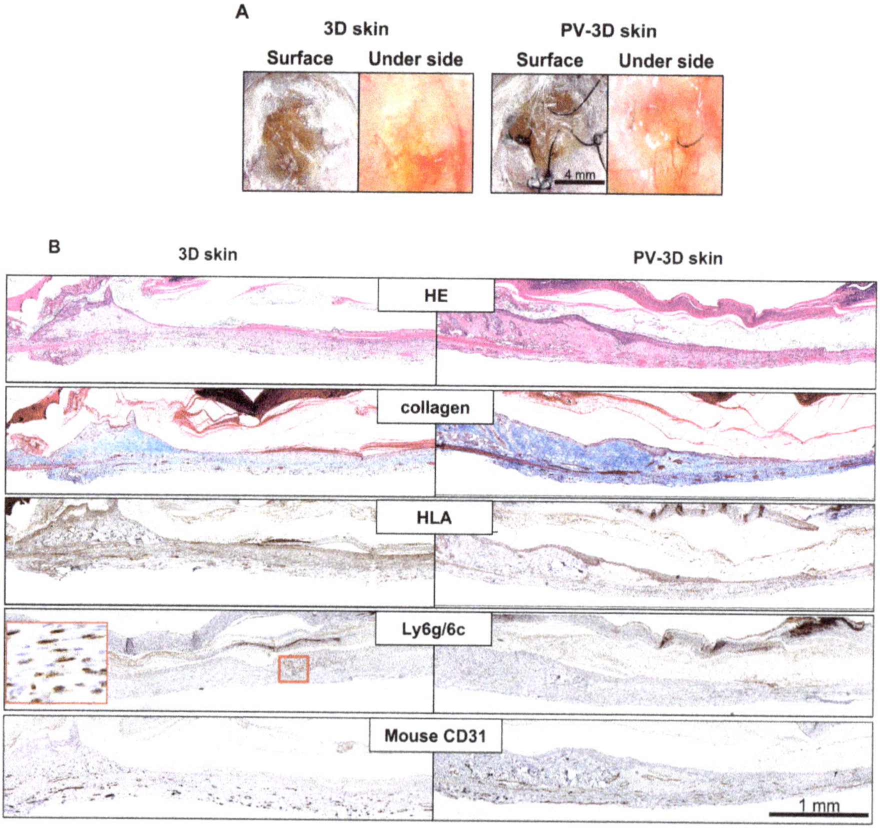

Figure 8. Wound site images on day 28. (**A**) Macroscopic images of the wound area and (**B**) Histological images at the wound site.

4. Discussion

This study demonstrates that PV-3D skin accelerates healing of mitomycin C (MMC)-induced refractory ulcers. Successful treatment of chronic wounds relies on improving local blood perfusion, as inadequate vascularization is a significant barrier to wound healing. Importantly, direct anastomosis was not observed between PV-derived and host vessels was confirmed, and human CD31$^+$ areas were scarcely detected on day 14. On the other hand, by day 14 post-transplantation, sufficient host-driven vascular structures within the grafted skin-like tissue indicated enhanced blood flow and successful healing. Therefore, the role of PV-3D skin in this model can be better characterized as creating a supportive microenvironment that enables and guides host-mediated repair, rather than as a functional graft vasculature.

Thermographic analysis revealed that commencing on day 3 post-transplantation, the PV-3D skin group exhibited temperatures comparable to those of the surrounding healthy tissue, similar to conventional skin grafts. This finding suggests the presence of an epidermal barrier that reduces evaporative water loss and heat dissipation [13,14]. Furthermore, the enhanced heat retention observed in the PV-3D skin group indicates augmented local blood flow perfusion, a critical factor in wound healing. Skin temperature is a well-established indicator of vascularization, reflecting inflammation or increased metabolic activity. Recent studies employing infrared thermographic imaging have indicated that wound temperature may serve as a predictive marker for graft integration and healing. Consequently, thermographic observations may assess the engraftment of transplanted tissues [15]. The findings of our study suggest that PV-3D skin was integrated, thus indicating its favorable engraftment.

The assessment of blood flow using laser speckle contrast imaging, a well-established method for evaluating microcirculatory dynamics in skin wounds [16–19], supported the finding of increased perfusion in the wound area following PV-3D skin transplantation. Histological analysis performed on day 7 post-transplantation indicated an increase in host-derived vascular structures around the PV-3D skin, and by day 14, blood vessel formation was evident within the wound site. These findings indicate that early enhancement of blood flow in both the graft and underlying tissue improved wound healing outcomes. This observation aligns with those of the previous report highlighting the role of vascularized grafts in facilitating granulation tissue formation and neovascularization [10]. Given that blood vessels derived from cultured PV-3D skin were barely expressed in the transplanted wound after 14 days, these findings suggest that they reflected increased blood flow around and beneath the graft up to day 7, but then reflected blood flow within the graft via newly formed murine vessels after day 10.

Fibroblast activation protein (FAP) is expressed by activated fibroblasts and epithelial cells during wound healing [20]. In this study, FAP staining on day 14 revealed activation of mouse-derived epidermal cells and fibroblasts in the PV-3D skin group, suggesting enhanced epithelial-dermal interactions that are essential for effective wound healing. In the mid-layer of the wound, regions lacking HLA and human FAP staining were detected. However, based on tissue morphology, these areas were likely derived from the cultured skin. Considering these points, histology along the time course supports a host-driven mechanism. Post-engraftment, the cultured skin may have served as a scaffold for recruiting and activating host-derived fibroblasts and endothelial cells, facilitating tissue regeneration.

The MMC model recapitulates key features of refractory ulcers through inhibition of DNA synthesis and consequent impairment of fibroblast and endothelial dynamics [12,21–28]. Within this stringent context, the PV-3D skin group showed the presence of fibroblasts and vascular structures of host-derived cells and collagen production, suggesting that PV-3D skin improved the refractory ulcer condition and wound healing by enhancing dermal remodeling and vascular integration. The observed benefits are consistent with reports that biomaterials containing endothelial components can enhance blood perfusion and healing [29–31]. It is known that angiogenesis can occur along decellularized vascular and luminal structures [32,33]. Likewise, despite the lack of direct vascular anastomosis with host vessels, residual vascular structures and endothelial cells within the PV-3D skin likely contributed to early tissue perfusion and neovascularization at the wound site. As healing progressed, partial degeneration of the PV-3D skin may have occurred, potentially influenced by the hostile microenvironment of the MMC-induced refractory ulcer. However, by day 14, the human endothelial structures were all plausibly shed or replaced in association with the progression of host-derived neovascularization. Moreover, cultured skin with a well-organized epidermal-dermal structure alters refractory ulcers from a non-healing stage to one conductive to regeneration [34,35]. These interpretations differ from out acute full-thickness wound model, in which early restoration of flow and anastomosis between graft-derived and host vessels were observed [10]. The present MMC model imposes a more hostile microenvironment for endothelial survival and proliferation, plausibly shifting the dominant mechanism from direct graft vascular function to indirect augmentation of host neovascularization and perfusion.

Angiogenesis is crucial in wound healing [7,8], with HIF-1α playing a central role by stimulating the production of pro-angiogenic factors such as FGF, VEGF, and cytokines like TNF-α [36–38]. However, in refractory ulcers, this hypoxia-driven response is often impaired, leading to delayed wound healing [37]. Since HIF-1α is activated under hypoxic conditions and regulates endothelial cell function, PV-3D skin transplantation likely helped normalize the hypoxic response at the wound site, thereby promoting HIF-1α expression and facilitating angiogenesis. This mechanism could explain the increased vascularization and improved wound healing in the PV-3D skin group. Although thermography and laser Doppler flowmetry demonstrated increased blood flow up to day 7, the expression of HIF-1α was also observed at day 7, which may appear contradictory at first glance. However, in the normal wound healing process, HIF-1α expression can persist for approximately 7 days after injury [36,39], and the present findings may therefore reflect the sustained therapeutic effect on refractory ulcers.

Recurrence is a major concern in chronic refractory ulcers, often attributed to the fragility of regenerated epidermis and prolonged inflammation [6,40]. Long-term histological observations of PV-3D skin suggested a low likelihood of recurrence and minimal inflammation. This is consistent with the maintenance of supporting tissue and proper wound architecture, as indicated by collagen staining. Regarding the human tissue, both PV-3D skin and 3D skin had largely disappeared by 4 weeks, likely reflecting replacement by host murine tissue, which was already evident by 2 weeks of observation. Although SCID mice are largely immunodeficient, xenografts are not completely free from rejection. If it occurs, reported timelines for graft rejection and loss in SCID mice generally begin at least 4–6 weeks after transplantation [41,42], supporting the idea that host tissue replacement predominates during the healing process. Additionally, the absence of functional T and B cells in SCID mice alters the inflammatory course compared to normal wound healing, although innate immune responses still occur [43–45]. Immunohistochemistry for Ly6g/6c, a marker of macrophages and neutrophils, revealed only few inflammatory cells in PV-3D skin over the long term, suggesting that this approach may be effective in preventing chronic inflammation associated with refractory ulcers.

This study has several limitations. Firstly, although this study utilized a faithful MMC-induced ulcer model, further investigation is required to determine whether similar mechanisms apply to other non-healing models, such as ischemic, congestive/hyperemic, and diabetic ulcers [46–49]. Further work should test the use of PV-3D skin in (i) ischemic flap/venous-stasis models and (ii) diabetic models (e.g., db/db or streptozotocin-induced), by integrating longitudinal perfusion imaging with assessments of graft-host anastomosis and multiplexed profiling of the wound bed. These studies should examine stromal–immune–endothelial crosstalk, focusing on HIF-1α dependent responses and FAP positive stromal activation, to define how PV-3D skin modulates non-healing microenvironments. While these factors are beyond the scope of the present study, they represent important directions for further investigation.

Overall, this study demonstrated that PV-3D skin transplantation promotes wound healing in a drug-induced refractory ulcer model by enhancing vascularization and tissue regeneration. In particular, the PV-3D skin accelerates early perfusion and angiogenic activity, as evidenced by increased HIF-1α expression and subsequent host CD31$^+$ vascular ingrowth, together with progressive matrix remodeling and tissue maturation. Given that inadequate blood perfusion is a major challenge in the treatment of chronic wounds, these findings highlight the potential of PV-3D skin as an advanced biomaterial capable of overcoming ischemia-related tissue repair failure.

5. Conclusions

In this study, we developed a novel biomaterial containing pre-formed blood vessels that could aid in restoring the blood flow necessary to treat refractory ulcers and evaluated its effectiveness. Although long-term validation and safety testing using large animal models are necessary next step, the PV-3D skin represents a promising therapeutic strategy for refractory ulcers, potentially contributing to the development of next-generation skin substitutes with enhanced regenerative capabilities.

Supplementary Materials: The following supporting information can be downloaded at https://www.mdpi.com/article/10.3390/jfb16110409/s1, Figure S1: Engineering of human pre-vascularized full-thickness skin substitute.

Author Contributions: Conceptualization, S.T. and H.M.; methodology, H.M.; software, S.T.; validation, H.M., S.N. and R.A.; formal analysis, S.T.; investigation, S.T. and T.S.; resources, Y.T.; data curation, S.T.; writing—original draft preparation, S.T.; writing—review and editing, S.T. and H.M.; visualization, S.T.; supervision, H.M.; project administration, S.N. and R.A.; funding acquisition, Y.T. and H.M. All authors have read and agreed to the published version of the manuscript.

Funding: This research was funded by JSPS KAKENHI, grant number 20K21912 (H.M.), 25K02906 (H.M.) and 22K18177 (Y.T.).

Institutional Review Board Statement: The animal study protocol was approved by the Animal Ethics Committee of the National Defense Medical College (Approval No. 22062, approval date: 14 March 2023).

Informed Consent Statement: Not applicable.

Data Availability Statement: The raw data supporting the conclusions of this article will be made available by the corresponding authors upon request.

Acknowledgments: The authors gratefully acknowledge M. Akashi (Osaka University) and S. Kawauchi (National Defense Medical College Research Institute) for their expert technical support and helpful suggestions.

Conflicts of Interest: The authors declare no conflicts of interest.

Abbreviations

The following abbreviations are used in this manuscript:

3D	three-dimensional
cAMP	cyclic adenosine monophosphate
DMEM	Dulbecco's modified Eagle's medium
FAP	Fibroblast activation protein
FGF	fibroblast growth factor
FN	fibronectin
G	gelatin
H&E	Hematoxylin and Eosin
HIF-1α	hypoxia-inducible factor 1-alpha
HLA	Human leukocyte antigen
HMVECs	Human dermal microvascular endothelial cells
IgG	immunoglobulin G
MMC	mitomycin C
NHDFs	Neonatal normal human dermal fibroblasts
NHEKs	Neonatal human epidermal keratinocytes
PBS	phosphate-buffered saline
PGs	prostaglandins
PV-3D skin	pre-vascularized three-dimensional skin substitutes
SCID	severe combined immunodeficiency

References

1. Mudge, E.; Price, P.; Walkley, N.; Harding, K.G. A randomized controlled trial of larval therapy for the debridement of leg ulcers: Results of a multicenter, randomized, controlled, open, observer blind, parallel group study. *Wound Repair Regen.* **2014**, *22*, 43–51. [CrossRef]
2. Wilcox, J.R.; Carter, M.J.; Covington, S. Frequency of debridements and time to heal: A retrospective cohort study of 312 744 wounds. *JAMA Dermatol.* **2013**, *149*, 1050–1058. [CrossRef]
3. Banakh, I.; Cheshire, P.; Rahman, M.; Carmichael, I.; Jagadeesan, P.; Cameron, N.R.; Cleland, H.; Akbarzadeh, S. A Comparative Study of Engineered Dermal Templates for Skin Wound Repair in a Mouse Model. *Int. J. Mol. Sci.* **2020**, *21*, 4508. [CrossRef]
4. Fu, H.; Zhang, D.; Zeng, J.; Fu, Q.; Chen, Z.; Sun, X.; Yang, Y.; Li, S.; Chen, M. Application of 3D-printed tissue-engineered skin substitute using innovative biomaterial loaded with human adipose-derived stem cells in wound healing. *Int. J. Bioprint* **2023**, *9*, 674. [CrossRef]
5. Romanelli, M.; Dini, V.; Bertone, M.S. Randomized comparison of OASIS wound matrix versus moist wound dressing in the treatment of difficult-to-heal wounds of mixed arterial/venous etiology. *Adv. Skin Wound Care* **2010**, *23*, 34–38. [CrossRef] [PubMed]
6. Burgess, J.L.; Wyant, W.A.; Abdo Abujamra, B.; Kirsner, R.S.; Jozic, I. Diabetic Wound-Healing Science. *Medicina* **2021**, *57*, 1072. [CrossRef] [PubMed]
7. Okonkwo, U.A.; DiPietro, L.A. Diabetes and Wound Angiogenesis. *Int. J. Mol. Sci.* **2017**, *18*, 1419. [CrossRef] [PubMed]
8. Catrina, S.B.; Zheng, X. Disturbed hypoxic responses as a pathogenic mechanism of diabetic foot ulcers. *Diabetes Metab. Res. Rev.* **2016**, *32* (Suppl. S1), 179–185. [CrossRef]
9. Powers, J.G.; Higham, C.; Broussard, K.; Phillips, T.J. Wound healing and treating wounds: Chronic wound care and management. *J. Am. Acad. Dermatol.* **2016**, *74*, 607–625, quiz 625-606. [CrossRef]
10. Miyazaki, H.; Tsunoi, Y.; Akagi, T.; Sato, S.; Akashi, M.; Saitoh, D. A novel strategy to engineer pre-vascularized 3-dimensional skin substitutes to achieve efficient, functional engraftment. *Sci. Rep.* **2019**, *9*, 7797. [CrossRef]
11. Matsusaki, M.; Kadowaki, K.; Nakahara, Y.; Akashi, M. Fabrication of cellular multilayers with nanometer-sized extracellular matrix films. *Angew. Chem. Int. Ed. Engl.* **2007**, *46*, 4689–4692. [CrossRef]
12. Nagano, H.; Suematsu, Y.; Takuma, M.; Aoki, S.; Satoh, A.; Takayama, E.; Kinoshita, M.; Morimoto, Y.; Takeoka, S.; Fujie, T.; et al. Enhanced cellular engraftment of adipose-derived mesenchymal stem cell spheroids by using nanosheets as scaffolds. *Sci. Rep.* **2021**, *11*, 14500. [CrossRef]
13. Seno, S.I.; Shimazu, H.; Kobayashi, H.; Kogure, E.; Watanabe, A.; Isoyama, T. Quantitative evaluation of skin barrier function using water evaporation time related to transepidermal water loss. *Skin Res. Technol.* **2023**, *29*, e13242. [CrossRef]

14. Surinchak, J.S.; Malinowski, J.A.; Wilson, D.R.; Maibach, H.I. Skin wound healing determined by water loss. *J. Surg. Res.* **1985**, *38*, 258–262. [CrossRef] [PubMed]
15. Hwang, Y.S.; Byeon, J.Y.; Lee, D.W.; Kim, J.H.; Lim, S.; Choi, H.J. Skin graft monitoring using forward-looking infrared thermal imaging. *Int. Wound J.* **2024**, *21*, e70107. [CrossRef] [PubMed]
16. Cracowski, J.L.; Gaillard-Bigot, F.; Cracowski, C.; Roustit, M.; Millet, C. Skin microdialysis coupled with laser speckle contrast imaging to assess microvascular reactivity. *Microvasc. Res.* **2011**, *82*, 333–338. [CrossRef]
17. Linkous, C.; Pagan, A.D.; Shope, C.; Andrews, L.; Snyder, A.; Ye, T.; Valdebran, M. Applications of Laser Speckle Contrast Imaging Technology in Dermatology. *JID Innov.* **2023**, *3*, 100187. [CrossRef] [PubMed]
18. Mahé, G.; Humeau-Heurtier, A.; Durand, S.; Leftheriotis, G.; Abraham, P. Assessment of skin microvascular function and dysfunction with laser speckle contrast imaging. *Circ. Cardiovasc. Imaging* **2012**, *5*, 155–163. [CrossRef]
19. Mahé, G.; Rousseau, P.; Durand, S.; Bricq, S.; Leftheriotis, G.; Abraham, P. Laser speckle contrast imaging accurately measures blood flow over moving skin surfaces. *Microvasc. Res.* **2011**, *81*, 183–188. [CrossRef]
20. Fitzgerald, A.A.; Weiner, L.M. The role of fibroblast activation protein in health and malignancy. *Cancer Metastasis Rev.* **2020**, *39*, 783–803. [CrossRef]
21. Nambu, M.; Ishihara, M.; Nakamura, S.; Mizuno, H.; Yanagibayashi, S.; Kanatani, Y.; Hattori, H.; Takase, B.; Ishizuka, T.; Kishimoto, S.; et al. Enhanced healing of mitomycin C-treated wounds in rats using inbred adipose tissue-derived stromal cells within an atelocollagen matrix. *Wound Repair Regen.* **2007**, *15*, 505–510. [CrossRef] [PubMed]
22. Porter, G.T.; Gadre, S.A.; Calhoun, K.H. The effects of intradermal and topical mitomycin C on wound healing. *Otolaryngol. Head Neck Surg.* **2006**, *135*, 56–60. [CrossRef]
23. Ribeiro, F.e.A.; Guaraldo, L.; Borges, J.e.P.; Vianna, M.R.; Eckley, C.A. Study of wound healing in rats treated with topical and injected mitomycin C. *Ann. Otol. Rhinol. Laryngol.* **2008**, *117*, 786–790. [CrossRef]
24. Takikawa, M.; Ishihara, M.; Takabayashi, Y.; Sumi, Y.; Yoshida, R.; Nakamura, S.; Hattori, H.; Yanagibayashi, S.; Yamamoto, N.; Kiyosawa, T. Enhanced healing of mitomycin C-treated healing-impaired wounds in rats with PRP-containing fragmin/protamine microparticles (PRP&F/P MPs). *J. Plast. Surg. Hand Surg.* **2015**, *49*, 268–274. [CrossRef]
25. Dachlan, I.; Kurniawan, H.S.; Wicaksana, A.; Fauzi, A.R.; Makrufardi, F.; Seswandhana, R. The effect of platelet-rich fibrin on normal dermal fibroblast proliferation after mitomycin-c treatment: An in vitro study. *Ann. Med. Surg.* **2021**, *62*, 473–476. [CrossRef]
26. Bradner, W.T. Mitomycin C: A clinical update. *Cancer Treat. Rev.* **2001**, *27*, 35–50. [CrossRef]
27. McKenna, E.; Traganos, F.; Zhao, H.; Darzynkiewicz, Z. Persistent DNA damage caused by low levels of mitomycin C induces irreversible cell senescence. *Cell Cycle* **2012**, *11*, 3132–3140. [CrossRef]
28. Wang, Y.W.; Ren, J.H.; Xia, K.; Wang, S.H.; Yin, T.F.; Xie, D.H.; Li, L.H. Effect of mitomycin on normal dermal fibroblast and HaCat cell: An in vitro study. *J. Zhejiang Univ. Sci. B* **2012**, *13*, 997–1005. [CrossRef] [PubMed]
29. Frueh, F.S.; Spater, T.; Korbel, C.; Scheuer, C.; Simson, A.C.; Lindenblatt, N.; Giovanoli, P.; Menger, M.D.; Laschke, M.W. Prevascularization of dermal substitutes with adipose tissue-derived microvascular fragments enhances early skin grafting. *Sci. Rep.* **2018**, *8*, 10977. [CrossRef] [PubMed]
30. Kim, K.L.; Song, S.H.; Choi, K.S.; Suh, W. Cooperation of endothelial and smooth muscle cells derived from human induced pluripotent stem cells enhances neovascularization in dermal wounds. *Tissue Eng. Part A* **2013**, *19*, 2478–2485. [CrossRef]
31. Rufaihah, A.J.; Huang, N.F.; Jamé, S.; Lee, J.C.; Nguyen, H.N.; Byers, B.; De, A.; Okogbaa, J.; Rollins, M.; Reijo-Pera, R.; et al. Endothelial cells derived from human iPSCS increase capillary density and improve perfusion in a mouse model of peripheral arterial disease. *Arterioscler. Thromb. Vasc. Biol.* **2011**, *31*, e72–e79. [CrossRef]
32. Chiu, A.; Sharma, D.; Zhao, F. Tissue Engineering-Based Strategies for Diabetic Foot Ulcer Management. *Adv. Wound Care* **2023**, *12*, 145–167. [CrossRef]
33. Treadwell, T.; Sabolinski, M.L.; Skornicki, M.; Parsons, N.B. Comparative Effectiveness of a Bioengineered Living Cellular Construct and Cryopreserved Cadaveric Skin Allograft for the Treatment of Venous Leg Ulcers in a Real-World Setting. *Adv. Wound Care* **2018**, *7*, 69–76. [CrossRef] [PubMed]
34. Stone, R.C.; Stojadinovic, O.; Rosa, A.M.; Ramirez, H.A.; Badiavas, E.; Blumenberg, M.; Tomic-Canic, M. A bioengineered living cell construct activates an acute wound healing response in venous leg ulcers. *Sci. Transl. Med.* **2017**, *9*, eaaf8611. [CrossRef]
35. Stone, R.C.; Stojadinovic, O.; Sawaya, A.P.; Glinos, G.D.; Lindley, L.E.; Pastar, I.; Badiavas, E.; Tomic-Canic, M. A bioengineered living cell construct activates metallothionein/zinc/MMP8 and inhibits TGFβ to stimulate remodeling of fibrotic venous leg ulcers. *Wound Repair Regen.* **2020**, *28*, 164–176. [CrossRef]
36. Hong, W.X.; Hu, M.S.; Esquivel, M.; Liang, G.Y.; Rennert, R.C.; McArdle, A.; Paik, K.J.; Duscher, D.; Gurtner, G.C.; Lorenz, H.P.; et al. The Role of Hypoxia-Inducible Factor in Wound Healing. *Adv. Wound Care* **2014**, *3*, 390–399. [CrossRef]

37. Mace, K.A.; Yu, D.H.; Paydar, K.Z.; Boudreau, N.; Young, D.M. Sustained expression of Hif-1alpha in the diabetic environment promotes angiogenesis and cutaneous wound repair. *Wound Repair Regen.* **2007**, *15*, 636–645. [CrossRef] [PubMed]
38. Botusan, I.R.; Sunkari, V.G.; Savu, O.; Catrina, A.I.; Grünler, J.; Lindberg, S.; Pereira, T.; Ylä-Herttuala, S.; Poellinger, L.; Brismar, K.; et al. Stabilization of HIF-1alpha is critical to improve wound healing in diabetic mice. *Proc. Natl. Acad. Sci. USA* **2008**, *105*, 19426–19431. [CrossRef]
39. Albina, J.E.; Mastrofrancesco, B.; Vessella, J.A.; Louis, C.A.; Henry, W.L.; Reichner, J.S. HIF-1 expression in healing wounds: HIF-1alpha induction in primary inflammatory cells by TNF-alpha. *Am. J. Physiol. Cell Physiol.* **2001**, *281*, C1971–C1977. [CrossRef]
40. Charles, J.; Harrison, C.; Britt, H. Chronic skin ulcers. *Aust. Fam. Physician* **2014**, *43*, 587. [PubMed]
41. Kaufmann, R.; Mielke, V.; Reimann, J.; Klein, C.E.; Sterry, W. Cellular and molecular composition of human skin in long-term xenografts on SCID mice. *Exp. Dermatol.* **1993**, *2*, 209–216. [CrossRef]
42. Soria, A.; Boccara, D.; Chonco, L.; Yahia, N.; Dufossée, M.; Cardinaud, S.; Moris, A.; Liard, C.; Joulin-Giet, A.; Julithe, M.; et al. Long-term maintenance of skin immune system in a NOD-Scid IL2rγ(null) mouse model transplanted with human skin. *Exp. Dermatol.* **2014**, *23*, 850–852. [CrossRef] [PubMed]
43. Bancroft, G.J.; Kelly, J.P. Macrophage activation and innate resistance to infection in SCID mice. *Immunobiology* **1994**, *191*, 424–431. [CrossRef]
44. Jimi, S.; Kimura, M.; De Francesco, F.; Riccio, M.; Hara, S.; Ohjimi, H. Acceleration Mechanisms of Skin Wound Healing by Autologous Micrograft in Mice. *Int. J. Mol. Sci.* **2017**, *18*, 1675. [CrossRef]
45. Kim, S.W.; Zhang, H.Z.; Guo, L.; Kim, J.M.; Kim, M.H. Amniotic mesenchymal stem cells enhance wound healing in diabetic NOD/SCID mice through high angiogenic and engraftment capabilities. *PLoS ONE* **2012**, *7*, e41105. [CrossRef]
46. Karim, A.S.; Liu, A.; Lin, C.; Uselmann, A.J.; Eliceiri, K.W.; Brown, M.E.; Gibson, A.L.F. Evolution of ischemia and neovascularization in a murine model of full thickness human wound healing. *Wound Repair Regen.* **2020**, *28*, 812–822. [CrossRef]
47. Li, G.; Ko, C.N.; Li, D.; Yang, C.; Wang, W.; Yang, G.J.; Di Primo, C.; Wong, V.K.W.; Xiang, Y.; Lin, L.; et al. A small molecule HIF-1α stabilizer that accelerates diabetic wound healing. *Nat. Commun.* **2021**, *12*, 3363. [CrossRef] [PubMed]
48. Mohd Hilmi, A.B.; Hassan, A.; Halim, A.S. A Bilayer Engineered Skin Substitute for Wound Repair in an Irradiation-Impeded Healing Model on Rat. *Adv. Wound Care* **2015**, *4*, 312–320. [CrossRef]
49. Zhu, N.N.; Lu, M.J.; Chen, Y.Q.; Jin, X.J.; Zhou, X.; Wei, H.W.; Liu, X.Q.; Duan, L.S.; Yin, L.; Guo, J.R. Autologous blood transfusion stimulates wound healing in diabetic mice through activation of the HIF-1α pathway by improving the blood preservation solution. *FASEB J.* **2020**, *34*, 6038–6054. [CrossRef] [PubMed]

Disclaimer/Publisher's Note: The statements, opinions and data contained in all publications are solely those of the individual author(s) and contributor(s) and not of MDPI and/or the editor(s). MDPI and/or the editor(s) disclaim responsibility for any injury to people or property resulting from any ideas, methods, instructions or products referred to in the content.

Journal of
Functional Biomaterials

MDPI

Article

Chlorogenic Acid–Strontium-Containing Dual-Functional Bioresorbable External Stent Suppresses Venous Graft Restenosis via Hippo-YAP Signaling Pathway

Ge Zhu [1,†], Su Wang [1,†], Zhang Liu [1], Shengji Gu [1], Feng Chen [2,*] and Wangfu Zang [1,*]

[1] Department of Cardio-Thoracic Surgery, Shanghai Tenth People's Hospital, School of Medicine, Tongji University, Shanghai 200072, China; zg01k39@rjh.com.cn (G.Z.); ws7202@shtrhospital.com (S.W.); 1910858@tongji.edu.cn (Z.L.); 2022110739@stu.njmu.edu.cn (S.G.)
[2] Shanghai Key Laboratory of Craniomaxillofacial Development and Diseases, Shanghai Stomatological Hospital & School of Stomatology, Fudan University, Shanghai 201102, China
* Correspondence: chen_feng@fudan.edu.cn (F.C.); zangwf@tongji.edu.cn (W.Z.)
† These authors contributed equally to this work and share first authorship.

Abstract: Vein graft restenosis remains a major complication following coronary artery bypass grafting (CABG), mainly due to the abnormal proliferation of vascular smooth muscle cells (VSMCs) and impaired endothelial repair. While external stents (eStents) can provide mechanical support and limit adverse remodeling, traditional metallic stents are non-degradable and may induce chronic inflammation and fibrosis. In contrast, many bioresorbable materials degrade too quickly or lack mechanical strength. These challenges highlight the need for external stents that combine sufficient mechanical strength with biodegradability to support long-term graft patency. This is the first study that develops a chlorogenic acid–strontium (SrCA)-loaded polycaprolactone bioresorbable eStent that inhibits VSMC proliferation and enhances endothelial repair via Hippo–Yes-associated protein (YAP) signaling, addressing vein graft restenosis post-CABG. Combining mechanical support and biodegradability, it overcomes the limitations of non-degradable stents and rapidly degrading biomaterials, elucidates the potential of natural polyphenol–metal ion complexes in vascular remodeling, and offers an innovative strategy for the prevention of vein graft restenosis.

Keywords: vein graft failure; external stent; chlorogenic acid; neointimal hyperplasia; hippo signaling pathway

1. Introduction

Coronary artery disease (CAD) remains a leading global health burden [1,2]. Coronary artery bypass grafting (CABG) is the gold standard treatment for multivessel disease and complex or severe anatomical stenoses [3,4], with saphenous vein grafts (SVGs) widely used due to their availability and length [5,6]. However, vein graft failure—primarily driven by restenosis from intimal hyperplasia—significantly compromises long-term outcomes [7]. This pathological remodeling, exacerbated by arterial pressure and cyclic stretch, leads to uncontrolled vascular smooth muscle cell (VSMC) proliferation and luminal narrowing [8,9]. Effective strategies to prevent vein graft restenosis are urgently needed to improve graft durability and clinical prognosis after CABG.

Several therapeutic strategies have been developed to mitigate vein graft remodeling, including gene therapy, modifications in vein harvesting techniques, and the use of external vein graft stents. Among these, external vein graft stents have demonstrated promising outcomes in enhancing graft patency rates and suppressing intimal hyperplasia [10–13]. Several studies have highlighted that the efficacy of stents largely depends on the materials and structural design used in their fabrication. For instance, Murphy et al. reported that polyester-based stents could contribute to vein graft failure, owing to their stiffness and incomplete conduit design [14]. The prolonged presence of foreign materials significantly heightens the risk of infection and immune rejection [15]. Consequently, bare-metal stents alone have shown limited efficacy in preventing vein graft failure [16,17]. To tackle these difficulties, the use of biodegradable materials has been suggested as a promising alternative. Polycaprolactone (PCL), a biodegradable and biocompatible polymer, is widely employed in tissue engineering, owing to its advantageous characteristics. Extrusion-based 3D printing has emerged as a promising method for fabricating customized stents tailored to the needs of personalized medicine [18,19].

The exact mechanisms by which external stents suppress intimal hyperplasia remain poorly understood. Recent studies have identified Yes-associated protein (YAP), a downstream effector of the Hippo signaling pathway, as a key player in vascular remodeling [20]. Additionally, YAP expression has been strongly linked to VSMCs proliferation and migration, key contributors to intimal hyperplasia and graft restenosis [21,22]. These insights suggest that modulating the Hippo-YAP axis may provide a novel strategy to mitigate pathological remodeling in vein grafts.

In recent years, biomaterials have garnered significant attention in medical research. Chlorogenic acid (CGA), for instance, has demonstrated promising effects, such as inhibiting platelet aggregation [23], reducing heart failure [24], and preventing myocardial infarction [25–27]. However, its low absolute bioavailability significantly limits its clinical utility. Incorporating strontium ions (Sr^{2+}) into biomaterials has emerged as an effective strategy to enhance the biological performance of implants [28]. Notably, recent findings suggest that Sr^{2+} may protect myocardial tissue from ischemia/reperfusion injury [29]. These insights highlight the potential for novel biomaterial designs in addressing challenges related to vein grafts and cardiovascular health.

In this study, we developed a novel chlorogenic acid–strontium (SrCA) dual-functional bioresorbable eStent using electrospinning technology to fabricate custom-woven PCL tubular stents with optimized radial strength, axial flexibility, and luminal patency. These mechanical properties ensure adequate early-stage support while reducing long-term foreign body reactions during biodegradation [30]. To our knowledge, this is the first SrCA-loaded eStent designed to inhibit VSMC proliferation and promote endothelial repair via modulation of the Hippo-YAP signaling pathway, thereby targeting the pathological mechanisms underlying vein graft restenosis after CABG. Our findings not only elucidate the mechanistic role of this biofunctional eStent in suppressing vein graft remodeling but also highlight its clinical potential as a next-generation therapeutic strategy for improving long-term graft patency in vascular surgery.

2. Materials and Methods

2.1. SrCA Fabrication

The preparation method for SrCA is as follows (Figure 1A): Dissolve 26 mg of CGA and 10 mg of strontium chloride in 15 mL of deionized water under magnetic stirring for 10 min until fully dissolved. Add 5 mg of potassium hydroxide to the mixed solution, adjusting the pH to 8. Transfer the solution into a 25 mL polytetrafluoroethylene (PTFE) hydrothermal reaction vessel using a pipette. Place the vessel in a preheated oven at 120 °C and allow the reaction to proceed for 24 h. After completion, cool the reaction mixture naturally to room temperature. Collect the reaction product via centrifugation and wash it 3–5 times with deionized water and absolute ethanol, respectively. Finally, disperse the product in absolute ethanol and store it at 4 °C for future use.

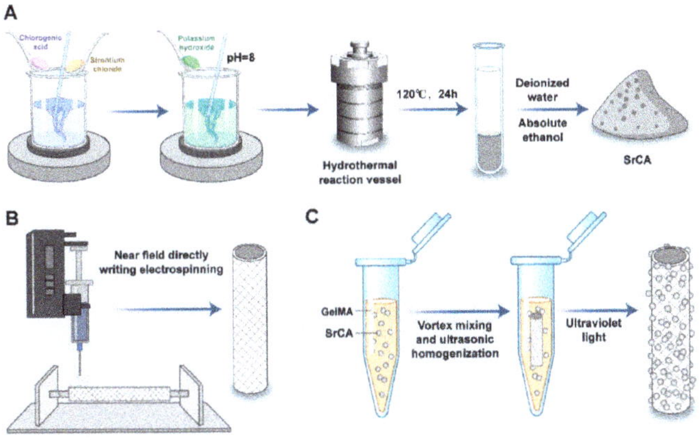

Figure 1. Schematic representation of the fabrication of (**A**) SrCA, (**B**) PCL stent, and (**C**) SrCA-loaded eStent.

2.2. SrCA eStent Fabrication

The preparation method for the PCL stent is as follows (Figure 1B): Dissolve 3.5 g of PCL in 10 mL of trifluoroethanol under magnetic stirring in a 50 °C thermostatic water bath for 12 h to ensure complete dissolution, forming a 20% (w/w) PCL solution. Load the prepared PCL solution into a 10 mL (21-gauge) syringe and employ a near-field directly writing electrospinning platform (NFDWE) (Changsha Nayi Instrument Technology Co., Changsha, China) for stent fabrication. During the process, set the electrostatic direct writing rate at 0.1 mL/h, maintain a voltage of 2.5 kV, adjust the height to 3 mm, and control the collection speed at 10 mm/s. The resulting PCL stent is maintained at room temperature for future application.

The preparation method for the porous gelatin methacryloyl (GelMA) hydrogel loaded with SrCA is as follows (Figure 1C): Add an appropriate amount of phosphate-buffered saline (PBS) to a bottle containing porous GelMA to prepare a 6% (w/v) GelMA hydrogel precursor solution. Stir the prepared solution in a 37 °C light-protected water bath using a magnetic stirrer for 1 h. Immediately filter the solution utilizing a 0.22 μm sterile needle filter to ensure sterility and set it aside for further use. Weigh pre-prepared SrCA using centrifugation and mix it with the porous GelMA precursor solution in a 37 °C thermostatic water bath. Prepare mixtures at concentrations of 20% (w/v), 10% (w/v), 5% (w/v), and 2.5% (w/v) by weight. Use vortex mixing and ultrasonic homogenization to ensure uniform distribution. Additionally, prepare a blank GelMA group with no additives. Transfer the SrCA /GelMA mixtures into four separate EP tubes. Submerge the pre-fabricated PCL stents into these solutions. After immersion, remove the stents and expose them to ultraviolet light for 3 min at room temperature in the absence of light to induce crosslinking and self-curing. The final products include GelMA-coated eStents with SrCA concentrations of 20% (w/v), 10% (w/v), 5% (w/v), and 2.5% (w/v), as well as a blank GelMA hydrogel eStent for control experiments.

2.3. Characterization of SrCA and SrCA eStents

The morphology and surface structure of SrCA and the eStents were examined using scanning electron microscopy (SEM, ZEISS, Gemini 300, Oberkochen, Germany) and transmission electron microscopy (TEM, FEI, Tecnai G2 F20, Hillsboro, OR, USA). TEM was specifically employed to analyze the size and morphology of SrCA. Selected area electron diffraction (SAED, FEI, Tecnai G2 F20, Hillsboro, OR, USA) was performed to determine whether SrCA exhibits a crystalline structure. Energy-dispersive X-ray spectroscopy (EDS, Xplore, Oxford, UK) elemental mapping was conducted to analyze the elemental composition of SrCA and eStents. Fourier-transform infrared spectroscopy (FTIR, Thermo Fisher, Waltham, MA, USA) was used to assess SrCA chemical properties. The tensile stress–strain mechanical performance of the eStents was evaluated using an electronic tensile testing machine (HY-940FS, Shanghai, China).

2.4. Cell Culture and SrCA eStent Extract Preparation

Rat aortic smooth muscle cells (A7r5, ATCC® CRL-1444™) were cultured in Dulbecco's Modified Eagle Medium (DMEM, Gibco, Waltham, MA, USA) supplemented with 10% fetal bovine serum (FBS, Gibco), 100 U/mL penicillin, and 100 μg/mL streptomycin. Cells were maintained at 37 °C in a humidified 5% CO_2 atmosphere and passaged at 80–90% confluence using 0.25% trypsin–EDTA (Gibco).

SrCA eStents were sterilized under UV light (254 nm, 30 min per side) and preconditioned in serum-free DMEM for 24 h at 37 °C. SrCA eStent extracts were collected by incubating SrCA eStents in serum-free medium (1 cm^2 stent/mL medium) for 72 h.

2.5. Cell Counting Kit-8 (CCK-8) Assay

The CCK-8 assay was conducted in accordance with the manufacturer's guidelines to evaluate the cytotoxicity of SrCA eStents on A7r5 cells. A7r5 cells were plated into 96-well plates at a density of 2000 cells per well and maintained on tissue culture plates (TCPs) for 24 and 48 h with various concentrations of SrCA eStent extract. Subsequently, 10 μL of CCK-8 reagent was added to each well, followed by a two-hour incubation in a cell culture incubator. Finally, the absorbance of each well was measured at 450 nm with a microplate reader (Bio Tek Synergyz, Winooski, VT, USA).

2.6. Live/Dead Assay

Cells were seeded onto 24-well plates containing SrCA eStent extracts at a density of 5×10^4 cells/cm^2 and cultured for 72 h. Cells were washed twice with PBS and incubated with 2 μM Calcein-AM (live cells, green fluorescence) and 4 μM propidium iodide (PI, dead cells, red fluorescence) (Thermo Fisher, Waltham, MA, USA) in PBS for 30 min at 37 °C. Fluorescent images were obtained using a confocal laser scanning microscope (LEICA DMI6000B, Wetzlar, Germany).

2.7. Cell Migration Assay

Cells were seeded in 6-well plates (2×10^5 cells/well) and grown to 100% confluence. A standardized scratch was created using a 200 μL sterile pipette tip. Debris was removed by washing with PBS. Cells were treated with (a) SrCA eStent extracts, (b) SrCA-free eStent extracts, or (c) DMEM (blank control group). Scratch closure was monitored at 0, 24, and 48 h using an inverted phase-contrast microscope (Olympus CKX31SF, Tokyo, Japan). ImageJ software (version 1.52v) was adopted for the quantitative analysis of the cell migration amount.

2.8. Quantitative Real-Time PCR

Total RNA was extracted using TRIzol®. YAP/TAZ target genes (CTGF, CYR61) and contractile markers (α-SMA, SM22α) were amplified with SYBR Green primers (sequences provided in Table S1).

As per the manufacturer's guidelines, total RNA was extracted from A7r5 cells co-cultured with the SrCA eStent extract using TRIzol reagent (Shanghai Shenggong Biotechnology Co., Shanghai, China). Complementary DNA (cDNA) was synthesized using the PrimeScript™ RT Reagent Kit with gDNA Eraser (TaKaRa, Kusatsu, Japan). Quantitative real-time PCR (RT-qPCR) was performed using the TB Green Premix Ex Taq™ II kit (TaKaRa, Kusatsu, Japan) on a fluorescence real-time PCR system (CFX96™, Bio-Rad, Hercules, CA, USA), with GAPDH used as the internal control. Gene expression levels were determined utilizing the $2^{-\Delta\Delta Ct}$ method. Primer sequences are provided in Supplementary Materials Table S1.

2.9. Western Blot

The proteins from A7r5 co-cultured with SrCA eStent extract were extracted utilizing RIPA lysis buffer (Gibco, Waltham, MA, USA), which included a combination of protease and phosphatase inhibitors. Equal quantities of protein were separated by 10% sodium dodecyl–sulfate polyacrylamide gel electrophoresis (SDS-PAGE) and subsequently transferred onto polyvinylidene difluoride (PVDF) membranes (Beyotime, Shanghai, China). After blocking, the membranes were incubated with a primary antibody against YAP (Absin, Shanghai, China), followed by incubation performed with the relevant HRP-conjugated secondary antibody. The expression level of glyceraldehyde 3-phosphate dehydrogenase (GAPDH) served as an internal control.

2.10. Surgical Procedure

This research and all animal procedures received approval from the Ethics Committee of Shanghai Tenth People's Hospital, Tongji University School of Medicine, and adhered to the National Institutes of Health Guide for the Care and Use of Laboratory Animals (NIH Publications No. 85-23 [31], revised 1996).

Sprague–Dawley rats (male, 300–350 g, $n = 6$/group) were anesthetized via intraperitoneal injection of 3% pentobarbital sodium at a dosage of 0.1 mL/100 g. The right external jugular vein was meticulously isolated, dissected, and immersed in a heparin saline solution. Following the immobilization and transection of the ipsilateral common carotid artery, a vein graft was implanted utilizing the "cuff" technique. Both ends of the dissected artery were everted over a custom-made cuff (1.5 mm), created from the cannula of a 20G indwelling needle and secured using 8-0 Prolene sutures. The cuffed artery was then inserted into the harvested jugular vein. Prior to ligation at the anastomotic junction, the grafts in the stent group were externally wrapped with SrCA eStents matching the vein length. Non-stented vein grafts served as controls. Animals in each group were euthanized at 2 and 4 weeks post-implantation before vein graft harvesting.

2.11. Histomorphology

The harvested vein grafts underwent fixation in 4% paraformaldehyde, were embedded in paraffin, and subsequently sectioned into 5 μm thick slices. Hematoxylin and eosin (H&E) staining was performed following standard protocols to characterize morphological changes. To identify the smooth muscle layer surrounding neointimal hyperplasia, elastic van Gieson (EVG) staining was conducted according to the manufacturer's instructions.

A quantitative evaluation of vessel wall thickness was performed using ImageJ software (version 1.52v).

2.12. Statistical Analysis

Data were analyzed with SPSS v22.0 software (IBM, Armonk, NY, USA). A one-way analysis of variance was employed to assess the differences among groups. An unpaired Student's *t*-test was employed to evaluate the differences between the two groups. All measurements were performed three times independently, and the results are expressed as the mean ± SD. $p < 0.05$ was considered significant.

3. Results

3.1. Fabrication and Characterization of SrCA

The SrCA complex (Figure 2A) displayed uniform spherical morphology (avg. 500 nm, Figure S1) with smooth surfaces (Figure 2B). SAED patterns (Figure 2B inset) displayed diffuse rings, indicating the amorphous nature of the SrCA. EDS elemental mapping (Figure 2C) demonstrated a uniform spatial distribution of carbon (C), oxygen (O), and strontium (Sr) within the composite. FTIR spectroscopy (Figure 2D) identified characteristic absorption bands, with 1619 cm^{-1} and 1386 cm^{-1} corresponding to Sr-Cl vibrational modes from the strontium chloride precursor and 1290 cm^{-1} attributable to aromatic C-C skeletal vibrations of CGA. The presence of these distinct peaks in the SrCA spectrum, along with the absence of isolated precursor signatures, confirmed successful coordination between CGA and strontium ions through hydrothermal synthesis.

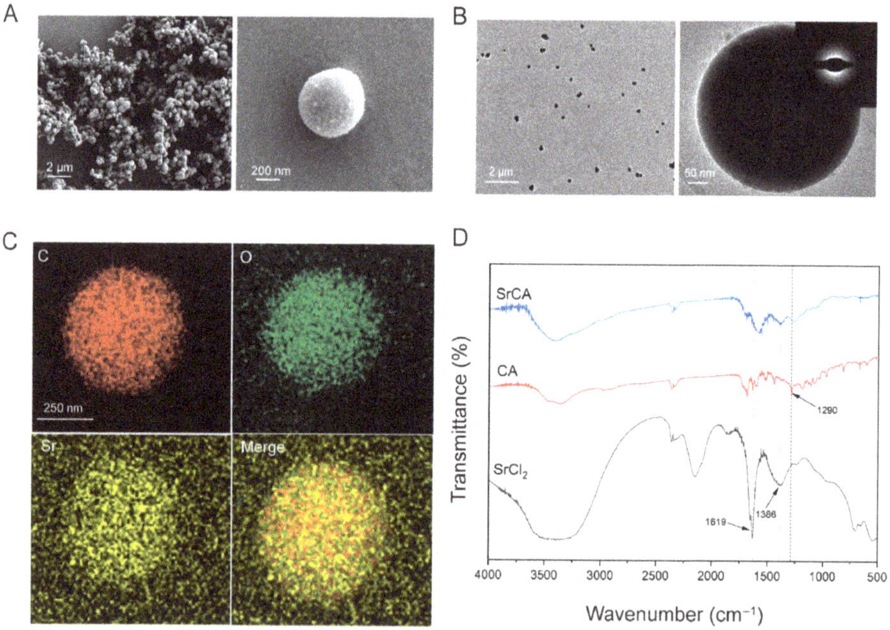

Figure 2. (**A**) SEM, (**B**) TEM, and SAED. (**C**) Elemental mappings of SrCA. (**D**) FTIR of SrCA, CA, and $SrCl_2$.

3.2. Fabrication and Characterization of SrCA eStents

The PCL stent exhibited a 3D porous network (Figure 3A–C) with smooth fibers. Following the GelMA photocrossloading of SrCA nanoparticles onto the stents, SEM imaging (Figure 3D) showed retained three-dimensional porosity with distinct surface roughening. SrCA nanoparticles (indicated by white arrows) were uniformly distributed along the fiber surfaces. EDS elemental mapping (Figure 3E) verified a homogeneous dispersion of carbon (C), oxygen (O), and strontium (Sr) across the fibrous network, confirming successful SrCA immobilization. Mechanical tests demonstrated that the stents have excellent radial ductility and cyclic resilience (Figure S2).

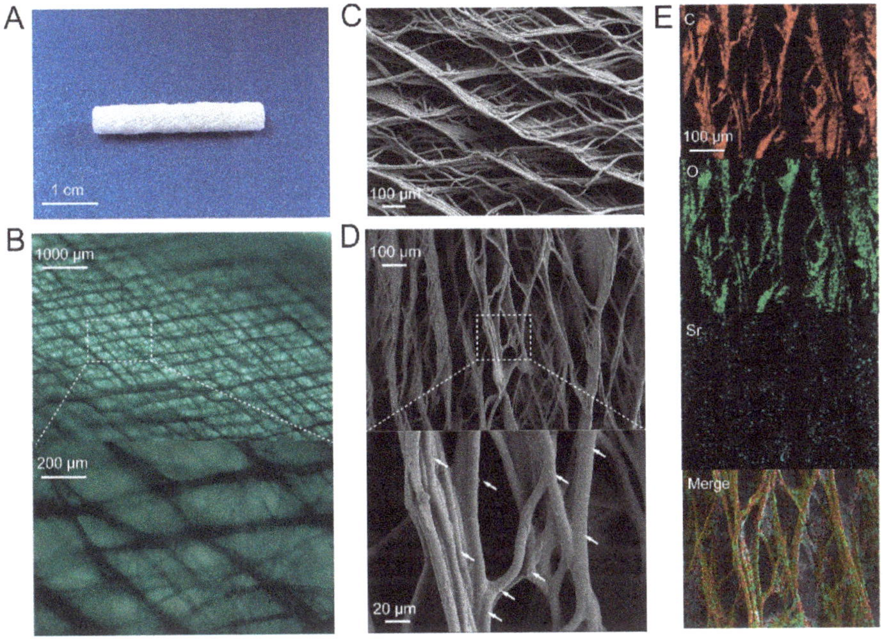

Figure 3. (**A**,**B**) Optical photograph; (**C**) SEM PCL eStent. (**D**) SEM. (**E**) Elemental mappings of SrCA eStent. White arrows refer to SrCA nanoparticles.

3.3. eStent Inhibited Intimal Hyperplasia of Vein Grafts and Regulated Vascular Remodeling

The CCK-8 assay (Figure 4A) demonstrated a concentration-dependent inhibition of VSMC proliferation by the eStents. At 48 h, viability decreased to 75.55% (5% SrCA), 58.86% (10% SrCA), and 48.50% (20% SrCA) compared to the blank group ($p < 0.0001$). Notably, 0% of the SrCA stents exhibited no cytotoxicity (94.14% viability), confirming the biocompatibility of GelMA/PCL composites. Conversely, 20% SrCA significantly reduced VEC viability (63.39%, Figure S3).

The LIVE/DEAD assay results (Figure 4B) demonstrated that both 5% and 10% SrCA effectively inhibited the growth of VSMCs. A cell migration assay (Figure 4C) revealed a dose-dependent suppression of VSMC migration. Wound closure rates at 48 h decreased from 90.50% (blank) to 43.22% (5% SrCA) and 23.45% (10% SrCA) ($p < 0.0001$).

Molecular analyses demonstrated SrCA-mediated Hippo pathway activation. RT-qPCR (Figure 4D) showed an upregulation of Yap1 (6.66-fold), Lats1 (2.74-fold), and Lats2 (5.91-fold) in 10% SrCA-treated VSMCs ($p < 0.001$ vs. blank). Western blotting (Figure 4E) confirmed corresponding YAP downregulation, indicating a pathway-specific suppression of VSMC proliferation.

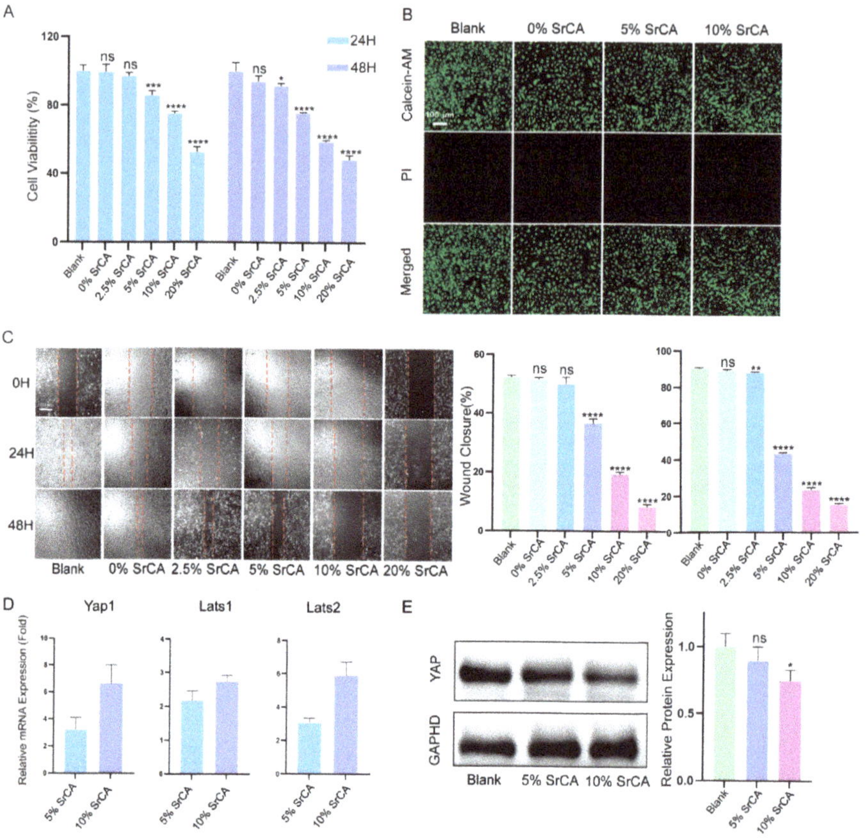

Figure 4. (**A**) CCK-8 assay results of VSMCs after co-culture with different concentrations of SrCA eStent extracts for 24 and 48 h. (**B**) LIVE/DEAD assay results of VSMCs after co-culturing with different concentrations of SrCA eStent extracts for 24 h. (**C**) Cell migration assay results of VSMCs after co-culturing with different concentrations of SrCA eStent extracts for 24 and 48 h, and a statistical diagram of the corresponding wound healing rate percentage. (**D**) RT-qPCR results of VSMCs after co-culturing with different concentrations of SrCA eStent extracts for 24 h. (**E**) Western blot analysis of VSMCs after co-culturing with different concentrations of SrCA eStent extracts for 24 h. (ns indicates not significant, * $p < 0.05$, ** $p < 0.01$, *** $p < 0.001$, **** $p < 0.0001$. Green fluorescence: Calcein-AM represents live cells; red fluorescence: PI represents dead cells).

3.4. In Vivo Efficacy of EStents in Vein Grafts

H&E staining (Figure 5) demonstrated that the stent-implanted groups (5% and 10% SrCA) maintained endothelial integrity at both 2-week and 4-week time points, with continuous and intact luminal surfaces. EVG staining (Figure 6) further revealed a significant reduction in smooth muscle layer thickness in the stent groups. At 4 weeks post-implantation, the average smooth muscle layer thickness in the 5% SrCA and 10% SrCA groups decreased to 82.87 μm and 41.61 μm, respectively, compared to 195.77 μm in the blank group ($p < 0.0001$) (Figure 6C). These findings indicate that SrCA eStents effectively suppress neointimal hyperplasia and vascular remodeling in vivo.

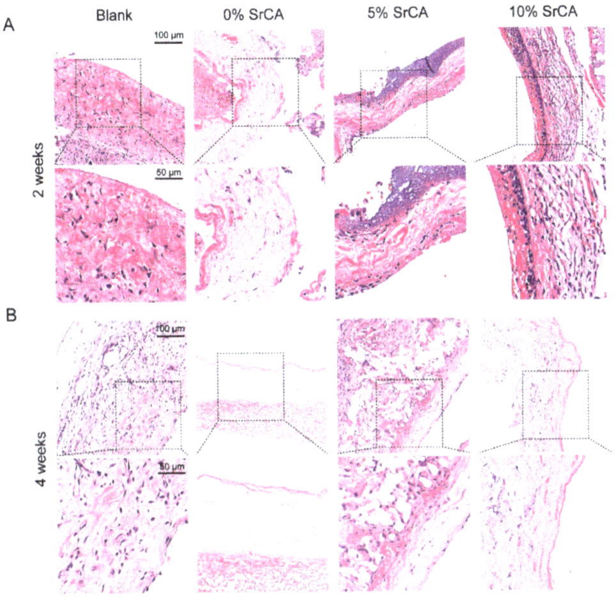

Figure 5. H&E staining results of vein graft samples at (**A**) 2 weeks and (**B**) 4 weeks after placement of the SrCA eStents.

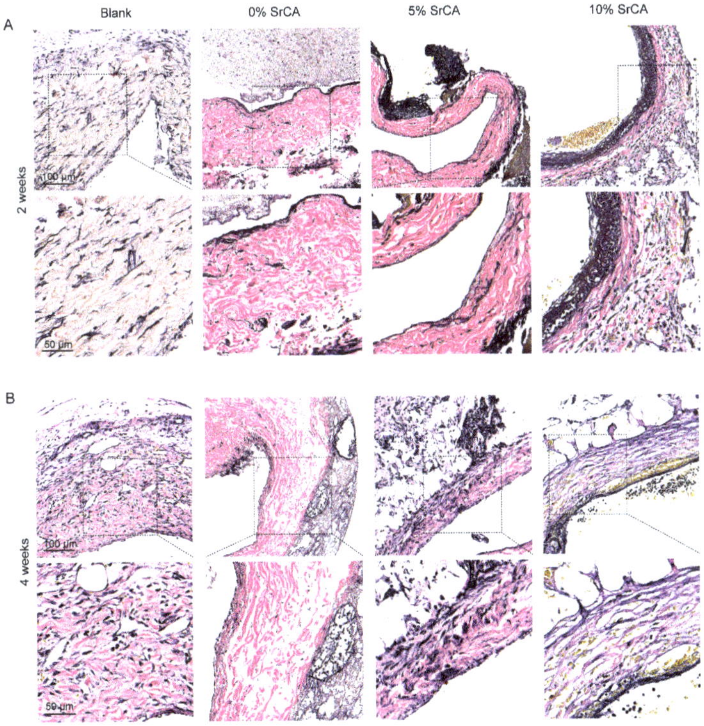

Figure 6. *Cont.*

C

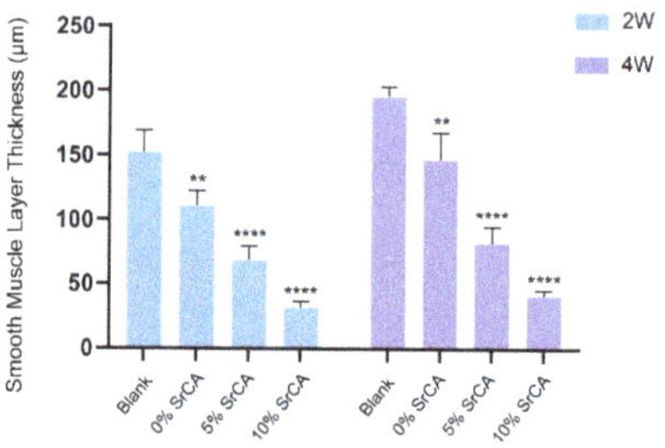

Figure 6. EVG staining results of vein graft samples at (**A**) 2 weeks and (**B**) 4 weeks after placement of the SrCA eStents. (**C**) Statistical diagram of smooth muscle layer thickness. (** $p < 0.01$, **** $p < 0.0001$).

4. Discussions

CABG remains a primary treatment strategy for severe CAD [32–34]. However, the long-term efficacy of CABG is notably constrained by vein graft restenosis [6]. When exposed to the arterial high-pressure environment, vein grafts—characterized by thin vessel walls and a lack of elastic fibers—are prone to excessive dilation and endothelial injury. This initiates a cascade of pathological responses, including VSMC proliferation and migration, along with abnormal extracellular matrix (ECM) deposition, ultimately leading to and exacerbating neointimal hyperplasia (NIH) [8,9,35]. Therefore, minimizing mechanical stress is an effective strategy for enhancing vein graft patency. In recent years, external vascular stents have garnered significant attention due to their mechanical support properties. By limiting graft overexpansion and providing radial stabilization, eStents reduce wall stress and have been shown to lower the incidence of NIH [10–12]. However, purely mechanical intervention cannot block the molecular mechanisms driving VSMC proliferation. Moreover, traditional metallic eStents are non-degradable, often inducing chronic inflammation and fibrotic remodeling of the vessel wall [15,16]. On the other hand, many bioresorbable materials degrade too rapidly or lack sufficient mechanical strength to provide long-term support [36]. Against this backdrop, there is an urgent necessity to develop external stents that integrate mechanical adaptability, biodegradability, and targeted regulatory functions to address both biomechanical and biological challenges in vein graft failure.

PCL has been widely utilized in clinical applications, owing to its superior biocompatibility, stiffness, mechanical elasticity, biodegradability, non-toxicity, thermal stability, and its favorable rheological and viscoelastic properties [37]. Moreover, polycaprolactone (PCL) has been demonstrated to undergo complete metabolic degradation and excretion, with no associated systemic toxicity [38]. Importantly, its controlled degradation rate (approximately two years for complete resorption), low elastic modulus (300–400 MPa), and minimal inflammatory response make it an ideal candidate material for external vein graft stents [37,39]. Compared to the brittleness of polylactic acid (PLA) and the rapid degradation of polyglycolic acid (PGA), the gradual degradation profile of PCL enables a progressive transfer of mechanical load to the healing vessel, facilitating adaptive remodeling during the vascular recovery process [37]. In this study, a three-dimensional porous PCL stent was fabricated using electrostatic direct writing technology. The eStent's architecture was designed not only to provide mechanical support but also to create a permissive microenvironment for cellular infiltration and vascular regeneration.

From a pharmacological perspective, although the local delivery of rapamycin has been shown to partially suppress VSMC activity, its non-selective cytotoxicity often results in delayed endothelial repair and late-stage thrombosis [40]. The synergistic effect of CGA and strontium ions (Sr^{2+}) offers a novel strategy for the inhibition of NIH. As a natural polyphenol, CGA has been demonstrated to effectively counteract adrenaline-induced vascular injury by scavenging reactive oxygen species (ROS), inhibiting DNA damage signaling pathways, and maintaining VSMC homeostasis [41]. This highlights its potential as a natural therapeutic candidate for vascular diseases such as atherosclerosis. However, its application in vein grafts has not yet been reported. Strontium ions, on the other hand, exert pro-angiogenic effects by activating signaling pathways such as vascular endothelial growth factor (VEGF), significantly upregulating the expression of CD31 (an endothelial marker) and α-smooth muscle actin (α-SMA), thereby promoting neovascularization [42]. Additionally, Sr^{2+} modulates the local microenvironment and reduces the release of inflammatory cytokines [42]. Recent studies have shown that methacrylated gelatin (GelMA), which combines the advantages of natural materials and synthetic polymers, exhibits excellent biocompatibility while retaining bioactivity and possessing favorable mechanical properties [43]. The presence of matrix metalloproteinase (MMP)-sensitive sites and the arginine–glycine–aspartic acid (RGD) sequences in gelatin facilitates cell adhesion, migration, proliferation, and differentiation [43]. Furthermore, due to its highly porous structure, GelMA provides efficient channels for nutrient and waste exchange, as well as ample space for cell growth, significantly enhancing cellular proliferation. The RGD motifs in gelatin offer anchoring sites for cell adhesion, while Sr^{2+} activates intracellular signaling pathways to further support endothelial cell function [42]. This is the first study that reports the integration of CGA and Sr^{2+} into a composite complex (SrCA), which is incorporated into the PCL eStent via GelMA-based photocrosslinking, achieving dual-functional synergistic release.

The 10% SrCA group showed 58.9% VSMC proliferation inhibition (Figure 4A) and promoted endothelial cell growth (Figure S3). A scratch assay revealed 23.4% wound closure at 48h (Figure 4C), indicating the potent suppression of VSMC migration. The concentration-dependent dual effect of the SrCA eStent—suppressing VSMCs while supporting endothelial cells—highlights its therapeutic window advantage. Previous studies also support this conclusion, demonstrating that CGA possesses excellent biocompatibility [44,45]. Similarly, Sr^{2+} has been shown to exhibit favorable biocompatibility and low cytotoxicity [46,47]. This bifunctional property overcomes the limitations of conventional agents such as rapamycin, which inhibit neointimal hyperplasia but impair endothelial repair [48].

The Hippo signaling pathway is an evolutionarily conserved regulatory mechanism initially discovered in Drosophila, where it is essential for suppressing cell proliferation and controlling organ size [49,50]. In mammals, the core components of this pathway include the serine/threonine kinases Lats and their downstream effectors YAP (Yes-associated protein) and TAZ (transcriptional co-activator with PDZ-binding motif) [51]. When activated, YAP and TAZ translocate into the nucleus, where they associate with TEA domain (TEAD) family transcription factors, driving the expression of genes related to cell proliferation, apoptosis, and migration [51]. Conversely, phosphorylated YAP is rendered inactive and is retained in the cytoplasm, thereby exerting growth-suppressive effects. Recent studies have demonstrated a strong association between dysregulation of the Hippo pathway and cardiovascular diseases, particularly in the context of VSMC phenotypic switching and vascular remodeling [21]. Wang et al. reported that YAP overexpression promotes VSMC proliferation and migration, whereas genetic knockout or pharmacological inhibition of YAP—such as with verteporfin—significantly attenuates neointimal formation following vascular injury [52,53].

RT-qPCR analysis in this study revealed a concentration-dependent upregulation of Hippo pathway-related gene expression in VSMCs treated with the extract of the SrCA eStents. Compared with the blank control group, the mRNA expression levels of Yap1, Lats1, and Lats2 were elevated by 3.26-fold/6.66-fold, 2.18-fold/2.74-fold, and 3.07-fold/5.91-fold, respectively, in the 5% SrCA and 10% SrCA groups (Figure 4D). This finding was further validated at the protein level by Western blotting, which showed a dose-dependent decrease in nuclear YAP expression (Figure 4E). The inverse correlation between increased Yap1 mRNA levels and decreased nuclear YAP expression suggests that SrCA effectively inhibits downstream pro-proliferative signaling by enhancing LATS-mediated YAP phosphorylation and subsequent degradation.

A histopathological evaluation further confirmed the therapeutic efficacy of the SrCA eStents. H&E staining revealed well-preserved endothelial continuity in the eStent-implanted groups, with a smaller area of endothelial denudation compared to the blank control group (Figure 5). EVG staining showed that at 4 weeks post-implantation, neointimal thickness in the 5% SrCA and 10% SrCA groups was reduced relative to the control group (Figure 6). Importantly, the mechanical stability of the eStents, combined with the pharmacological effects of SrCA, provided a dual-mode intervention: mechanical support reduced vessel wall tension, while Hippo pathway activation suppressed VSMC proliferation and preserved endothelial function. Collectively, these effects contributed to a marked reduction in neointimal hyperplasia in the 10% SrCA group.

Although this study demonstrated the efficacy of the SrCA eStents in inhibiting venous graft restenosis, several limitations should be acknowledged. First, while rodent vein graft models are cost-effective and suitable for preliminary screening, they fail to fully replicate the hemodynamic stresses and chronic inflammatory responses seen in human CABG. Second, the 4-week observation period may be insufficient to evaluate long-term outcomes. Third, although our findings suggest that the activation of the Hippo-YAP/TAZ pathway is a key mechanism, proteomic or phosphoproteomic analyses were not conducted to comprehensively identify potential off-target effects or synergistic signaling networks. Fourth, we did not yet assess the systemic toxicity or detailed cellular responses in non-vascular healthy cells such as fibroblasts, which are essential for ensuring biocompatibility and biosafety. Lastly, clinical translation will require validation in large animal models, such as porcine coronary vein graft models. Addressing these limitations will strengthen the therapeutic rationale for the use of SrCA eStents in vein graft interventions.

5. Conclusions

In this study, a biodegradable eStent loaded with SrCA was successfully developed, offering an effective strategy to prevent vein graft restenosis through a unique "mechanical–chemical" synergistic mechanism. Experimental results demonstrated that the SrCA eStent significantly inhibited VSMC proliferation by activating the Hippo-YAP/TAZ signaling pathway while simultaneously promoting VEC growth. This dual functionality is attributed to the synergistic effects of CGA, known for its antioxidant and anti-inflammatory properties, and Sr^{2+}, which enhances endothelial repair. Despite the limitations associated with rodent models and short-term observation, this study is the first to reveal the targeted regulatory potential of natural polyphenol–metal ion complexes in vascular remodeling, providing a theoretical foundation for the development of next-generation "pro-healing" vascular stents. In future studies, we plan to employ large animal models to further validate our findings and facilitate clinical translation.

Supplementary Materials: The following supporting information can be downloaded at: https://www.mdpi.com/article/10.3390/jfb16070259/s1, Figure S1: DLS Analysis of SrCA; Figure S2: Mechanical Properties of EStent; Figure S3: CCK-8 Assay of VECs; Table S1: Forward and reverse primer sequences used for each target gene synthesis in the RT-qPCR reaction.

Author Contributions: G.Z.: conceptualization, formal analysis, investigation, writing—original draft. S.W.: conceptualization, validation, resources, writing—original draft. Z.L.: formal analysis, investigation. S.G.: investigation, data curation. F.C.: methodology, resources, writing—review and editing, supervision. W.Z.: methodology, writing—review and editing, supervision, project administration. All authors have read and agreed to the published version of the manuscript.

Funding: This research was funded by the National Key R&D Program of China (2022YFE0123500) and the Shanghai Science and Technology Committee (no. 16140901600).

Institutional Review Board Statement: The animal study protocol was approved by the Ethics Committee of Shanghai Tenth People's Hospital, Tongji University School of Medicine (SHDSYY-2023-Y0634, approval date 3 January 2023).

Informed Consent Statement: Not applicable.

Data Availability Statement: The original contributions presented in this study are included in the article and Supplementary Materials. Further inquiries can be directed to the corresponding author.

Conflicts of Interest: The authors declare no conflicts of interest.

References

1. Dalen, J.E.; Alpert, J.S.; Goldberg, R.J.; Weinstein, R.S. The epidemic of the 20(th) century: Coronary heart disease. *Am. J. Med.* **2014**, *127*, 807–812. [CrossRef]
2. Naghavi, M.; Wang, H.; Lozano, R.; Davis, A.; Liang, X.; Zhou, M.; Vollset, S.E.; Abbasoglu Ozgoren, A.; Abdalla, S.; Abd-Allah, F.; et al. Global, regional, and national age-sex specific all-cause and cause-specific mortality for 240 causes of death, 1990–2013: A systematic analysis for the Global Burden of Disease Study 2013. *Lancet* **2015**, *385*, 117–171. [CrossRef]
3. Mohr, F.W.; Morice, M.-C.; Kappetein, A.P.; E Feldman, T.; Ståhle, E.; Colombo, A.; Mack, M.J.; Holmes, D.R.; Morel, M.-A.; Van Dyck, N.; et al. Coronary artery bypass graft surgery versus percutaneous coronary intervention in patients with three-vessel disease and left main coronary disease: 5-year follow-up of the randomised, clinical SYNTAX trial. *Lancet* **2013**, *381*, 629–638. [CrossRef]
4. Brener, S.J.; Lytle, B.W.; Casserly, I.P.; Schneider, J.P.; Topol, E.J.; Lauer, M.S. Propensity Analysis of Long-Term Survival After Surgical or Percutaneous Revascularization in Patients With Multivessel Coronary Artery Disease and High-Risk Features. *Circulation* **2004**, *109*, 2290–2295. [CrossRef] [PubMed]
5. Gaudino, M.; Benedetto, U.; Fremes, S.; Biondi-Zoccai, G.; Sedrakyan, A.; Puskas, J.D.; Angelini, G.D.; Buxton, B.; Frati, G.; Hare, D.L.; et al. Radial-Artery or Saphenous-Vein Grafts in Coronary-Artery Bypass Surgery. *N. Engl. J. Med.* **2018**, *378*, 2069–2077. [CrossRef] [PubMed]
6. Mounsey, C.A.; Mawhinney, J.A.; Werner, R.S.; Taggart, D.P. Does Previous Transradial Catheterization Preclude Use of the Radial Artery as a Conduit in Coronary Artery Bypass Surgery? *Circulation* **2016**, *134*, 681–688. [CrossRef]
7. De Vries, M.R.; Simons, K.H.; Jukema, J.W.; Braun, J.; Quax, P.H.A. Vein graft failure: From pathophysiology to clinical outcomes. *Nat. Rev. Cardiol.* **2016**, *13*, 451–470. [CrossRef]
8. Harskamp, R.E.; Lopes, R.D.; Baisden, C.E.; de Winter, R.J.; Alexander, J.H. Saphenous Vein Graft Failure After Coronary Artery Bypass Surgery. *Ann. Surg.* **2013**, *257*, 824–833. [CrossRef]
9. Yahagi, K.; Kolodgie, F.D.; Otsuka, F.; Finn, A.V.; Davis, H.R.; Joner, M.; Virmani, R. Pathophysiology of native coronary, vein graft, and in-stent atherosclerosis. *Nat. Rev. Cardiol.* **2015**, *13*, 79–98. [CrossRef]
10. Zilla, P.; Human, P.; Wolf, M.; Lichtenberg, W.; Rafiee, N.; Bezuidenhout, D.; Samodien, N.; Schmidt, C.; Franz, T. Constrictive external nitinol meshes inhibit vein graft intimal hyperplasia in nonhuman primates. *J. Thorac. Cardiovasc. Surg.* **2008**, *136*, 717–725. [CrossRef]
11. Sato, A.; Kawamoto, S.; Watanabe, M.; Suzuki, Y.; Takahashi, G.; Masaki, N.; Kumagai, K.; Saijo, Y.; Tabayashi, K.; Saiki, Y. A novel biodegradable external mesh stent improved long-term patency of vein grafts by inhibiting intimal–medial hyperplasia in an experimental canine model. *Gen. Thorac. Cardiovasc. Surg.* **2015**, *64*, 1–9. [CrossRef]

12. Chu, T.; Li, Q.; Dai, C.; Li, X.; Kong, X.; Fan, Y.; Yin, H.; Ge, J. A novel Nanocellulose-Gelatin-AS-IV external stent resists EndMT by activating autophagy to prevent restenosis of grafts. *Bioact. Mater.* **2023**, *22*, 466–481. [CrossRef] [PubMed]
13. Ben-Gal, Y.; Taggart, D.P.; Williams, M.R.; Orion, E.; Uretzky, G.; Shofti, R.; Banai, S.; Yosef, L.; Bolotin, G. Expandable external support device to improve Saphenous Vein Graft Patency after CABG. *J. Cardiothorac. Surg.* **2013**, *8*, 122. [CrossRef] [PubMed]
14. Murphy, G.J.; Newby, A.C.; Jeremy, J.Y.; Baumbach, A.; Angelini, G.D. A randomized trial of an external Dacron sheath for the prevention of vein graft disease: The Extent study. *J. Thorac. Cardiovasc. Surg.* **2007**, *134*, 504–505. [CrossRef] [PubMed]
15. Gharamti, A.; Kanafani, Z.A. Vascular Graft Infections. *Infect. Dis. Clin. N. Am.* **2018**, *32*, 789–809. [CrossRef]
16. Taggart, D.P.; Ben Gal, Y.; Lees, B.; Patel, N.; Webb, C.; Rehman, S.M.; Desouza, A.; Yadav, R.; De Robertis, F.; Dalby, M.; et al. A Randomized Trial of External Stenting for Saphenous Vein Grafts in Coronary Artery Bypass Grafting. *Ann. Thorac. Surg.* **2015**, *99*, 2039–2045. [CrossRef]
17. Taggart, D.P.; Gavrilov, Y.; Krasopoulos, G.; Rajakaruna, C.; Zacharias, J.; De Silva, R.; Channon, K.M.; Gehrig, T.; Donovan, T.J.; Friedrich, I.; et al. External stenting and disease progression in saphenous vein grafts two years after coronary artery bypass grafting: A multicenter randomized trial. *J. Thorac. Cardiovasc. Surg.* **2022**, *164*, 1532–1541.e2. [CrossRef]
18. Grigoryan, B.; Paulsen, S.J.; Corbett, D.C.; Sazer, D.W.; Fortin, C.L.; Zaita, A.J.; Greenfield, P.T.; Calafat, N.J.; Gounley, J.P.; Ta, A.H.; et al. Multivascular networks and functional intravascular topologies within biocompatible hydrogels. *Science* **2019**, *364*, 458–464. [CrossRef]
19. Lee, A.; Hudson, A.R.; Shiwarski, D.J.; Tashman, J.W.; Hinton, T.J.; Yerneni, S.; Bliley, J.M.; Campbell, P.G.; Feinberg, A.W. 3D bioprinting of collagen to rebuild components of the human heart. *Science* **2019**, *365*, 482–487. [CrossRef]
20. He, J.; Bao, Q.; Yan, M.; Liang, J.; Zhu, Y.; Wang, C.; Ai, D. The role of Hippo/yes-associated protein signalling in vascular remodelling associated with cardiovascular disease. *Br. J. Pharmacol.* **2017**, *175*, 1354–1361. [CrossRef]
21. Ruiter, M.S.; Pesce, M. Mechanotransduction in Coronary Vein Graft Disease. *Front. Cardiovasc. Med.* **2018**, *5*, 20. [CrossRef] [PubMed]
22. Xie, C.; Guo, Y.; Zhu, T.; Zhang, J.; Ma, P.X.; Chen, Y.E. Yap1 Protein Regulates Vascular Smooth Muscle Cell Phenotypic Switch by Interaction with Myocardin. *J. Biol. Chem.* **2012**, *287*, 14598–14605. [CrossRef]
23. Amin, R.P.; Kunaparaju, N.; Kumar, S.; Taldone, T.; Barletta, M.A.; Zito, S.W. Structure elucidation and inhibitory effects on human platelet aggregation of chlorogenic acid from Wrightia tinctoria. *J. Complement. Integr. Med.* **2013**, *10*, 97–104. [CrossRef] [PubMed]
24. Li, Y.; Shen, D.; Tang, X.; Li, X.; Wo, D.; Yan, H.; Song, R.; Feng, J.; Li, P.; Zhang, J.; et al. Chlorogenic acid prevents isoproterenol-induced hypertrophy in neonatal rat myocytes. *Toxicol. Lett.* **2014**, *226*, 257–263. [CrossRef] [PubMed]
25. Akila, P.; Vennila, L. Chlorogenic acid a dietary polyphenol attenuates isoproterenol induced myocardial oxidative stress in rat myocardium: An in vivo study. *Biomed. Pharmacother.* **2016**, *84*, 208–214. [CrossRef]
26. Li, Y.; Ren, X.; Lio, C.; Sun, W.; Lai, K.; Liu, Y.; Zhang, Z.; Liang, J.; Zhou, H.; Liu, L.; et al. A chlorogenic acid-phospholipid complex ameliorates post-myocardial infarction inflammatory response mediated by mitochondrial reactive oxygen species in SAMP8 mice. *Pharmacol. Res.* **2018**, *130*, 110–122. [CrossRef]
27. Akila, P.; Asaikumar, L.; Vennila, L. Chlorogenic acid ameliorates isoproterenol-induced myocardial injury in rats by stabilizing mitochondrial and lysosomal enzymes. *Biomed. Pharmacother.* **2017**, *85*, 582–591. [CrossRef]
28. Gu, Z.; Xie, H.; Li, L.; Zhang, X.; Liu, F.; Yu, X. Application of strontium-doped calcium polyphosphate scaffold on angiogenesis for bone tissue engineering. *J. Mater. Sci. Mater. Med.* **2013**, *24*, 1251–1260. [CrossRef]
29. Xing, M.; Jiang, Y.; Bi, W.; Gao, L.; Zhou, Y.-L.; Rao, S.-L.; Ma, L.-L.; Zhang, Z.-W.; Yang, H.-T.; Chang, J. Strontium ions protect hearts against myocardial ischemia/reperfusion injury. *Sci. Adv.* **2021**, *7*, eabe0726. [CrossRef]
30. Yang, Y.; Lei, D.; Huang, S.; Yang, Q.; Song, B.; Guo, Y.; Shen, A.; Yuan, Z.; Li, S.; Qing, F.-L.; et al. Elastic 3D-Printed Hybrid Polymeric Scaffold Improves Cardiac Remodeling after Myocardial Infarction. *Adv. Healthc. Mater.* **2019**, *8*, e1900065. [CrossRef]
31. *No. 85-23*; Guide for the Care and Use of Laboratory Animals. National Academies Press (US) Copyright 1996 by the National Academy of Sciences: Washington, DC, USA, 1996.
32. Fihn, S.D.; Blankenship, J.C.; Alexander, K.P.; Bittl, J.A.; Byrne, J.G.; Fletcher, B.J.; Fonarow, G.C.; Lange, R.A.; Levine, G.N.; Maddox, T.M.; et al. 2014 ACC/AHA/AATS/PCNA/SCAI/STS focused update of the guideline for the diagnosis and management of patients with stable ischemic heart disease. *J. Thorac. Cardiovasc. Surg.* **2015**, *149*, e5–e23. [CrossRef]
33. Bates, E.R. Review: In CAD, CABG reduced 5-year mortality more than PCI in multivessel but not left main disease. *Ann. Intern. Med.* **2018**, *168*, Jc66. [CrossRef] [PubMed]
34. Gaudino, M.; Taggart, D.; Suma, H.; Puskas, J.D.; Crea, F.; Massetti, M. The Choice of Conduits in Coronary Artery Bypass Surgery. *J. Am. Coll. Cardiol.* **2015**, *66*, 1729–1737. [CrossRef]
35. Wadey, K.; Lopes, J.; Bendeck, M.; George, S. Role of smooth muscle cells in coronary artery bypass grafting failure. *Cardiovasc. Res.* **2018**, *114*, 601–610. [CrossRef] [PubMed]

36. Cheng, H.; Yang, X.; Che, X.; Yang, M.; Zhai, G. Biomedical application and controlled drug release of electrospun fibrous materials. *Mater. Sci. Eng. C* **2018**, *90*, 750–763. [CrossRef]
37. Arif, Z.U.; Khalid, M.Y.; Noroozi, R.; Sadeghianmaryan, A.; Jalalvand, M.; Hossain, M. Recent advances in 3D-printed polylactide and polycaprolactone-based biomaterials for tissue engineering applications. *Int. J. Biol. Macromol.* **2022**, *218*, 930–968. [CrossRef] [PubMed]
38. Sun, H.; Mei, L.; Song, C.; Cui, X.; Wang, P. The in vivo degradation, absorption and excretion of PCL-based implant. *Biomaterials* **2006**, *27*, 1735–1740. [CrossRef]
39. Yang, Q.; Lei, D.; Huang, S.; Yang, Y.; Jiang, C.; Shi, H.; Chen, W.; Zhao, Q.; You, Z.; Ye, X. A novel biodegradable external stent regulates vein graft remodeling via the Hippo-YAP and mTOR signaling pathways. *Biomaterials* **2020**, *258*, 120254. [CrossRef]
40. Giacoppo, D.; Alfonso, F.; Xu, B.; Claessen, B.E.; Adriaenssens, T.; Jensen, C.; Pérez-Vizcayno, M.J.; Kang, D.-Y.; Degenhardt, R.; Pleva, L.; et al. Drug-Coated Balloon Angioplasty Versus Drug-Eluting Stent Implantation in Patients with Coronary Stent Restenosis. *J. Am. Coll. Cardiol.* **2020**, *75*, 2664–2678. [CrossRef]
41. Wang, J.; Li, J.; Liu, J.; Xu, M.; Tong, X.; Wang, J. Chlorogenic acid prevents isoproterenol-induced DNA damage in vascular smooth muscle cells. *Mol. Med. Rep.* **2016**, *14*, 4063–4068. [CrossRef]
42. Chen, Y.; Ye, M.; Wang, X.; Li, W.; Chen, W. Functionalized gelatin/strontium hydrogel bearing endothelial progenitor cells for accelerating angiogenesis in wound tissue regeneration. *Biomater. Adv.* **2022**, *136*, 212803. [CrossRef] [PubMed]
43. Reynolds, A.R.; Hart, I.R.; Watson, A.R.; Welti, J.C.; Silva, R.G.; Robinson, S.D.; Da Violante, G.; Gourlaouen, M.; Salih, M.; Jones, M.C.; et al. Stimulation of tumor growth and angiogenesis by low concentrations of RGD-mimetic integrin inhibitors. *Nat. Med.* **2009**, *15*, 392–400. [CrossRef] [PubMed]
44. Chen, X.; Yu, T.; Kong, Q.; Xu, H.; Zhao, Z.; Li, G.; Fan, H.; Wang, Y. A chlorogenic acid functional strategy of anti-inflammation, anti-coagulation and promoted endothelial proliferation for bioprosthetic artificial heart valves. *J. Mater. Chem. B* **2023**, *11*, 2663–2673. [CrossRef] [PubMed]
45. Fang, J.; Jin, X.; Xu, B.; Nan, L.; Liu, S.; Wang, J.; Niu, N.; Wu, Z.; Chen, F.; Liu, J. Chlorogenic acid releasing microspheres enhanced electrospun conduits to promote peripheral nerve regeneration. *Biomater. Sci.* **2023**, *11*, 7909–7925. [CrossRef]
46. Dutta, S.; Khan, R.; Prakash, N.S.; Gupta, S.; Ghosh, D.; Nandi, S.K.; Roy, M. In Vitro Degradation and In Vivo Biocompatibility of Strontium-Doped Magnesium Phosphate-Reinforced Magnesium Composites. *ACS Biomater. Sci. Eng.* **2022**, *8*, 4236–4248. [CrossRef]
47. Sahoo, S.; Sinha, A.; Das, M. Synthesis, characterization and in vitro biocompatibility study of strontium titanate ceramic: A potential biomaterial. *J. Mech. Behav. Biomed. Mater.* **2020**, *102*, 103494. [CrossRef]
48. Chen, S.-Y.; Wang, J.; Jia, F.; Shen, Z.-D.; Zhang, W.-B.; Wang, Y.-X.; Ren, K.-F.; Fu, G.-S.; Ji, J. Bioinspired NO release coating enhances endothelial cells and inhibits smooth muscle cells. *J. Mater. Chem. B* **2022**, *10*, 2454–2462. [CrossRef] [PubMed]
49. Liu, B.; Zheng, Y.; Yin, F.; Yu, J.; Silverman, N.; Pan, D. Toll Receptor-Mediated Hippo Signaling Controls Innate Immunity in Drosophila. *Cell* **2016**, *164*, 406–419. [CrossRef]
50. Rauskolb, C.; Sun, S.; Sun, G.; Pan, Y.; Irvine, K.D. Cytoskeletal tension inhibits Hippo signaling through an Ajuba-Warts complex. *Cell* **2014**, *158*, 143–156. [CrossRef]
51. Meng, Z.; Moroishi, T.; Guan, K.-L. Mechanisms of Hippo pathway regulation. *Genes Dev.* **2016**, *30*, 1–17. [CrossRef]
52. Wang, X.; Hu, G.; Gao, X.; Wang, Y.; Zhang, W.; Harmon, E.Y.; Zhi, X.; Xu, Z.; Lennartz, M.R.; Barroso, M.; et al. The induction of yes-associated protein expression after arterial injury is crucial for smooth muscle phenotypic modulation and neointima formation. *Arterioscler. Thromb. Vasc. Biol.* **2012**, *32*, 2662–2669. [CrossRef] [PubMed]
53. Kimura, T.E.; Duggirala, A.; Smith, M.C.; White, S.; Sala-Newby, G.B.; Newby, A.C.; Bond, M. The Hippo pathway mediates inhibition of vascular smooth muscle cell proliferation by cAMP. *J. Mol. Cell Cardiol.* **2016**, *90*, 1–10. [CrossRef] [PubMed]

Disclaimer/Publisher's Note: The statements, opinions and data contained in all publications are solely those of the individual author(s) and contributor(s) and not of MDPI and/or the editor(s). MDPI and/or the editor(s) disclaim responsibility for any injury to people or property resulting from any ideas, methods, instructions or products referred to in the content.

Journal of
Functional Biomaterials

MDPI

Review

Peripheral Nerve Protection Strategies: Recent Advances and Potential Clinical Applications

Weronika Radecka [1,2], Wiktoria Nogalska [1] and Maria Siemionow [1,*]

1 Department of Orthopaedics, University of Illinois at Chicago, Chicago, IL 60607, USA; radecka.weronika@gmail.com (W.R.); nogalskawiktoria@gmail.com (W.N.)
2 Department of Anatomy, Institute of Medical Sciences, University of Opole, 45040 Opole, Poland
* Correspondence: siemiom@uic.edu or siemiom@hotmail.com; Tel.: +1-312-996-2578

Abstract: Peripheral nerve injuries (PNIs) are a significant clinical challenge, often resulting in persistent sensory and motor deficits despite surgical repair. Autologous nerve grafts remain the gold standard for repair; however, outcomes are frequently suboptimal due to donor site morbidity and inconsistent functional recovery. A major obstacle in nerve regeneration is the formation of postoperative adhesions and fibrosis, which impede healing and necessitate revision surgeries. Nerve protectors from biological, synthetic, and hybrid materials offer a promising tissue engineering strategy to enhance nerve regeneration. These protectors are applied as a protective barrier when a nerve is severed without the gap, allowing for direct repair. They provide mechanical support and reduce scarring. Biocompatible biological wraps, including vascularized fat flaps, vein wraps, collagen-based materials, human amniotic membrane (hAM), porcine small intestinal submucosa (PSIS), and chitosan, modulate immune responses and promote vascularization. Synthetic alternatives, like polycaprolactone (PCL), provide mechanical stability with controlled degradation. Hybrid wraps, such as PCL-amnion, combine the benefits of both. Despite optimistic results, the heterogeneity of study methodologies hinders direct comparisons and standardization. This review highlights the latest developments in nerve wraps, their clinical applications, limitations, and future potential, guiding clinicians in selecting the most appropriate materials for peripheral nerve repair.

Keywords: nerve protection; nerve protector; nerve regeneration; nerve injuries; regenerative medicine; nerve repair; nerve wrap; synthetic nerve wrap; biological nerve wrap; tissue engineering

1. Introduction

Peripheral nerve injuries (PNIs) represent a significant clinical challenge affecting millions worldwide, often resulting in long-term disability, including chronic neuropathic pain, sensory loss, and motor deficits [1]. Effective management of PNI is essential for the restoration of function and improved quality of life of the affected patients [2]. However, despite advancements in surgical techniques and the introduction of biomaterials for support of nerve regeneration, managing PNI still remains complex, with patient recovery outcomes varying significantly [3].

Historically, the understanding of PNI was limited; by the late 18th century, it was widely believed that regeneration of damaged peripheral nerves was impossible [4]. Modern microsurgical techniques improve nerve regeneration after trauma; however, still, only half of the patients report satisfactory recovery, whereas one-third of patients may see little or no improvement [5]. Various nerve repair techniques offer distinct advantages and potential complications. This highlights the ongoing need for innovation and optimization in therapeutic strategies for nerve injuries [6].

Scar tissue formation is a common consequence of PNI and surgical interventions, significantly impeding nerve regeneration and predisposing patients to complications such as traction neuropathy. Understanding the mechanisms behind scar formation is crucial for improving treatment outcomes and advancing regenerative strategies. Effective prevention of scar tissue formation remains a key focus in the development of biomaterials and surgical techniques [2].

One of the most promising innovations in this field is the use of a nerve wrap—a bioengineered barrier that surrounds the nerve, providing a protective interface [7]. Nerve wraps are used when a nerve is transected without a resulting gap, where direct repair is feasible, and grafting is not required (Figure 1). An ideal nerve protector should be biocompatible, minimize the risk of fibrosis and scarring, and ensure sufficient nerve mobility [8,9]. The concept of nerve wrapping dates back to 1989, when Masear first proposed using an autologous vein as a protective sheath [10]. Since then, a diverse array of biological, synthetic, and hybrid nerve protectors has been introduced, each designed to support the natural regeneration process [9].

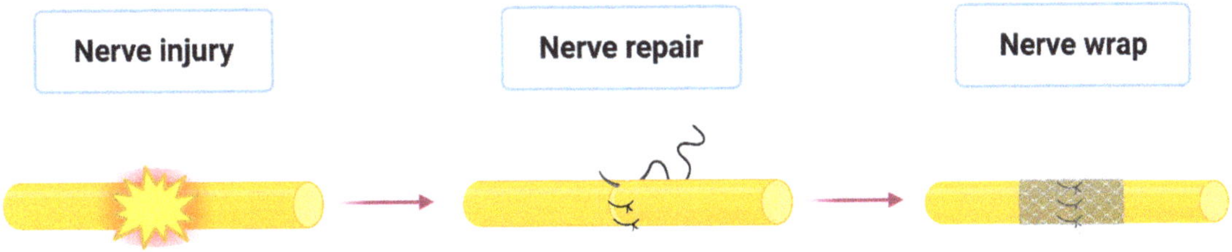

Figure 1. Application of the nerve protector. The figure illustrates the application of nerve-protecting wrap after nerve injury and repair.

This article presents a narrative review of current strategies and materials used to protect peripheral nerves from scar tissue formation, fibrosis, and inflammation following nerve injuries or surgical procedures. Relevant literature was identified through structured searches of PubMed, Scopus, and Web of Science using keyword combinations such as "peripheral nerve protection," "nerve protector," "nerve wrap," "nerve injury," "nerve regeneration," and "regenerative medicine" (Figure 2). Original experimental and clinical studies published in English were included, with priority given to articles from 2020 to 2025 to ensure coverage of the most recent advancements in the field. Additional studies were identified by screening the reference lists of the selected articles. The review is strictly focused on peripheral nerve protection; studies addressing the central nervous system or biomaterials developed exclusively for other tissue types were excluded.

This review provides a comprehensive analysis of the current progress in PNI management, emphasizing cutting-edge surgical techniques and biomaterials. It evaluates their efficacy, limitations, and clinical potential while highlighting their impact on patient outcomes. Additionally, it identifies remaining challenges and suggests future research directions for peripheral nerve repair.

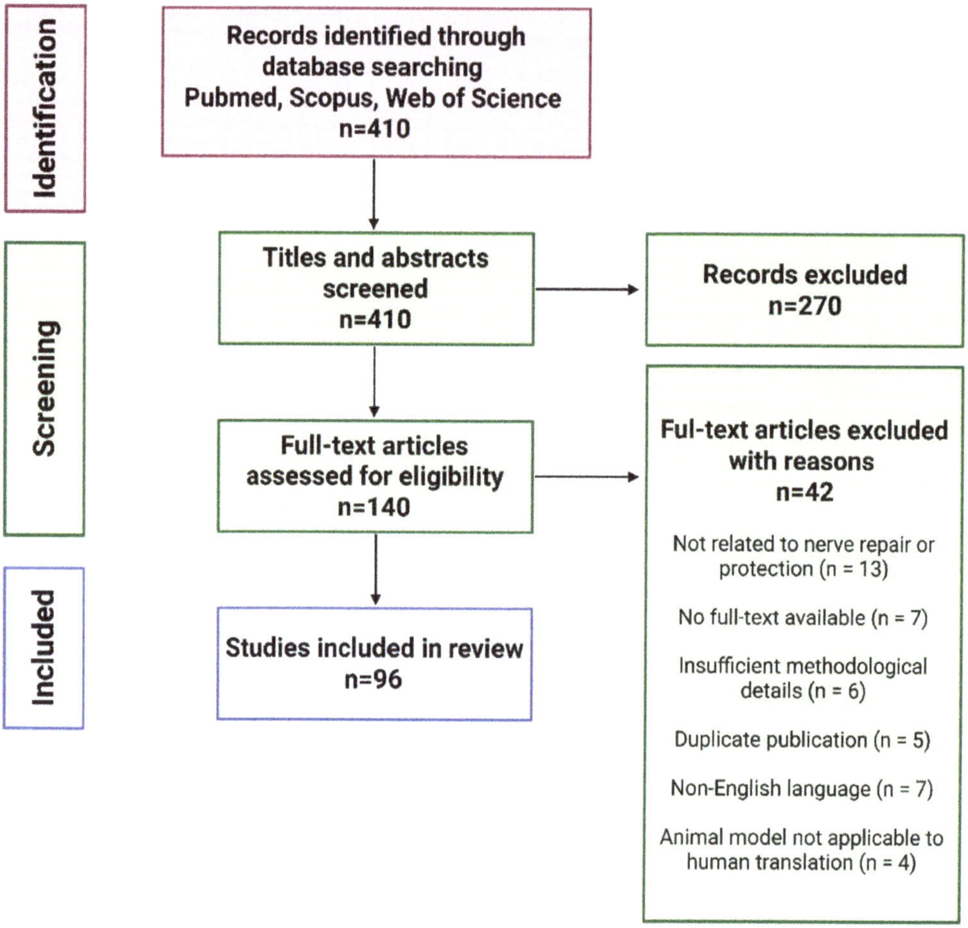

Figure 2. Flow chart of article preselection, selection, and exclusion for nerve wrapping materials review.

2. Nerve Protection Strategies: Biological vs. Synthetic Nerve Wraps/Protectors

Following the initial exploration of autologous tissues for nerve protection, research has expanded to include other biological materials with regenerative potential [11] (Figure 3) (Table 1). Recognizing that natural tissues might provide an optimal environment for nerve regeneration, researchers have developed biological nerve wraps. By integrating with the surrounding tissue and utilizing intrinsic repair mechanisms, these wraps promote axonal regrowth and regeneration. As the complexity of nerve injuries became more evident, the focus shifted from tissue derivatives to engineered scaffolds, resulting in the development of biological and synthetic nerve wraps [5,12,13]. Despite differences in composition and degradation profiles, both types of materials function as physical barriers, shielding regenerating nerves from external compression and scar tissue formation [11]. The choice between biological versus synthetic wraps depends on various factors, including clinical indications, extent of injury, anatomical location, patient-specific considerations, and surgeon preference [14].

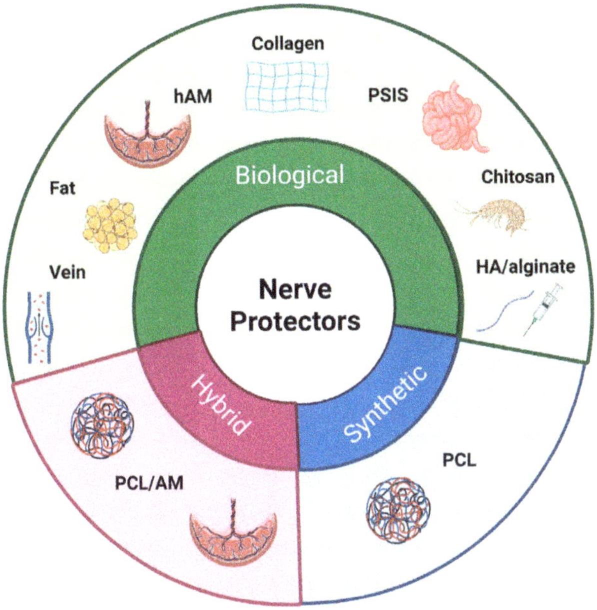

Figure 3. Classification of nerve protectors. The figure presents a classification of nerve protectors into three main types: biological, synthetic, and hybrid materials.

Table 1. Comparison of nerve wrapping materials based on key characteristics.

Material	Success Rate	Biological Response	Immunogenicity	Resorption Rate	Clinical Applicability
Vein	Moderate	Promotes tissue regeneration, low scar formation	Low	N/A	Recurrent and persistent compressive neuropathies
Hypothenar Fat Flap	Moderate	Supports regeneration, reduces scarring	Low	N/A	Recurrent and persistent compressive neuropathies
Human Amniotic Membrane (hAM)	High	Strong regenerative properties, anti-inflammatory, antifibrotic, antibacterial	Low	1–3 months	Recurrent and persistent compressive neuropathies
Collagen Type I	Moderate	Supports cellular growth and collagen formation	Moderate	4–8 months 36–48 months	Recurrent and persistent compressive neuropathies
Porcine Small Intestinal Submucosa	High	Supports tissue regeneration, some risk of immune response	Moderate	3 months	Recurrent and persistent compressive neuropathies
Chitosan	Limited data	Biodegradable, supports tissue growth, low inflammation	Low	3–18 months	Experimental/ preclinical stage
Hyaluronic acid	High	Enhances tissue repair, anti-inflammatory	Low	3–6 months	Recurrent and persistent compressive neuropathies
Human Epineural Patch (hEP)	Limited data	Promotes tissue regeneration, low scar formation, low fibrosis, anti-inflammatory	Low	12 months	Experimental/ preclinical stage
Poly-ε-caprolactone (PCL)	Limited data	Biocompatible, supports cellular attachment	Moderate	18–36 months	Experimental/ preclinical stage
Poly-ε-caprolactone (PCL)/Amnion	Limited data	Anti-inflammatory, antifibrotic, anti-adhesion properties, promotion of nerve generation	Low	Limited data	Experimental/ preclinical stage

3. Biological Nerve Protectors

Pursuing effective nerve repair strategies has long driven advancements in regenerative medicine, with biological nerve protectors emerging as a cornerstone in this field. Derived from human or animal tissue, they share several key characteristics that contribute to their effectiveness. Their inherent biocompatibility significantly reduces the risk of immune rejection and enhances the likelihood of successful integration into the body [15]. Additionally, their resorbable nature ensures they naturally degrade over time, eliminating the need for surgical removal [16]. Many biological protectors, such as autologous fat flaps and vein wraps, facilitate nerve gliding and reduce adhesions. Others, like collagen-based wraps and human amniotic membrane (hAM), contribute to vascularization and immune modulation, further enhancing nerve recovery. Their anti-inflammatory and antifibrotic properties mitigate perineural scarring. However, they come with certain disadvantages, such as rapid degradation, which may lead to premature loss of structural support before the nerve has fully regenerated [17]. Seven nerve-protecting wraps made of porcine submucosa, collagen, and calcium alginate have received FDA approval for clinical use in the United States [18,19] (Table 2).

Table 2. FDA-approved nerve protectors.

510K Approval	Product Name	Material	Company	Address	FDA Approval Year
K041620	NeuraWrap™	Type I bovine collagen and chondroitin-6-sulfate	Integra LifeSciences, Corp.	Plainsboro, NJ, USA	2004
K060952	NeuroMend™	Type I bovine collagen and chitosan	Styker, Corp.	Kalamazoo, MI, USA	2006
K132660	AxoGuard® Nerve Protector	Porcine small intestinal submucosa (SIS) extracellular collagen matrix (ECM)	Axogen, Inc.	Alachua, FL, USA	2014
K190246	NeuroShield™	Chitosan	Checkpoint Surgical, Inc.	Cleveland, OH, USA	2019
K210665	Nerve Tape®	Porcine small intestinal submucosa (SIS) extracellular collagen matrix (ECM)	BioCircuit Technologies, Inc.	Atlanta, GA, USA	2022
K223640	Axoguard® HA+ Nerve Protector	Porcine small intestinal submucosa (SIS) extracellular collagen matrix (ECM), sodium hyaluronate, and sodium alginate	Axogen, Inc.	Alachua, FL, USA	2023
K232029	Versawrap nerve protector	Calcium alginate and hyaluronic acid	Alafair Biosciences, Inc.	Austin, TX, USA	2023

3.1. *Vein Wraps for Enhancement of Nerve Regeneration*

Since Masear first introduced vein wrapping (VW) for nerve protection, autologous veins have been utilized to support nerve repair techniques [10] (Table 3). This method employs freshly harvested vein tissue as a protective barrier around nerves, using the vein's smooth intimal layer to create a gliding surface [20]. VW is accessible and cost effective, making it suitable for addressing both traumatic nerve injuries and chronic compressive neuropathies [21]. Studies suggest that VW may contribute to neuroprotection by promoting the release of basic fibroblast growth factor (bFGF), which triggers the induction of heme oxygenase-1 (HO-1), an enzyme with immunomodulatory and antinociceptive effects. In rat sciatic nerve injury models, this mechanism has shown promise in managing pain [22,23]. Furthermore, vein-derived cytokines such as interleukin-4 (IL-4) and interleukin-10 (IL-10) enhance the expression of M2 macrophage markers like CD206 and Arginase-1 (ARG1), contributing to an anti-inflammatory environment and facilitating tissue regeneration while reducing neuropathic pain [24]. Clinically, VW has been shown to be an effective technique for preventing complications like perineural scarring and neuromas. Application of VW following nerve repair demonstrated improved motor and sensory recovery compared to standard neurorrhaphy [20]. However, the clinical application of VW can present challenges such as donor site morbidity, longer surgery time, and technical difficulties, which must be carefully weighed when considering the application of VW for enhancement of peripheral nerve regeneration [8].

Table 3. First reports on the use of various materials as nerve protecting wraps.

Researcher	Year	Nerve Wrap Material	Animal
		Biological Nerve Wraps	
Masear et al. [10]	1989	Vein	Rat
Strickland JW et al. [25]	1969	Hypothenar Fat Flap	Rat
O' Neill AC et al. [26]	2009	Human Amniotic Membrane (hAM)	Rat
Kim PD et al. [27]	2010	Collagen Type I	Rat
Kokkalis ZT et al. [28]	2011	Porcine Small Intestinal Submucosa	Rat
Hones KM [29]	2023	Hyaluronic acid	Rat
		Synthetic Nerve Wraps	
Sahrane et al. [30]	2019	Poly-ε-caprolactone (PCL)	Rat
		Hybrid Nerve Wraps	
Liu et al. [31]	2023	Poly-ε-caprolactone (PCL)/Amnion	Rat

3.2. *Vascularized Fat Flaps for Nerve Protection*

Different materials for nerve protection have emerged over the last few decades. Both vascularized and free-fat flaps have been used for peripheral nerve protection. However, the vascularized pedicle flaps have become more popular in recent years. These flaps offer several advantages, such as revascularizing nerve tissue, thereby reducing fibrosis, preventing nerve adherence to surrounding structures, and allowing nerve gliding within its natural surroundings [32]. They have gained attention due to their easy availability and abundance. Adipose tissue is rich in adipose-derived stem cells (ADSCs), which can differentiate into various cell types, particularly Schwann-like cells, and guide regenerating axons [33–35]. The rich capillary network within adipose tissue helps sustain Schwann cells and other supporting cellular components, ensuring a steady influx of growth factors. The procedure is minimally invasive, safe, and effective in ameliorating persistent neuropathic pain and promoting nerve recovery, particularly in cases of chronic nerve injury or recalcitrant carpal tunnel syndrome [36]. As a result, there is a reduced need for antineuropathic medications, thus enhancing patients' quality of life while minimizing drug side effects [37]. Clinical studies have shown that the application of a vascularized fat flap during the anterior subcutaneous transposition of the ulnar nerve reduces perineural scarring and supports nerve regeneration by mimicking the natural fatty surroundings of peripheral nerves [38]. However, despite many advantages, there are still challenges, such as donor site morbidity and extended operative time due to the fat flap harvest, which must be considered when choosing fat flaps as the protective barrier against scar and adhesion formation [39].

3.3. *Human Amniotic Membrane as the Protective Barrier*

The human amniotic membrane (hAM), well recognized for its wound-healing properties, has emerged as an effective biomaterial for the enhancement of peripheral nerve regeneration. Derived from the placenta, this non-immunogenic tissue is rich in collagen, growth factors, and glycoproteins [40]. Its extracellular matrix is designed to promote wound healing and modulate cellular processes. hAM's regenerative potential is further enhanced by its anti-inflammatory cytokines, along with its antifibrotic and antibacterial properties that facilitate nerve healing and reduce the incidence of postoperative adhesions [41,42]. One of the key advantages of hAM is its lack of human leukocyte antigen (HLA) expression, significantly reducing the risk of immune rejection and enabling broad clinical applicability without complications typically associated with foreign tissue materials [43]. In a rat model of complete common peroneal nerve transection, the effect of amniotic membrane (AM) wrapping was evaluated by comparing two groups: the suture group (nerve transection with end-to-end repair) and the AM wrap group (nerve transection with end-to-end repair followed by two-layer AM wrapping). AM application significantly enhanced functional recovery, as evidenced by improved nerve conduction velocity and axonal transport. In contrast, the suture group without wrapping showed limited muscle regeneration and persistent perineural fibrosis [44]. Similar benefits were observed in clinical settings, where patients undergoing cubital tunnel decompression supported by wrapping the ulnar nerve with hAM experienced a reduced recurrence of paresthesias compared to those without hAM application [45]. The widespread availability, cost-effectiveness, and potent biological activity of hAM make it a highly advantageous tool in neuroprotection strategies. As research continues to support its clinical applications, hAM is positioned as a leading choice for enhancing nerve recovery and reducing post-surgical complications following PNI [46].

3.4. *Bioresorbable Collagen for Nerve Regeneration*

Once regarded as a simple structural protein, collagen has now become one of the key materials used to support peripheral nerve repair due to its biocompatibility and low immunogenicity [47]. As a natural component of the extracellular matrix (ECM), collagen provides a conducive scaffold for cellular attachment, Schwann cell adhesion, proliferation, and migration, supporting the alignment of regenerating axons [48]. Unlike synthetic materials, which may provoke chronic foreign body reactions, purified collagen exhibits minimal antigenicity, particularly when enzymatically treated to remove immunogenic telopeptides or chemically crosslinked to modulate degradation kinetics [49]. The bioresorbable nature of collagen ensures gradual degradation, reducing the risks associated with foreign body materials and allowing for controlled healing without the need for surgical removal [50,51]. Preclinical studies using rat sciatic nerve models confirmed that the selective permeability of the collagen Type 1 wrap allowed the diffusion of essential micronutrients and growth factors while preventing infiltration by fibroblasts [52]. Clinically, collagen nerve wraps are effective in preserving nerve tissue homeostasis in conditions such as carpal and cubital tunnel syndrome [53,54]. Commercially distributed, FDA-approved products, such as NeuraWrap™ (Integra LifeSciences Corporation, Plainsboro, NJ, USA) and NeuroMend™ Wrap (Stryker Corp., Kalamazoo, MI, USA), are designed to unroll and naturally conform to the shape of the injured nerve, ensuring an optimal fit without compression [55,56]. NeuraWrap™ and NeuroMend™ mainly differ in their resorption rates, with NeuraWrap™ breaking down over 36 to 48 months, while NeuroMend™ degrades within 4 to 8 months [57]. Despite their advantages, the differences in resorption rates, influenced by crosslinking and host enzymatic activity factors, may lead to nerve compression [58]. Furthermore, while inflammatory responses are generally mild, occasional foreign body reactions have been reported, necessitating careful material selection and post-implantation monitoring [59].

3.5. Porcine-Derived Biomaterials for Nerve Regeneration

In the complex landscape of nerve repair, where meticulous surgical technique and biocompatibility of the applied materials dictate the success of regeneration, porcine small intestine submucosa (PSIS) has emerged as a compelling biomaterial. With its structural resemblance to the native human extracellular matrix (ECM), PSIS provides a supportive microenvironment for nerve healing. In addition, it fully degrades within 3 months following application [60,61]. Advances in decellularization techniques have optimized the safety profile of these matrices by reducing immunogenicity while preserving their bioactive components, such as collagen (Types 1 and 3), laminin, and fibronectin, key factors in axonal growth, Schwann cell migration, and myelination [62]. Pre-clinical studies have demonstrated encouraging results, including revascularization and remodeling into connective tissue similar to the nerve epineurium after implantation [28]. Clinically, PSIS wraps have gained attention, particularly in recurrent and persistent carpal tunnel syndrome and cubital tunnel syndrome, where chronic nerve compression is a major concern [63,64]. Their application in upper extremity surgeries has been associated with reduced pain and improved functional outcomes [65]. Investigations into PSIS use in ulnar nerve revisions and other peripheral nerve conditions highlight its ability to reduce inflammatory reactions [66]. This anti-inflammatory effect has been attributed to the material's ability to modulate the local immune response, including the downregulation of pro-inflammatory cytokines, such as TNF-α and IL-1β, and the promotion of a macrophage phenotypic shift from the pro-inflammatory M1 to the pro-regenerative M2 subtype, thereby creating a microenvironment favorable for nerve healing and axonal regeneration [67]. In the commercial space, three variants of PSIS nerve repair solutions are available, including Axoguard® Nerve Protector (Axogen, Inc. Alachua, FL, USA), the more advanced Axoguard® HA+ Nerve Protector (Axogen, Inc. Alachua, FL, USA), which incorporates a resorbable hyaluronate–alginate gel layer, and Nerve Tape® (BioCircuit Technologies, Inc, Atlanta, GA, USA), a suture-less coaptation device [68]. Nerve Tape utilizes microhooks embedded in a flexible, biocompatible material to enable quick and precise alignment of nerve ends without sutures, providing an effective alternative to traditional microsuture neurorrhaphy [69,70]. The presented innovations reflect a paradigm shift in peripheral nerve repair techniques, where the application of biomaterials bridges the gap between surgical intervention and natural regeneration. While PSIS-based products are generally well tolerated, rare immune responses have been reported and should be taken into consideration when making choices between different biomaterials used for the enhancement of nerve regeneration [71].

3.6. Chitosan-Based Nerve Protectors

As the quest for innovative biomaterials in nerve repair continues, chitosan nerve sheets emerge as an alternative in regenerative medicine. Derived from chitin, a natural polysaccharide from the exoskeletons of crustaceans, chitosan is a biodegradable and biocompatible material [72]. It is rich in deacetylated chitin and has shown potential in inhibiting extraneural scarring and promoting neural regeneration, as demonstrated in experimental studies on rat tibial nerves [73]. Research suggests it effectively supports nerve regeneration in sheets and as a drug carrier, increasing its clinical potential. In particular, its ability to reduce scarring and support Schwann cell migration indicates its possible use as an alternative to currently used approaches [74]. Chitosan's inherent antimicrobial properties further reduce the risk of infection [75,76]. However, the degradation time of chitosan-based wraps can vary depending on factors such as the degree of acetylation and the specific formulation used. Some studies have reported that chitosan nerve wraps did not show significant degradation after extended periods, suggesting that the material's persistence may offer prolonged support during nerve regeneration [77]. Currently, NeuroShield® (Checkpoint Surgical, Inc. Cleveland, OH, USA) is the only commercially available chitosan-based nerve wrap, and there is a limited number of preclinical and clinical studies evaluating its efficacy [78].

3.7. Human Epineural Patch, a Novel Strategy for Nerve Protection

One of the novel approaches explored in our laboratory is the human Epineural Patch (hEP), a nerve protector derived from the human epineurium—the outermost layer of peripheral nerves [79]. In this study, we investigated the efficacy of the hEP in a complex rat sciatic nerve injury model that included crush injury, transection, and end-to-end repair, conditions designed to replicate clinically relevant nerve trauma. Unlike other biological wraps, which often degrade prematurely or lack sufficient structural integrity, our findings demonstrate that the hEP provides a durable protective effect supporting nerve healing without inducing an immune response [79]. Following sciatic nerve repair, the application of the hEP to the injury site enhanced nerve regeneration and vascularization, as evidenced by increased expression of vasculogenic (VEGF and vFV) and neurogenic (laminin B, NGF, and S 100) markers. These results confirm the hEP's potential to promote nerve regeneration. Furthermore, compared to the human amnion membrane (hAM), hEP application significantly improved long-term functional outcomes, as validated by standard functional tests. Thus, the hEP presents a novel and valuable protective option for peripheral nerve reconstruction, offering an alternative tool in regenerative medicine [79].

In our previous research, we have explored the various applications of the human epineurium, including epineural jackets, sheaths, and conduits. These studies have confirmed the neuroregenerative, anti-inflammatory, and immunomodulatory potential of epineurium-based products [80–85]. Given its unique neuroregenerative properties, the hEP represents a valuable addition to the repertoire of biological nerve wraps, making it a strong candidate for clinical translation. These findings are promising; however, the clinical potential of the hEP should be further validated through clinical studies.

3.8. A New Perspective on Biological Nerve Protection

Nerve protectors are not limited to traditional wrap formats. New products, such as the VersaWrap (Alafair Biosciences, Inc, Austin, TX, USA), initially designed for tendon, ligament, and skeletal muscle protection, represent a bioresorbable plant-based hydrogel composed of hyaluronic acid (HA) and alginate [86]. To expand these applications, recent studies have explored its potential in enhancing nerve regeneration following PNI. VersaWrap can be implanted as either a sheet or gel via a syringe, offering an innovative approach to nerve protection [29,87]. It features a nano-porous structure that allows the passage of small molecules, including growth factors while acting as a barrier to larger proteins and fibroblasts. The hydrogel gradually resorbs within 3 to 6 months following implantation, though residual biopolymer traces may persist for up to two years, ensuring extended protection during the critical phases of nerve regeneration and healing [29]. Preliminary clinical data support its safety and efficacy in patients undergoing revision surgery for upper-extremity compressive neuropathies. Despite the study's limitations—such as a small sample size and retrospective design—all patients experienced symptomatic improvement, expressed postoperative satisfaction, and did not require further surgical intervention within the observed postoperative period [87].

4. Synthetic Nerve Protectors

While biological wraps leverage nature's regenerative mechanisms, the development of synthetic nerve protectors arose from the need to overcome the limitations of biological wraps, marking the beginning of research on synthetic products to enhance nerve regeneration [88]. In recent years, biodegradable polymers have been increasingly utilized in the fabrication of scaffolds for neural tissue engineering, with polycaprolactone (PCL) being the most extensively studied for nerve wrap applications [89]. In contrast, polymers, such as polyglycolic acid (PGA), polylactic acid (PLA), and their copolymer poly(lactic-co-glycolic) acid (PLGA), degrade into acidic byproducts, which can induce localized inflammatory responses and create a suboptimal microenvironment, ultimately impairing peripheral nerve regeneration. PCL-based wraps offer controlled degradation rates, allowing them to gradually degrade over time as the nerve undergoes recovery [90]. Furthermore, their mechanical properties can be tailored to match the physiological needs of the regenerating nerve, providing a scaffold that guides axonal outgrowth while minimizing perineural fibrosis.

Polycaprolactone (PCL) Wraps for Nerve Protection

Initially developed for controlled drug delivery and tissue engineering, poly-ε-caprolactone (PCL) has found a vital role in nerve repair [30]. Its biodegradable nature, along with its ability to mimic the extracellular matrix and integrate with various coatings and polymers, improves biocompatibility and mechanical strength [91]. Its widespread use is further driven by its affordability, natural resorption over several months, and nonimmunogenic properties, making it a practical choice for biomedical applications. However, despite advantages it should be noted that synthetic polymers such as PCL are inherently hydrophobic, which can impair cellular adhesion and potentially trigger inflammatory or immune responses. Recent advances in electrospinning techniques have enabled the creation of PCL scaffolds with interconnected macropores, forming nerve wraps that reduce scarring and promote healing with minimal inflammation [92]. In a study by Sarhane et al., an innovative nanofiber PCL wrap was evaluated in a rat model involving primary nerve transection followed by direct repair. At five weeks post-surgery, the PCL wrap demonstrated efficacy in enhancing neuroregeneration. The macroporous structure of the wrap facilitated controlled tissue ingrowth and preserved nerve continuity. Histological assessments revealed minimal scarring and smooth integration with surrounding tissues, with no signs of nerve damage or adverse inflammatory responses. In contrast, the control group exhibited significant fibrosis, compromising neural integrity [30]. The effectiveness of PCL can be further improved when it is coated with different materials. A study by Harley-Troxe et al. showed that PCL fibers coated with graphene oxide promote axonal growth in a rat model of sciatic nerve injury, emphasizing the promise of these composite materials in nerve regeneration applications [89]. However, clinical evidence remains limited, and further research is necessary to validate their efficacy in human applications.

5. Hybrid Synthetic-Biological Wraps as the Protective Barrier

Hybrid wraps represent a bridge between biological and synthetic nerve protection options, addressing unfavorable outcomes of peripheral nerve injury, such as axonal escape and excessive scarring, which often compromise functional recovery. Currently, the only reported hybrid constructs are nanofibrous membranes composed of polycaprolactone (PCL) and amniotic membrane (AM), which have shown significant potential in addressing the neurobiology of an injury by modulating macrophage polarization and regulating the inflammatory microenvironment [93]. PCL/AM hybrid wraps are known to reduce fibrosis and preserve the regenerative pathway. PCL provides mechanical stability, while AM fosters biological integration, together creating an optimal framework for axonal regrowth and Schwann cell migration. Evidence from preclinical studies using rat sciatic nerve injury models further supports the therapeutic potential of these materials. PCL-amnion nanofibrous membranes decrease the expression of pro-inflammatory cytokines such as

IL-6 and TNF-α while increasing levels of anti-inflammatory cytokines like IL-10 and IL-13, thereby promoting an M2 macrophage-dominant healing response. This immunomodulatory effect contributes to a reduction in fibrotic tissue formation by suppressing the excessive extracellular matrix deposition, ultimately preventing scar-related inhibition of axonal regrowth. [31]. At 16 weeks post-implantation, these scaffolds have been shown to enhance functional recovery by facilitating Schwann cell proliferation, accelerating axon regeneration, and reducing muscle denervation. Moreover, they contribute to higher axon numbers, increased myelin sheath thickness, and improved overall nerve maturity [94].

6. Limitations of Commercial Nerve Wraps

Although significant advancements have been made in the development of biological and synthetic wraps and materials for nerve protection and regeneration, no standardized guidelines exist for their optimal application in acute versus chronic nerve injuries. While nerve protectors serve as a barrier shielding injured nerves from surrounding tissues, they also present certain limitations (Table 4). Biocompatibility concerns, particularly with animal-derived materials, may trigger immune reactions or be unsuitable for patients due to ethical or religious considerations. Even biocompatible options can induce low-grade inflammatory responses over time, and their biodegradation may be slower than expected, potentially leading to fibrosis or foreign body reactions [95,96]. Additionally, standardized sizes may not accommodate all nerve repairs, resulting in mismatches that can cause iatrogenic constriction or slippage, thereby reducing effectiveness [11]. Proper placement of nerve wraps and protectors is crucial, as excessive tension can hinder nerve regeneration and increase the risk of scarring, fibrosis, and adhesion formation [2].

Table 4. Advantages and disadvantages of nerve wrapping materials.

Material	Advantages	Disadvantages
Vein	Cost-effective; biologically inert	Donor site morbidity; limited mechanical strength; extended surgery time
Hypothenar Fat Flap	Easy local application; well-vascularized	Donor site morbidity; extended surgery time; limited coverage
Human Amniotic Membrane (hAM)	No donor site morbidity; anti-inflammatory and non-immunogenic properties; low incidence of reported complications; relatively affordable; available in various sizes and shapes	Potential risk of disease transmission; requires specific regulatory approvals; limited mechanical strength and handling properties, rapid degradation rate
Collagen Type I	No donor site morbidity; off-the-shelf availability; reduces the risk of long-term foreign body response; selective permeability	Degradation rate can vary and might not match nerve regeneration timing; may elicit mild immune response depending on the source
Porcine Small Intestinal Submucosa	No donor site morbidity; durable; flexible; easy application; full degradation within 4 months; non-immunogenic when decellularized	Ethical/religious concerns for some patients; may elicit mild immune responses depending on processing
Chitosan	Transparency; flexibility and resistance to collapse	Variable degradation rate; low mechanical strength; limited clinical data; limited commercial availability
Hyaluronic acid	No donor site morbidity	Rapid degradation rate
Human Epineural Patch (hEP)	No donor site morbidity; neurogenic, vasculogenic, anti-inflammatory; non-immunogenic properties	Requires regulatory approval; lack of clinical data
Poly-ε-caprolactone (PCL)	Excellent mechanical properties; easy production	Poor cell affinity; poor hydrophilicity; risk of inflammation; potential for immune rejection
Poly-ε-caprolactone (PCL)/Amnion	Improved mechanical strength of AM through the incorporation of PCL nanofibers; high porosity; sustained release of exogenous growth factors	Poor cell affinity; poor hydrophilicity

7. Future Perspectives

Looking ahead, the next generation of nerve protectors must move beyond the limitations of current materials. A comprehensive, multilevel intervention can substantially improve functional recovery and, consequently, the long-term quality of life of patients with peripheral nerve injuries. Future strategies may involve the integration of nerve-protecting wraps with advanced therapeutic approaches such as stem cell therapies and controlled delivery of the growth factors. Incorporation of stem cells into nerve wraps may enhance nerve regeneration leading to cell differentiation into neural lineages, including motor neurons and Schwann cells. Moreover, stem cells can promote axonal growth by secreting paracrine factors, modulating the inflammatory milieu, facilitating tissue remodeling, reducing fibrosis, and enhancing repair quality. Furthermore, the release of neurogenic and angiogenic growth factors, such as NGF, BDNF, VEGF, and PDGF, respectively, in a controlled and sustained manner will further promote axonal growth and enhance neuronal survival while mitigating trauma-induced muscle atrophy. To be effective, therapeutic constructs must be both durable and capable of adapting to the dynamic nature of nerve regeneration. Therefore, future innovative approaches should prioritize the development of personalized and/or customizable nerve wraps and protectors that would address these challenges effectively.

8. Conclusion

Nerve protectors, including biological and synthetic wraps, have significantly advanced the management of peripheral nerve injuries (PNIs). They serve as an adjuvant treatment option in severe cases, where creating a barrier between the repaired nerves and surrounding tissues is essential to prevent fibrosis and tissue adhesions. Depending on the type and severity of the PNI, the application of nerve protective materials should be tailored to the individual patient's needs to achieve optimal outcomes. While nerve protectors enhance nerve regeneration, challenges such as biocompatibility and immune reactions remain. Further research is required to assess long-term neuroregenerative effects and optimize clinical applications of nerve wraps. Continued innovation and rigorous clinical evaluations are essential to ensuring their effectiveness and achieving consistent, reliable outcomes.

Author Contributions: Conceptualization, M.S.; data curation, W.R. and W.N.; writing—original draft preparation, W.R. and W.N.; writing—review and editing, M.S.; visualization, M.S.; supervision, M.S.; project administration, M.S.; funding acquisition, M.S. All authors have read and agreed to the published version of the manuscript.

Funding: This research received no external funding.

Institutional Review Board Statement: Not applicable.

Informed Consent Statement: Not applicable.

Data Availability Statement: No new data were created or analyzed in this study. Data sharing is not applicable to this article.

Acknowledgments: The authors acknowledge the use of BioRender software (RRID: SCR_018361, BioRender, Toronto, ON, Canada) to create Figures 1–3.

Conflicts of Interest: The authors declare no conflicts of interest.

Abbreviations

The following abbreviations are used in this manuscript:

PNIs	Peripheral nerve injuries
hAM	Human amniotic membrane
PSIS	Porcine small intestinal submucosa
PCL	Poly-ε-caprolactone
VW	Vein wrapping
bFGF	Basic fibroblast growth factor
HO-1	Heme oxygenase-1
IL-4	Interleukin-4
IL-10	Interleukin-10
ARG1	Arginase-1
ADSCs	Adipose-derived stem cells
HLA	Human leukocyte antigen
ECM	Extracellular matrix
VEGF	Vascular endothelial growth factor
NGF	Nerve growth factor
FDA	Food and Drug Administration
PGA	Polyglycolic acid
PLA	Polylactic acid
PLGA	Poly(lactic-co-glycolic) acid
TNF-α	Tumor necrosis factor-alpha
hEP	Human Epineural Patch
HA	Hyaluronic acid
NGF	Nerve growth factor
BDNF	Brain-derived neurotrophic factor
VEGF	Vascular endothelial growth factor
PDGF	Platelet-derived growth factor

References

1. Aman, M.; Zimmermann, K.S.; Thielen, M.; Thomas, B.; Daeschler, S.; Boecker, A.H.; Stolle, A.; Bigdeli, A.K.; Kneser, U.; Harhaus, L. An Epidemiological and Etiological Analysis of 5026 Peripheral Nerve Lesions from a European Level I Trauma Center. *J. Pers. Med.* **2022**, *12*, 1673. [CrossRef]
2. Siemionow, M.; Brzezicki, G. Current techniques and concepts in peripheral nerve repair. *Int. Rev. Neurobiol.* **2009**, *87*, 141–172. [CrossRef] [PubMed]
3. Bateman, E.A.; Pripotnev, S.; Larocerie-Salgado, J.; Ross, D.C.; Miller, T.A. Assessment, management, and rehabilitation of traumatic peripheral nerve injuries for non-surgeons. *Muscle Nerve* **2024**, *71*, 696–714. [CrossRef]
4. Dellon, E.S.; Dellon, A.L. The first nerve graft, Vulpian, and the nineteenth-century neural regeneration controversy. *J. Hand Surg. Am.* **1993**, *18*, 369–372. [CrossRef]
5. Thomson, S.E.; Ng, N.Y.; Riehle, M.O.; Kingham, P.J.; Dahlin, L.B.; Wiberg, M.; Hart, A.M. Bioengineered nerve conduits and wraps for peripheral nerve repair of the upper limb. *Cochrane Database Syst. Rev.* **2022**, *12*, CD012574. [CrossRef]
6. Rath, S.; Green, C.J. Selectivity of distal reinnervation of regenerating mixed motor and sensory nerve fibers across muscle grafts in rats. *Br. J. Plast. Surg.* **1991**, *44*, 215–218. [CrossRef] [PubMed]
7. Supra, R.; Agrawal, D.K. Peripheral nerve regeneration: Opportunities and challenges. *J. Spine Res. Surg.* **2023**, *5*, 10–18. [CrossRef] [PubMed]
8. Dy, C.J.; Aunins, B.; Brogan, D.M. Barriers to epineural scarring: Role in treatment of traumatic nerve injury and chronic compressive neuropathy. *J. Hand Surg. Am.* **2018**, *43*, 360–367. [CrossRef]
9. Thakker, A.; Sharma, S.C.; Hussain, N.M.; Devani, P.; Lahiri, A. Nerve wrapping for recurrent compression neuropathy: A systematic review. *J. Plast. Reconstr. Aesthetic Surg.* **2021**, *74*, 549–559. [CrossRef]
10. Masear, V. Venous wrapping of nerves to prevent scarring. In Proceedings of the 44th Annual Meeting of the American Society for Surgery of the Hand, Seattle, WA, USA, 9 September 1989.

11. Mayrhofer-Schmid, M.; Klemm, T.T.; Aman, M.; Kneser, U.; Eberlin, K.R.; Harhaus, L.; Boecker, A.H. Shielding the Nerve: A Systematic Review of Nerve Wrapping to Prevent Adhesions in the Rat Sciatic Nerve Model. *J. Pers. Med.* **2023**, *13*, 1431. [CrossRef]
12. Kehoe, S.; Zhang, X.F.; Boyd, D. FDA-approved guidance conduits and wraps for peripheral nerve injury: A review of materials and efficacy. *Injury* **2012**, *43*, 553–572. [CrossRef] [PubMed]
13. Fakhr, M.J.; Kavakebian, F.; Ababzadeh, S.; Rezapour, A. Challenges and Advances in Peripheral Nerve Tissue Engineering Critical Factors Affecting Nerve Regeneration. *J. Tissue Eng. Regen. Med.* **2024**, *2024*, 8868411. [CrossRef] [PubMed]
14. Taylor, C.S.; Haycock, J.W. Biomaterials and scaffolds for repair of the peripheral nervous system. In *Peripheral Nerve Tissue Engineering and Regeneration*; Phillips, J.B., Hercher, D., Hausner, T., Eds.; Springer: Berlin/Heidelberg, Germany, 2022. [CrossRef]
15. Fornasari, B.E.; Carta, G.; Gambarotta, G.; Raimondo, S. Natural-based biomaterials for peripheral nerve injury repair. *Front. Bioeng. Biotechnol.* **2020**, *8*, 554257. [CrossRef] [PubMed]
16. Liu, X.; Duan, X. Mechanisms and treatments of peripheral nerve injury. *Ann. Plast. Surg.* **2023**, *91*, 313–318. [CrossRef]
17. Crabtree, J.R.; Mulenga, C.M.; Tran, K.; Feinberg, K.; Santerre, J.P.; Borschel, G.H. Biohacking Nerve Repair: Novel Biomaterials, Local Drug Delivery, Electrical Stimulation, and Allografts to Aid Surgical Repair. *Bioengineering* **2024**, *11*, 776. [CrossRef]
18. Lam, T.C.; Leung, Y.Y. Innovations in peripheral nerve regeneration. *Bioengineering* **2024**, *11*, 444. [CrossRef]
19. Arkansas Blue Cross Blue Shield. Peripheral Nerve Repair and Reconstruction. Available online: https://journals.lww.com/jbjsjournal/abstract/2013/12040/peripheral_nerve_repair_and_reconstruction.9.aspx (accessed on 19 March 2025).
20. Mukai, M.; Uchida, K.; Hirosawa, N.; Murakami, K.; Inoue, G.; Miyagi, M.; Shiga, Y.; Sekiguchi, H.; Inage, K.; Orita, S.; et al. Frozen vein wrapping for chronic nerve constriction injury reduces sciatic nerve allodynia in a rat model. *BMC Neurosci.* **2022**, *23*, 1–6. [CrossRef]
21. Papatheodorou, L.K.; Sotereanos, D.G. Vein wrapping of peripheral nerves: Surgical technique. In *Compressive Neuropathies of the Upper Extremity*; Sotereanos, D.G., Papatheodorou, L.K., Eds.; Springer International Publishing: Berlin/Heidelberg, Germany, 2020; pp. 247–252. [CrossRef]
22. Hirosawa, N.; Uchida, K.; Kuniyoshi, K.; Murakami, K.; Inoue, G.; Miyagi, M.; Matsuura, Y.; Orita, S.; Inage, K.; Suzuki, T.; et al. Vein wrapping facilitates basic fibroblast growth factor-induced heme oxygenase-1 expression following chronic nerve constriction injury. *J. Orthop. Res.* **2017**, *36*, 898–905. [CrossRef]
23. Mukai, M.; Uchida, K.; Hirosawa, N.; Murakami, K.; Kuniyoshi, K.; Inoue, G.; Miyagi, M.; Sekiguchi, H.; Shiga, Y.; Inage, K.; et al. Wrapping With Basic Fibroblast Growth Factor-Impregnated Collagen Sheet Reduces Rat Sciatic Nerve Allodynia. *J. Orthop. Res.* **2019**, *37*, 2258–2263. [CrossRef]
24. Hirosawa, N.; Uchida, K.; Kuniyoshi, K.; Murakami, K.; Inoue, G.; Miyagi, M.; Matsuura, Y.; Orita, S.; Inage, K.; Suzuki, T.; et al. Vein wrapping promotes M2 macrophage polarization in a rat chronic constriction injury model. *J. Orthop. Res.* **2018**, *36*, 2210–2217. [CrossRef]
25. Strickland, J.W.; Idler, R.S.; Lourie, G.M.; Plancher, K.D. The hypothenar fat pad flap for management of recalcitrant carpal tunnel syndrome. *J. Hand Surg.* **1996**, *21*, 840–848. [CrossRef] [PubMed]
26. O'Neill, A.C.; Randolph, M.A.; Bujold, K.E.; Kochevar, I.E.; Redmond, R.W.; Winograd, J.M. Photochemical Sealing Improves Outcome Following Peripheral Neurorrhaphy. *J. Surg. Res.* **2009**, *151*, 33–39. [CrossRef] [PubMed]
27. Kim, P.D.; Hayes, A.; Amin, F.; Akelina, Y.; Hays, A.P.; Rosenwasser, M.P. Collagen nerve protector in rat sciatic nerve repair: A morphometric and histological analysis. *Microsurgery* **2010**, *30*, 392–396. [CrossRef]
28. Kokkalis, Z.T.; Pu, C.; Small, G.; Weiser, R.W.; Venouziou, A.; Sotereanos, D.G. Assessment of Processed Porcine Extracellular Matrix as a Protective Barrier in a Rabbit Nerve Wrap Model. *J. Reconstr. Microsurg.* **2010**, *27*, 019–028. [CrossRef]
29. Hones, K.M.; Nichols, D.S.; Barker, H.; Cox, E.; Hones, J.A.; Chim, H. Outcomes following use of VersaWrap nerve protector in treatment of patients with recurrent compressive neuropathies. *Front. Surg.* **2023**, *10*, 1123375. [CrossRef]
30. Sarhane, K.A.; Ibrahim, Z.; Martin, R.; Krick, K.; Cashman, C.R.; Tuffaha, S.H.; Broyles, J.M.; Prasad, N.; Yao, Z.-C.; Cooney, D.S.; et al. Macroporous nanofiber wraps promote axonal regeneration and functional recovery in nerve repair by limiting fibrosis. *Acta Biomater.* **2019**, *88*, 332–345. [CrossRef]
31. Liu, C.; Liu, D.; Zhang, X.; Hui, L.; Zhao, L. Nanofibrous polycaprolactone/amniotic membrane facilitates peripheral nerve regeneration by promoting macrophage polarization and regulating inflammatory microenvironment. *Int. Immunopharmacol.* **2023**, *121*, 110507. [CrossRef] [PubMed]
32. Langdell, H.C.; Zeng, S.L.; Pidgeon, T.S.; Mithani, S.K. Recalcitrant Neuropathies in the Upper Extremity. *J. Hand Surg. Glob. Online* **2023**, *5*, 503–509. [CrossRef]
33. Oh, S.H.; Chung, J.I. Oblique axis hypothenar free flaps: Tips for harvesting larger flaps with minimal donor site morbidity. *Arch. Plast. Surg.* **2023**, *50*, 279–287. [CrossRef]

34. Shin, A.; Saffari, T.; Saffari, S.; Vyas, K.; Mardini, S. Role of adipose tissue grafting and adipose-derived stem cells in peripheral nerve surgery. *Neural Regen. Res.* **2022**, *17*, 2179–2184. [CrossRef]
35. Danoff, J.R.; Lombardi, J.M.; Rosenwasser, M.P. Use of a pedicled adipose flap as a sling for anterior subcutaneous transposition of the ulnar nerve. *J. Hand Surg. Am.* **2014**, *39*, 552–555. [CrossRef]
36. Dibbs, R.P.; Ali, K.; Sarrami, S.M.; Koshy, J.C. Revision peripheral nerve surgery of the upper extremity. *Semin. Plast. Surg.* **2021**, *35*, 119–129. [CrossRef] [PubMed]
37. Kingham, P.J.; Kalbermatten, D.F.; Mahay, D.; Armstrong, S.J.; Wiberg, M.; Terenghi, G. Adipose-derived stem cells differentiate into a Schwann cell phenotype and promote neurite outgrowth in vitro. *Exp. Neurol.* **2007**, *207*, 267–274. [CrossRef] [PubMed]
38. Ching, R.C.; Wiberg, M.; Kingham, P.J. Schwann cell-like differentiated adipose stem cells promote neurite outgrowth via secreted exosomes and RNA transfer. *Stem Cell Res. Ther.* **2018**, *9*, 266. [CrossRef] [PubMed]
39. Riccio, M.; Gravina, P.; Pangrazi, P.P.; Cecconato, V.; Gigante, A.; De Francesco, F. Ulnar nerve anteposition with adipofascial flap, an alternative treatment for severe cubital syndrome. *BMC Surg.* **2023**, *23*, 268. [CrossRef]
40. Mamede, A.M.A.; Botelho, A.C. *Amniotic Membrane: Origin, Characterization, and Medical Applications*; Springer: Berlin/Heidelberg, Germany, 2015.
41. Wolfe, E.M.; Mathis, S.A.; Muñoz, N.d.l.O.; Ovadia, S.A.; Panthaki, Z.J. Comparison of human amniotic membrane and collagen nerve wraps around sciatic nerve reverse autografts in a rat model. *Biomater. Biosyst.* **2022**, *6*, 100048. [CrossRef]
42. Leal-Marin, S.; Kern, T.; Hofmann, N.; Pogozhykh, O.; Framme, C.; Börgel, M.; Figueiredo, C.; Glasmacher, B.; Gryshkov, O. Human Amniotic Membrane: A review on tissue engineering, application, and storage. *J. Biomed. Mater. Res. Part B Appl. Biomater.* **2020**, *109*, 1198–1215. [CrossRef]
43. McClendon, D.C.; Su, J.; Smith, D.W. Human amniotic allograft in hand surgery. *J. Hand Surg. Am.* **2023**, *48*, 388–395. [CrossRef]
44. Yu, L.-M.; Yu, C.-Y.; Zhang, Z.-Y.; Yang, J.; Fan, Z.-H.; Wang, D.-L.; Wang, Y.-Y.; Zhang, T. Fresh human amniotic membrane effectively promotes the repair of injured common peroneal nerve. *Neural Regen. Res.* **2019**, *14*, 2199–2208. [CrossRef]
45. Mirzayan, R.; Russo, F.; Yang, S.-J.T.; Lowe, N.; Shean, C.J.; Harness, N.G. Human Amniotic Membrane Wrapping of the Ulnar Nerve During Cubital Tunnel Surgery Reduces Recurrence of Symptoms. *Arch. Bone Jt. Surg.* **2022**, *10*, 969–975. [CrossRef]
46. Fénelon, M.; Catros, S.; Meyer, C.; Fricain, J.-C.; Obert, L.; Auber, F.; Louvrier, A.; Gindraux, F. Applications of Human Amniotic Membrane for Tissue Engineering. *Membranes* **2021**, *11*, 387. [CrossRef] [PubMed]
47. Li, X.; Zhang, X.; Hao, M.; Wang, D.; Jiang, Z.; Sun, L.; Gao, Y.; Jin, Y.; Lei, P.; Zhuo, Y. The application of collagen in the repair of peripheral nerve defect. *Front. Bioeng. Biotechnol.* **2022**, *10*, 973301. [CrossRef] [PubMed]
48. Jiang, M.; Chen, M.; Liu, N. Interactions between Schwann cell and extracellular matrix in peripheral nerve regeneration. *Front. Neurol.* **2024**, *15*, 1372168. [CrossRef] [PubMed]
49. Hardin-Young, J.; Carr, R.M.; Downing, G.J.; Condon, K.D.; Termin, P.L. Modification of native collagen reduces antigenicity but preserves cell compatibility. *Biotechnol. Bioeng.* **2000**, *49*, 675–682. [CrossRef]
50. Eleftheriadou, D.; Phillips, J.B. Collagen biomaterials for nerve tissue engineering. In *Peripheral Nerve Tissue Engineering and Regeneration*; Phillips, J.B., Hercher, D., Hausner, T., Eds.; Springer: Berlin/Heidelberg, Germany, 2022. [CrossRef]
51. Abedi, M.; Shafiee, M.; Afshari, F.; Mohammadi, H.; Ghasemi, Y. Collagen-Based Medical Devices for Regenerative Medicine and Tissue Engineering. *Appl. Biochem. Biotechnol.* **2023**, *196*, 5563–5603. [CrossRef]
52. Chen, S.; Gao, Y.-B.; Liu, Z.-G.; Lin, G.-D.; Guo, Y.; Chen, L.; Huang, B.-T.; Yin, Y.-B.; Yang, C.; Sun, L.-Y.; et al. Safety and efficacy of a nerve matrix membrane as a collagen nerve wrapping: A randomized, single-blind, multicenter clinical trial. *Neural Regen. Res.* **2021**, *16*, 1652–1659. [CrossRef]
53. Wong, G.C.; Chung, K.C. Bioengineered nerve conduits and wraps. *Hand Clin.* **2024**, *40*, 379–387. [CrossRef]
54. Spielman, A.F.; Sankaranarayanan, S.; Skowronski, P.; Lessard, A.-S.; Panthaki, Z. Recurrent and persistent carpal tunnel syndrome: "Triple-therapy approach". *J. Orthop.* **2020**, *22*, 431–435. [CrossRef]
55. Integra LifeSciences Corporation. NeuraWrap™ Nerve Protector. Available online: https://www.integranerve.com/nerve-protector (accessed on 19 March 2025).
56. Stryker Corporation. NeuroMend™ Wrap. Available online: https://www.stryker.com/us/en/trauma-and-extremities/products/neuromend.html (accessed on 19 March 2025).
57. Downey, M.S. A guide to nerve wrapping for tarsal tunnel surgery. *Podiatry Today* **2015**, *28*, 32.
58. Rault, I.; Frei, V.; Herbage, D.; Abdul-Malak, N.; Huc, A. Evaluation of different chemical methods for cros-linking collagen gel, films and sponges. *J. Mater. Sci. Mater. Med.* **1996**, *7*, 215–221. [CrossRef]
59. Mariani, E.; Lisignoli, G.; Borzì, R.M.; Pulsatelli, L. Biomaterials: Foreign Bodies or Tuners for the Immune Response? *Int. J. Mol. Sci.* **2019**, *20*, 636. [CrossRef]

60. Hanwright, P.J.; Rath, J.B.; von Guionneau, N.; Slavin, B.; Pinni, S.; Zlotolow, D.; Shores, J.; Dellon, A.L.; Tuffaha, S.H. The Effects of a Porcine Extracellular Matrix Nerve Wrap as an Adjunct to Primary Epineurial Repair. *J. Hand Surg.* **2021**, *46*, 813.e1–813.e8. [CrossRef] [PubMed]
61. Yi, J.-S.; Lee, H.-J.; Lee, H.-J.; Lee, I.-W.; Yang, J.-H. Rat Peripheral Nerve Regeneration Using Nerve Guidance Channel by Porcine Small Intestinal Submucosa. *J. Korean Neurosurg. Soc.* **2013**, *53*, 65–71. [CrossRef]
62. Li, T.; Javed, R.; Ao, Q. Xenogeneic decellularized extracellular matrix-based biomaterials for peripheral nerve repair and regeneration. *Curr. Neuropharmacol.* **2021**, *19*, 2152–2163. [CrossRef]
63. Grandizio, L.C.; Maschke, S.; Evans, P.J. The management of persistent and recurrent cubital tunnel syndrome. *J. Hand Surg. Am.* **2018**, *43*, 933–940. [CrossRef] [PubMed]
64. Imran, R.; George, S.; Jose, R.; Shirley, C.; Power, D.M. Clinical outcomes following neurolysis and porcine collagen extracellular matrix wrapping of scarred nerves in revision carpal tunnel decompression. *J. Plast. Reconstr. Aesthetic Surg.* **2022**, *75*, 2802–2808. [CrossRef] [PubMed]
65. Fones, L.; DePascal, M.; Ilyas, A.M. Use of nerve wraps in the upper extremity. *SurgiColl* **2024**, *2*. [CrossRef]
66. Burahee, A.S.; Duraku, L.S.; Bosman, R.; Shirley, C.; van der Oest, M.J.; Zuidam, M.J.; Power, D.M. Porcine submucosal extracellular matrix wrapping of the ulnar nerve in revision cubital tunnel surgery. *J. Plast. Reconstr. Aesthetic Surg.* **2024**, *98*, 176–183. [CrossRef]
67. Fujii, M.; Tanaka, R. Porcine Small Intestinal Submucosa Alters the Biochemical Properties of Wound Healing: A Narrative Review. *Biomedicines* **2022**, *10*, 2213. [CrossRef]
68. Jordaan, P.; Uhiara, O.; Power, D. Management of the scarred nerve using porcine submucosa extracellular matrix nerve wraps. *J. Musculoskelet. Surg. Res.* **2019**, *3*, 128. [CrossRef]
69. Eberlin, K.R.; Safa, B.; Buntic, R.; Rekant, M.S.; Richard, M.J.; Styron, J.F.; Bendale, G.; Isaacs, J. Usability of Nerve Tape: A Novel Sutureless Nerve Coaptation Device. *J. Hand Surg.* **2024**, *49*, 346–353. [CrossRef]
70. Bendale, G.S.; Sonntag, M.; Clements, I.P.; Isaacs, J.E. Biomechanical Testing of a Novel Device for Sutureless Nerve Repair. *Tissue Eng. Part C Methods* **2022**, *28*, 469–475. [CrossRef] [PubMed]
71. Zeng, W.; Osterman, M.; Stern, P.J. Inflammatory reactions to xenogenic nerve wraps: A report of three cases. *JBJS Case Connect.* **2019**, *9*, e0302. [CrossRef]
72. Zhang, M.; An, H.; Zhang, F.; Jiang, H.; Wan, T.; Wen, Y.; Han, N.; Zhang, P. Prospects of Using Chitosan-Based Biopolymers in the Treatment of Peripheral Nerve Injuries. *Int. J. Mol. Sci.* **2023**, *24*, 12956. [CrossRef] [PubMed]
73. Yao, P.; Li, P.; Jiang, J.J.; Li, H.Y. Anastomotic stoma coated with chitosan film as a betamethasone dipropionate carrier for peripheral nerve regeneration. *Neural Regen. Res.* **2018**, *13*, 309–316. [CrossRef]
74. Bąk, M.; Gutlowska, O.N.; Wagner, E.; Gosk, J. The role of chitin and chitosan in peripheral nerve reconstruction. *Polym. Med.* **2017**, *47*, 43–47. [CrossRef] [PubMed]
75. Kołodziejska, M.; Jankowska, K.; Klak, M.; Wszoła, M. Chitosan as an Underrated Polymer in Modern Tissue Engineering. *Nanomaterials* **2021**, *11*, 3019. [CrossRef]
76. Ke, C.-L.; Deng, F.-S.; Chuang, C.-Y.; Lin, C.-H. Antimicrobial Actions and Applications of Chitosan. *Polymers* **2021**, *13*, 904. [CrossRef]
77. Aranaz, I.; Alcántara, A.R.; Civera, M.C.; Arias, C.; Elorza, B.; Caballero, A.H.; Acosta, N. Chitosan: An Overview of Its Properties and Applications. *Polymers* **2021**, *13*, 3256. [CrossRef]
78. Jain, N.; Murchinson, M.; Rounds, A.; Bourland, B. Hemoclip migration after revision carpal tunnel release: A case report. *J. Surg. Case Rep.* **2023**, *2023*, rjad548. [CrossRef]
79. Siemionow, M.; Radecka, W.; Kozlowska, K.; Chambily, L.; Brodowska, S.; Kuc, D.; Filipek, G.; Budzynska, K. Protective Effect of the Human Epineural Patch Application after Sciatic Nerve Crush Injury Followed by Nerve Transection and End-to-End Repair. *Arch. Immunol. Ther. Exp.* **2025**, *73*. [CrossRef] [PubMed]
80. Siemionow, M.; Uygur, S.; Madajka, M. Application of epineural sheath as a novel approach for fat volume maintenance. *Ann Plast Surg.* **2017**, *79*, 606–612. [CrossRef]
81. Siemionow, M.; Cwykiel, J.; Uygur, S.; Kwiecien, G.; Oztürk, C.; Szopinski, J.; Madajka, M. Application of epineural sheath conduit for restoration of 6-cm long nerve defects in a sheep median nerve model. *Microsurgery* **2018**, *39*, 332–339. [CrossRef] [PubMed]
82. Siemionow, M.; Demir, Y.; Mukherjee, A.L. Repair of peripheral nerve defects with epineural sheath grafts. *Ann. Plast. Surg.* **2010**, *65*, 546–554. [CrossRef] [PubMed]
83. Siemionow, M.; Duggan, W.; Brzezicki, G.; Klimczak, A.; Grykien, C.; Gatherwright, J.; Nair, D. Peripheral Nerve Defect Repair With Epineural Tubes Supported With Bone Marrow Stromal Cells: A preliminary report. *Ann. Plast. Surg.* **2011**, *67*, 73–84. [CrossRef]

84. Siemionow, M.; Strojny, M.M.; Kozlowska, K.; Brodowska, S.; Grau-Kazmierczak, W.; Cwykiel, J. Application of Human Epineural Conduit Supported with Human Mesenchymal Stem Cells as a Novel Therapy for Enhancement of Nerve Gap Regeneration. *Stem Cell Rev. Rep.* **2021**, *18*, 642–659. [CrossRef]
85. Siemionow, M.; Tetik, C.; Ozer, K.; Ayhan, S.; Siemionow, K.; Browne, E. Epineural sleeve neurorrhaphy: Surgical technique and functional results—A preliminary report. *Ann. Plast. Surg.* **2002**, *48*, 281–285. [CrossRef]
86. Wan, R.; Zhao, G.; Adam, E.A.; Selim, O.A.; Sarcon, A.K.; Reisdorf, R.L.; Meves, A.; Zhao, C.; Moran, S.L. Evaluating the Effectiveness of Commercially Available Antiadhesion Tendon Protector Sheets in Tendon Repair Surgery Versus Tendon Repair Surgery Alone: A Preclinical Model Study. *J. Hand Surg.* **2024**. [CrossRef]
87. Adu, Y.; Harder, J.; Cox, C.; Baum, G.; Hernandez, E.J.; MacKay, B.J. Evaluating the effect of VersaWrap tendon protector on functional outcomes in operative tendon repairs. *Front. Surg.* **2024**, *11*, 1447515. [CrossRef]
88. Zhang, M.; Li, C.; Zhou, L.P.; Pi, W.; Zhang, P.X. Polymer scaffolds for biomedical applications in peripheral nerve reconstruction. *Molecules* **2021**, *26*, 2712. [CrossRef]
89. Harley-Troxell, M.E.; Steiner, R.; Newby, S.D.; Bow, A.J.; Masi, T.J.; Millis, N.; Matavosian, A.A.; Crouch, D.; Stephenson, S.; Anderson, D.E.; et al. Electrospun PCL Nerve Wrap Coated with Graphene Oxide Supports Axonal Growth in a Rat Sciatic Nerve Injury Model. *Pharmaceutics* **2024**, *16*, 1254. [CrossRef] [PubMed]
90. Jiang, Y.; Tang, X.; Li, T.; Ling, J.; Yang, Y. The success of biomaterial-based tissue engineering strategies for peripheral nerve regeneration. *Front. Bioeng. Biotechnol.* **2022**, *10*, 1039777. [CrossRef] [PubMed]
91. Dias, J.R.; Sousa, A.; Augusto, A.; Bártolo, P.J.; Granja, P.L. Electrospun Polycaprolactone (PCL) Degradation: An In Vitro and In Vivo Study. *Polymers* **2022**, *14*, 3397. [CrossRef]
92. Lopez, J.; Xin, K.; Quan, A.; Xiang, S.; Barone, A.A.L.; Budihardjo, J.; Musavi, L.; Mulla, S.; Redett, R.; Martin, R.; et al. Poly(ε-Caprolactone) Nanofiber Wrap Improves Nerve Regeneration and Functional Outcomes after Delayed Nerve Repair. *Plast. Reconstr. Surg.* **2019**, *144*, 48e–57e. [CrossRef]
93. Bai, J.; Liu, C.; Kong, L.; Tian, S.; Yu, K.; Tian, D. Electrospun Polycaprolactone (PCL)-Amnion Nanofibrous Membrane Promotes Nerve Regeneration and Prevents Fibrosis in a Rat Sciatic Nerve Transection Model. *Front. Surg.* **2022**, *9*, 842540. [CrossRef]
94. Dong, R.; Tian, S.; Bai, J.; Yu, K.; Liu, C.; Liu, L.; Tian, D. Electrospun Polycaprolactone (PCL)-Amnion Nanofibrous Membrane Promotes Nerve Repair after Neurolysis. *J. Biomater. Appl.* **2022**, *36*, 1390–1399. [CrossRef] [PubMed]
95. Nicolas, C.F.; Corvi, J.J.; Zheng, Y.; Park, K.H.; Akelina, Y.; Engemann, A.; Strauch, R.J. Resorbable Nerve Wraps: Can They Be Overtightened? *J. Reconstr. Microsurg.* **2022**, *38*, 694–702. [CrossRef]
96. Koenig, Z.A.; Burns, J.C.; Hayes, J.D. Necrotic granulomatous inflammation after use of small intestine submucosa matrix for recurrent compression neuropathy. *Plast. Reconstr. Surg. Glob. Open.* **2022**, *10*, e4378. [CrossRef]

Disclaimer/Publisher's Note: The statements, opinions and data contained in all publications are solely those of the individual author(s) and contributor(s) and not of MDPI and/or the editor(s). MDPI and/or the editor(s) disclaim responsibility for any injury to people or property resulting from any ideas, methods, instructions or products referred to in the content.

MDPI AG
Grosspeteranlage 5
4052 Basel
Switzerland
Tel.: +41 61 683 77 34

Journal of Functional Biomaterials Editorial Office
E-mail: jfb@mdpi.com
www.mdpi.com/journal/jfb

Disclaimer/Publisher's Note: The title and front matter of this reprint are at the discretion of the Guest Editors. The publisher is not responsible for their content or any associated concerns. The statements, opinions and data contained in all individual articles are solely those of the individual Editors and contributors and not of MDPI. MDPI disclaims responsibility for any injury to people or property resulting from any ideas, methods, instructions or products referred to in the content.

www.ingramcontent.com/pod-product-compliance
Lightning Source LLC
LaVergne TN
LVHW070453061126
851858LV00011B/232
* 9 7 8 3 7 2 5 8 8 1 7 5 8 *